Douglas G. Brookins

Geochemical Aspects of Radioactive Waste Disposal

With 53 Figures

Springer-Verlag
New York   Berlin   Heidelberg   Tokyo

DOUGLAS G. BROOKINS
Department of Geology
University of New Mexico
Albuquerque, New Mexico 87131
U.S.A.

Library of Congress Cataloging in Publication Data
Brookins, Douglas G.
  Geochemical aspects of radioactive waste disposal.
  Bibliography: p.
  Includes index.
  1. Radioactive waste disposal. 2. Geochemistry.
I. Title.
TD898.B76   1984   621.48′38   83-16896

Typeset by Ampersand, Inc., Rutland, Vermont.
Printed and bound by R.R. Donnelley & Sons, Harrisonburg, Virginia.
Printed in the United States of America.

9 8 7 6 5 4 3 2 1

ISBN 0-387-90916-8 Springer-Verlag New York Berlin Heidelberg Tokyo
ISBN 3-540-90916-8 Springer-Verlag Berlin Heidelberg New York Tokyo

For Barbara

Preface

There is an extremely voluminous literature on radioactive waste and its disposal, much in the form of government-sponsored research reports. To wade through this mountain of literature is indeed a tedious task, and it is safe to speculate that very few, if any, individuals have the time to examine each report that has been issued during the preceding ten years. This book attempts to summarize much of this literature.

Further, many workers in the geosciences have not received training in the nuclear sciences, and many nuclear scientists could be better versed in geology. In this book an attempt is made to cover some background material on radioactive wastes and geotoxicity that may not be an integral part of a geologist's training, and background material on geology and geochemistry for the nuclear scientist. The geochemical material is designed for both the geoscientist and the nuclear scientist.

There is no specific level for this book. Certainly, it should be useful to advanced undergraduates and graduates studying geology and nuclear science. It does not pretend to cover a tremendous amount of detail in all subjects, yet the references cited provide the necessary source materials for follow-up study. It is my intention that the reader of this book will have a better, broader understanding of the geochemical aspects of radioactive waste disposal than is otherwise available in any one source.

The focus of this book is on geochemistry, with an emphasis on natural analogs wherever possible. The attempt is made herein to show that rocks can, have, and in all probability will, retain radioactive waste materials without posing a threat to public health.

Albuquerque, New Mexico   Douglas G. Brookins
November 1983

Contents

CHAPTER 1
Introduction

Radioactive waste management is one of the most pressing problems facing the world today. Politically, it is a nightmare. Scientifically, there are solutions to many parts of the problem and well-identified areas in which to conduct additional necessary research. Radioactive waste management involves the entire world community, since approximately 35 nations have nuclear power and/or defense projects in use. The laboratory aspects of radioactive waste studies are fairly well defined, but the necessarily short duration of most of these experiments requires that natural analogs also be examined. Natural experiments over periods of 10^3–10^5 years were carried out at least hundreds of millions of years ago, however.

One of the main objectives of this book is to present geochemical aspects of radioactive waste disposal. The audience for this book requires that some background, including some elementary material, be presented as well. Thus the reader will find descriptive text on radioactive waste management as well as some basic geological and geochemical principles.

Many elements and isotopes must be considered in a total planned waste disposal system. Fission elements, actinide and actinide-daughter elements, and numerous other elements are involved. Table 1-1 lists important radionuclides, their half-lives, elements as analogs in addressing their behavior, and the most common forms these elements assume in the upper-crustal and surface environment. The behavior of most of these radionuclides is best explained in conjunction with natural analogs, and the reader is referred to Chapter 11, in which this approach is described. In addition, radionuclide migration or retardation is the critical factor of waste disposal to be addressed here.

When radioactive waste is deeply buried, the only fluid that can possibly interact with it is groundwater. It is important to consider various ground-

Table 1–1. Some important radionuclides in the earth's crust.

Nuclide	Half-life (yr)	Analog	Common forms (surface crustal sites)
^{3}H	12.3	H	H_2O, H^+, OH^-, $H_2(g)$
^{14}C	5×10^3	C	HCO_3^-, H_2CO_3, CO_3^{2-}, CO_2, CH_4
^{36}Cl	3.1×10^5	Cl	Cl^-, $HCl\,(g)$
^{63}Ni	100	Ni	Ni^{2+}, $NiOH^+$, $NiCl_2^0$, $NiCO_3$, $Ni(OH)_2$
^{90}Sr	29	Sr, Ca	Sr^{2+}, $SrCO_3$
^{93}Zr	1.5×10^6	Zr	$Zr(OH)_2^{2+}$, $Zr(OH)_5^-$, $ZrSiO_4$
^{94}Nb	2×10^4	Nb	$Nb(OH)_5$
^{107}Pd	7×10^6	Pd,Pt	Pd^{2+}, $Pd(OH)_2$, $PdCl_4^{2-}$
^{129}I	2×10^7	I, Cl,Br	I^-, IO_3^-
^{135}Cs	3×10^6	Cs,Rb	Cs^+
^{137}Cs	30	Cs,Rb	Cs^+
^{154}Eu	8.2	Eu,REE	Eu^{3+}, $Eu(OH)_2^+$, $EuSO_4^+$, $Eu(OH)_3$
^{79}Se	6.5×10^4	Se,S	Se^{2-}, HSe^-, $Se(s)$, SeO_4^{2-}, SeO_3^{2-}
^{93}Mo	3.5×10^3	Mo	Mo^{4+}, MoO_4^{2-}, MoS_2
^{99}Tc	2×10^5	Re,Mn	TcO_4^-, Tc^{4+}, TcO_2, TcS_2, $Tc(OH)_4$
^{106}Ru	1	Ru,Mo	RuO_4^{2-}, RuO_2, RuS_2
^{126}Sn	1×10^5	Sn	Sn^{2+}, $Sn(OH)^+$, SnS, SnO_2
^{147}Sm	1.3×10^{11}	Sm,Nd	Sm^{3+}, $Sm(OH)_2^+$, $Sm(OH)_3$
^{151}Sm	90	Sm,Nd	Sm^{3+}, $Sm(OH)_2^+$, $Sm(OH)_3$
^{210}Pb	22	Pb	Pb^{2+}, $Pb(OH)^+$, PbS,$PbCl^+$
^{226}Ra	1.6×10^3	Ra, Ba, Sr, Ca	Ra^{2+}
^{227}Ac	22	REE	Ac^{3+}, $Ac(OH)_2^+$, $Ac(OH)_3$
^{230}Th	8×10^4	Th	Th^{4+}, $Th(OH)_3^+$, ThO_2, $Th(OH)_2$
^{232}Th	1.4×10^{10}	Th	Th^{4+}, $Th(OH)_3^+$, ThO_2, $Th(OH)_4$
^{231}Pa	3×10^4	Pa	$Pa(OH)_5$
^{234}U	2.5×10^5	U	U–O and U–C complexes
^{235}U	7×10^8	U	U–O and U–C complexes
^{238}U	4.5×10^9	U	U–O and U–C complexes
^{237}Np	2×10^6	Np,U	NpO_2^+, NpO_2, complexes
^{238}Pu	88	Pu,Th	Pu^{3+}, PuO_2^+, PuO_2
^{239}Pu	2.4×10^4	Pu,Th	Pu^{3+}, PuO_2^+, PuO_2
^{241}Am	432	Nd	$Am(OH)^{3+}$, AmO_2
^{243}Am	7×10^3	Nd	$Am(OH)_3^+$, AmO_2
^{243}Cm	28	Nd	Cm–O complexes, $Cm(OH)_3$

Source: Modified from Moody (1982).

Table 1-2. Some deep groundwater compositions.

	Basalt	Granite	Tuff
temp(°C)	45	30	30
pH	9–10	9.8	8.5
Eh(v)	−0.5	0.17	—
Ca[a]	1.3	59	13
Mg	0.04	0.5	2
Na	250	125	49
K	1.9	0.4	4.7
Cl	148	283	7.6
SO_4	108	19	21
Fe	—	0.02	0.16
ΣCO_2	46	3	—
SiO_2	145	11	50
B	—	—	—
Li	—	—	0.05
F	37	4	2.3
Sr	—	—	0.05
Ba	—	—	0.2
Br	—	—	—
I	—	—	—
Cs	—	—	—
Rb	—	—	—
V	0.01	—	—
PO_4	—	0.1	—

Source: Modified from Moody (1982).
[a]All data reported in mg/L.

waters, their chemistry, and other properties. Table 1-2 lists some average groundwater compositions, assumed temperatures (i.e., at depths where repositories in each medium are considered), Eh, and pH. These average values apply to the generic basalt, granite, bedded salt (i.e., the Waste Isolation Pilot Plant (WIPP) site, New Mexico), and for average tuffaceous rocks at the Nevada Test Site (NTS), and will be discussed in subsequent chapters.

The speciation of several key radionuclides in groundwaters is given in Table 1-3. In addition to the simple cations and anions, there are a number of complex ions formed by various simple cations with OH^-, Cl^-, HCO_3^- (or CO_3^{2-}), SO_4^{2-}, F^-, $H_2PO_4^-$ (or HPO_4^{2-}), and humic or fulvic acids. These will be discussed in Chapters 7, 11, 12, and 13.

There are two points to emphasize. First, there are many complex variables and parameters that may affect possible radionuclide migration/retardation in natural media. Second, there are natural analogs that address

Table 1–3. Probable aqueous species in groundwater.

Ion	Species
Simple anions	HCO_3^-, CO_3^{2-}
	I^-, IO_3^-
	HSe^-, SeO_3^{2-}, SeO_4^{2-}
	TcO_4^-, MoO_4^{2-}, RuO_4^{2-}
Simple cations	Cs^+
	Ni^{4+}, Sr^{2+}, Sn^{2+}, Pb^{2+}, Pd^{2+}, Th^{4+}, Pu^{4+}, Zr^{4+}, Pa^{4+}, U^{4+}, Np^{4+}, UO_2^{2+}, PuO_2^{2+}, NpO_2^+
Complexions	Simple cations forming mono- or polynuclear
Ligand	species with ligand to left
OH^-	Ni^{2+}, Sr^{2+}, Sn^{2+}, Am^{3+}, Eu^{3+}, Zr^{4+}, Th^{4+}, Tc^{4+}, U^{4+}, Pu^{4+}, Pa^{5+}, UO_2^{2+}
Cl^-	Ni^{2+}, Sn^{2+}, Pd^{2+}, Pb^{2+}, Pt^{2+}, U^{4+}, Pu^{4+}
HCO_3^-/CO_3^{2-}	UO_2^{2+}, PuO_2^{2+}, Th^{4+}, NpO_2^+, Ni^{2+}, Tc^{4+}
SO_4^{2-}	Eu^{3+}, U^{4+}, Pu^{4+}, Am^{3+}, UO_2^{2+}, PuO_2^{2+}
F^-	U^{4+}, Pu^{4+}, UO_2^{2+}, PuO_2^{2+}, NpO_2^+
$H_2PO_4^-/HPO_4^{2-}$	UO_2^{2+}, PuO_2^{2+}, NpO_2^+, U^{4+}, Pu^{4+}, Th^{4+}
humic & fulvic acids	U^{4+}, UO_2^{2+}, Pu^{4+}, PuO_2^{2+}, Ni^{2+}, Pb^{2+}, Th^{4+}

Source: Modified from Moody (1982).

virtually all of the complexities. Whenever possible, I will use natural examples to address specific problems. In places, this approach demands logical extrapolations of analogs to specific reactions or interactions with man-made materials for which there may not be an ideal analog. Yet in the entire waste program (see Chapter 2), the multibarrier concept now proposed does allow analogs to be commented on as part of the entire package. As an example, if a man-made waste form is totally foreign to the natural environment, then it can be argued that, in the absence of a natural analog, one cannot truly predict ultimate behavior of this waste form to, for instance, leaching. Yet in the multibarrier system, the waste form is only the first of several materials involved. For any radionuclide to escape the buried waste, it must pass through the canister, overpack, engineered backfill, and, ultimately, the repository rock—each of which can be addressed by sophisticated laboratory work for which there are numerous natural analogs.

The prime objective of waste disposal is to isolate the radioactive waste from the environment for as long as necessary so that there is no threat to public health and safety. As pointed out by Martin *et al.* (1981, p. 3-16)

" . . . the regulation and licensing of a high-level waste repository will occur in the absence of any prior experience with comparable systems." This is true if natural analogs are not considered and only man-generated nuclear waste is considered (see Chapter 11 for discussion). The regulatory and licensing process is affected by uncertainties in fully understanding basic physical properties and reactions, viz., hydrologic and geologic problems of waste in a repository and complex waste–rock interactions. Further, it is not possible to characterize fully a geologic site, and, still further, the possibility of unanticipated interactions in complex systems such as the total waste system (i.e., waste form, canister, overpack, backfill, host rock) exists. These problems severely limit one's ability to provide accurate input for modeling, and without accurate input, the results of the models may be challenged. As pointed out by Martin *et al.* (1981), very prestigious groups, such as the National Academy of Science and the Interagency Review Group on Nuclear Waste Management, have noted uncertainties in various modeling methods, incluidng erroneous approximation techniques, algorithm errors, and unexpected logic errors in complex computer codes.

Yet despite these problems, natural systems reveal reasonable limits for input to modeling codes, and models themselves continue to be refined to a high degree of sophistication. In this chapter, models themselves are not treated. The reader is referred to Kocher (1981) for an up-to-date treatment of models and modeling philosophy. In this book the problems of input data and limits for the models, rather than the models themselves, will be stressed.

After considering several alternatives for regulating the geologic disposal of radioactive wastes, the Nuclear Regulatory Commission (see Martin *et al.*, 1981) set minimum performance standards for each of the major system elements and required the overall system to conform to newly set Environmental Protection Agency standards. Three separate system elements are identified by the NRC: (1) length of time after closure during which radionuclides are contained; (2) the rate at which radionuclides are released to the geologic setting after containment failures; and (3) the travel time through the geologic setting for radionuclides to reach the biosphere.

To address these topics requires that one be able to predict the behavior of buried radioactive waste in rock. Laboratory experiments are, in terms of geologic time, infinitely short. It is my intent to draw on natural analogs wherever possible to provide information for these and other associated problems.

Near-field effects in the waste repository are dominated by temperature effects, since at (and relatively soon after) waste storage the temperature is at a maximum, and reactions due to the increased temperature are more likely in this setting. The temperature begins to fall within a few hundred years after emplacement, and the total radioactivity drops dramatically over the same period. To prevent radionuclide loss from the waste during this early period of waste storage, the waste package will be designed to perform to specifications for the expected near-field conditions. In addition, the engineered barriers (see Chapter 13) will also be designed to limit the amount and rate of radionuclide transport from waste package site after the initial containment period; this will also limit the release of material to the repository environment. The amount of material that can reach the repository environment will be limited initially by the waste package and then by the backfill. The chemical properties of the backfill and the waste package will retard or inhibit radionuclide transport.

There are basically two geologic parameters to consider: (1) transport time of groundwater from the underground repository to the accessible environment, and (2) if radionuclides escape the engineered backfill system, transport time of individual radionuclides from the underground repository to the accessible environment. Of these, transport time of the groundwater will ultimately limit all material transport, and the NRC has decided to focus on this parameter (see Martin *et al.*, 1981).

Waste package containment times can be considered for any time period desired, but the NRC (Martin *et al.*, 1981) has chosen alternatives of 300, 1000, and 10,000 years. The 300-year containment time will allow for most of the highly radioactive waste products to be removed by decay (e.g., ^{90}Sr, ^{137}Cs), but setting such a short containment time will, in theory, permit the containment package to start to fail when temperature and total radioactivity are high. Thus hydrothermal reactions and phase changes in the backfill and in the near-backfill host rock may be severe, and it would be extremely difficult to predict and model radionuclide transport through the system.

The 1000-year containment time would, in theory, prevent any release until most of the fission products have decayed and the decay heat effects have dropped by about three orders of magnitude, at which point the temperature would have dropped significantly so that the probability of hydrothermal reactions would be lessened.

For the 10,000-year containment time, the bulk of the fission products would have decayed and many of the intermediate half-life transuranics (^{241}Am) would also have decayed. Heat generation would have decreased by four orders of magnitude and thermal and temperature gradients would be nearly ambient. The main problem is the difficulty of using laboratory data for extrapolating to such a lengthy time. The NRC (Martin *et al.*, 1981) has decided to use the 1000-year containment time with as many precautions as possible in order to approach likely conditions for the 10,000-year containment time, becuase they realize that the 300-year containment time is simply too short. Thus their approach is to attempt to guarantee waste package integrity for a minimum of 1000 years.

With further deliberation, the NRC has proposed that a leach rate (see Chapter 12) of the waste inventory of 10^{-5}/yr can be achieved with existing technology for waste package and engineered backfill. Although a rate of 10^{-5}/yr for the package alone is not adequate to prevent possible releases to the environment, it is sufficient to guarantee the decay of most fission products, and further, the backfill will more easily cope with such a low feed of material dictated by the 10^{-5}/yr rate. For the engineered backfill, a desired rate of release of 10^{-7}/yr will prevent any significant amounts of radiation from reaching the environment so as to pose a threat to public health or safety, although the 10^{-5}/yr rate is currently proposed here as well. The groundwater travel time advocated by the NRC (Martin *et al.*, 1981) is 10,000 years, because this time is sufficient to allow for decay of most fission products and transuranics. For long-half-life elements (^{99}Tc, ^{239}Pu) there will be a greater reliance on chemical retardation, but this is deemed likely to be the case based on natural analogs, laboratory experiments, theory, and other factors (see discussion in Chapters 5, 6, 11, 12, and 13).

Retrievability of buried waste is desirable. A period of 50 years during which the buried waste can be retrieved is proposed (Martin *et al.*, 1981). During this 50-year period, no events are anticipated to affect the waste package, and, if necessary, access to the waste is possible. After 50 years, the waste package area may be sealed.

Models should be carefully evaluated in terms of the natural environment. As Cohen and Smith (1981) point out, the EPA model used for risk evaluation for a population from HLW indicates that release of 3 Ci of Ra would cause 10 cancers over a 10,000-year period. Yet the top 10 m of soil in the United States have been calculated to contain 2.6×10^8 Ci of naturally occurring Ra (Cohen and Smith, 1981). One percent of natural background radiation dosage is due to Ra, hence the EPA model would predict 8.7 million cases of cancer per year—yet all cancers from all causes are only 350,000; therefore the EPA model does not yield meaningful information in this case. Similarly, Cohen and Smith (1981) take the reports of iodine hazard to task. They point out that the estimation of a possible dose of 3.3 rem/yr to the thyroid from iodine released from a waste repository is totally wrong. Since there is a natural abundance of iodine in the earth, for radioiodine from a 1000-meter-deep repository, the ratio of ^{129}I/stable I is 1.3×10^{-7}, and since there is no known mechanism to segregate iodine isotopes, and the total number of iodine atoms in the thyroid is fixed, the calculated whole-body and thyroid doses due to ^{129}I are 9×10^{-8} rem/yr and 7.5×10^{-6} rem/yr, respectively (Cohen and Smith, 1981)—or a six-orders-of-magnitude difference from the "model" value. What Cohen and Smith (1981) emphasize is that one must consider natural abundances of elements and their distribution, and how the natural system behaves. When this is either known or can be discussed in terms of reasonable limits, then, and only then, can successful modeling be carried out.

Models have been extensively used in the study of radioactive wastes and their possible impact on the environment. In fact, some 300 models were available in early 1981 (Smith and Cohen, 1981). The discrepancies between models has been pointed out by Smith and Cohen (1981) and by Cohen and Smith (1981), and the unrestricted use of such models without full evaluation of proper input justifiably criticized [e.g., " . . . garbage in, garbage out (GIGO)" (Cohen and Smith, 1981)]. Yet attempts are constantly being made to improve both the quality and identification of input and to derive more suitable (not necessarily complex!) models. In this vein, it is important to comment on the models (not computer output) now advocated by the EPA to address various aspects of radioactive waste transport (Smith *et al.*, 1981). Specific repository release models include river, ocean, land surface, and air. The first three models are likely to be applied to an actual repository, whereas the air model would probably only apply for an event such as accompanies a volcanic eruption. Yet the intent here is not to focus on models that have given erroneous results. Sophisticated models have been developed and tested (see Kocher, 1981, and Brookins, 1983 for examples). Ultimately, the models will provide a means for integrating laboratory and field data with theoretical considerations to fully characterize any possible interactions between any part of the waste package with the surrounding rocks in the presence of fluid. In addition, the models will ultimately test ways in which fluid can possibly reach the waste package in the first place, as oposed to assuming some fluid in "instantaneous" contact with the waste form. Hopefully the day when the models reach this goal is not far off.

In subsequent chapters, general treatment of radioactive waste management, natural radation in the geochemical environment, and descriptive material on different types of radioactive wastes and the multibarrier concept are followed by treatment of geochronologic aspects of the radwaste problem, a generic treatment of geological sites under consideration, and a closer look at some specific sites. Alternate waste options are then treated, followed by a discussion of low-level radioactive waste, uranium milling and mill tailings, and an in-depth look at natural analogs. Special focus is given to waste-form options, followed by discussion of engineered barriers and candidate canister materials. Some aspects of the potential toxicity of radwaste, especially when compared to other hazardous materials, follow. The attempt is to focus on geochemical aspects of all subject matter treated.

The literature devoted to radwaste problems over the last 10 years or so is truly voluminous, and the references cited represent only a small number of total references canvassed in assembling this book. The reader may find some of his or her favorite articles or reports missing from the references cited, but this is to be expected considering the magnitude of the job of trying to cover the existing literature; therefore the cited entries are ones that I feel are necessary.

CHAPTER 2

Radioactive Waste Management

Introduction

For purposes of management, radioactive wastes are categorized as defense wastes or civilian wastes. Defense wastes include all government-generated nuclear waste, most of which come from the national weapons program, although a small part result from nondefense government research and development activities. Civilian wastes include high-level wastes (HLW) from power reactors, low-level (LLW) and other non-high-level wastes, and spent fuel. The types of wastes and their inventories are described in detail in Chapter 4.

In 1971, the Atomic Energy Commission (AEC) asked Congress to authorize facilities for waste disposal in an abandoned salt mine in Lyons, Kansas. This plan called for disposal of commercial HLW and defense transuranic wastes (TRU). Due to a combination of political factors and testing results controversies, the AEC abandoned work at the Lyons site in 1972. In 1974, a new, parallel program, which became known as the WIPP (Waste Isolation Pilot Plant), was started in southeastern New Mexico. The intent of the WIPP (see Chapter 7) is to receive defense TRU and test for HLW behavior in a salt medium. As part of its plan to inspect a variety of rock types for waste repository potential, the Department of Energy (DOE) has begun to evaluate a number of sites. These include the Hanford Reservation (basalt), Washington; the Nevada Test site (NTS) (granite–tuff–shale), Paradox Basin, Utah and part of Colorado and Texas (bedded salt), and Louisiana–east Texas and Mississippi (salt domes). Expansion of testing at the WIPP is also planned. In addition, bedded salt, shale, and granite from several broad areas have been inspected (see Chapter 6).

The DOE, through its Defense Nuclear Waste Activity Program, identifies five activities to be handled: (1) decontamination and decommissioning (2) interim waste operations, (3) long-term waste management technology, (4) terminal storage, and (5) transportation. The reader is referred to DOE (1980) for details on problems of decontamination, decommissioning, and transportation. Interim storage will provide safe handling and storage of DOE radioactive wastes pending utilization of the long-term waste management technology program, which, in turn, covers research and development on HLW, TRU, LLW, and air-borne wastes. Terminal storage has always been planned for defense TRU at WIPP, although the plans have been changed to include testing of HLW in bedded salt.

Commercial nuclear wastes are mainly generated by nuclear power plants, but medicine, industry, and research and development facilities generate some nuclear wastes as well. The AEC recognized early that geologic repositories are a viable approach to the disposal of nuclear wastes, and research in this area has been continuing for about the last 30 years. The DOE sees the problem handled in three activities: (1) terminal isolation research and development, (2) waste systems evaluation and public interaction, and (3) waste treatment technology.

Terminal isolation is being studied in a number of geologic media throughout the United States (see Chapter 6 for details) and by cooperative studies with Canada, France, Japan, Germany, and the United Kingdom.

Waste systems evaluation and public interaction are assured by a series of Environmental Impact Statements and supporting documents that address many problems, including health and safety, economics, risk assessment, and quality assurance. State and local governments actively participate in all steps of the process, from site selection to final waste program.

Waste treatment technology involves selection and testing of candidate materials for disposal of nuclear wastes. Candidate waste forms (see Chapter 12) include cement, glass, ceramics, calcines, and many other, waste-specific forms. These materials are subject to rigorous testing to determine how such materials will behave in a rock repository, as well as the interaction of the waste form with surrounding engineered barriers in the presence of water.

Strategic planning for the management of nuclear wastes was drastically improved by the formation of the Interagency Review Group (IRG) on Nuclear Waste Management (Presidential Order of March 1978). In a series of reports, the IRG proposed that the choice of a geologic site for the disposal of nuclear wastes be made in 1985 from four or five fully evaluated candidate sites, and that operation of the repository(ies) begin in the mid-1990s. Further, the IRG report states

Successful isolation of radioactive wastes from the biosphere appears feasible for periods of thousands of years provided that the systems view is utilized rigorously to evaluate the suitability of sites and designs, to minimize the influence of future human activities, and to select a waste form that is compatible with its host rock.

The supporting experimental data and analyses to document this are being provided by the DOE today.

The Government has been responsible for production of major amounts of nuclear wastes since the mid-1940s, mainly as products of separating plutonium for nuclear weapons. Formerly, all nuclear wastes with low levels of activity, including LLW and TRU, were disposed of in shallow land burial sites. Storage of TRU in this fashion was stopped in 1970; it will, as is the case with HLW, be stored in mined geologic repositories from which it can be retrieved.

In the case of commerical wastes, a major program, the National Waste Terminal Storage (NWTS) Program, was started in 1976. Today it comprises three major elements; the Office of Nuclear Waste Isolation (ONWI), the Basalt Waste Isolation Project (BWIP), and the Nevada Nuclear Waste Storage Investigations (NNWSI). The breakdown of responsibility of these three groups is as follows. The ONWI is managed for the DOE by the Battelle Memorial Institute, Columbus, Ohio; the BWIP is managed for the DOE by Rockwell International (Hanford Operations, Hanford, Washington); and the NNWSI is managed by the DOE.

Some of the geologic exploration efforts of the ONWI have been concentrated in the four geographic areas underlain by bedded or dome salt (see Fig. 6-1). In the Salina salt region, a literature review phase has been completed and no further work is planned at this time. In Paradox Basin, three or four areas have been identified for additional work (see Chapter 7). The Permian Basin salts include sites for repository consideration in the Palo Duro and Dalhart basins. In the Gulf Coast region several dome salts are being investigated for repository consideration.

The BWIP is responsible for determining if the basalts of the Hanford, Washington site are suitable for terminal waste isolation. Investigations include site evaluation and choosing a specific site for testing, technology development and application specific to geologic systems of basalt, and *in situ* testing in an underground facility for thermomechanical and radiation-related effects of basalt-waste systems. Testng for a variety of purposes (see DOE, 1980, p. 119) over a period of 5 years is underway (see also Chapter 7).

The NNWSI project is important because of the diversity of rock types being considered. Granite, argillite, and tuff are all potential candidates at the NTS, and all are being tested. Specific activities at the NTS include: (1) site evaluations for location of a repository, (2) technical support for determination of the characterization of available geologic media and weapons testing seismic effects, (3) *in situ* testing of thermomechanical and radiation effects in an underground facility. The tuffaceous rocks of Yucca Mountain are receiving the most attention at the present time, and a preliminary decision to table research on other rock types has been made (see DOE, 1980, p. 122).

Geologic tests at the NTS include studies of tectonics, hydrology, stratigraphy, natural seismicity, and volcanism. Nongeologic studies include those of radionuclide transport (Los Alamos National Laboratory) with granite, shale, and tuff as waste-migration barriers. Electrical heater tests have been made in shale and granite and are now under way in tuff. Emplacement of commercial spent fuel in granite has been made to test the rock response to natural radioactivity.

Objectives of the National Waste Terminal Storage Program

The objectives of the National Waste Terminal Storage (NWTS) Program provide interim guidance for the disposal of radioactive wastes. They are designed to supplement the regulations put forth by the Environmental Protection Agency (EPA) and the Nuclear Regulatory Commission (NRC).

The primary method of waste disposal being pursued by the NWTS program is that of geologic isolation. In theory, a geologic repository will consist of three parts: the site, the repository, and the waste package. These three parts will act as multibarriers to prevent any escape of radionuclides from the buried waste. The waste package comprises the waste form as well as a series of engineered barriers that are designed to minimize interactions between the waste form, rock, and groundwater. The barriers will include a filler (in the case of spent fuel), canister, and engineered backfill. While the repository is operational, the waste package will guarantee no release of radionuclides, as well as ensure retrievability. The thermal period is that in which the short-lived fission products dominate the waste package and thermal output and radiation are high. In this period the waste package and the repository will contain the radionuclides. Long-term waste isolation is guaranteed by the site proper.

There are similarities between the repository and conventional mines. Access to the underground is by shafts and corridors, and rooms are excavated just like mine tunnels. In the case of the repository, however, man-made barriers will be used to isolate and contain the stored wastes. Further, the goal of surface facilities and activities is preservation of the containment of the waste in the rocks. Repository sites will be chosen so that the host rock is suitable for building the repository. They must provide natural barriers to any possible escape of radionuclides by keeping the waste in its original location, retard radionuclide migration in the rock, and help in preventing human intrusion. In Fig. 2-1 a hypothetical waste isolation system is shown.

Site suitability for radioactive waste storage will be accomplished by interagency agreement. The NRC will define the requirements by which to judge the ultimate site suitability. The EPA has the job of providing standards to guide the NRC requirements, and the DOE is to handle the construction of the facility, including additional research and related work

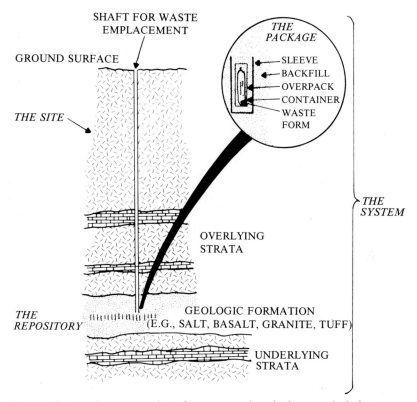

Fig. 2-1. Schematic cross section of a conceptual geologic waste isolation system. *Source:* Modified from DOE (1982).

parallel to NRC and EPA guidelines. In its search for suitable sites, the DOE has provided useful criteria to quickly focus in on favorable areas. The land first inspected in any area is deferred for some future assessment if it is less desirble than other lands, whether this is based on geologic evidence, lack of data, isolation factors, or other factors.

The siting process is guided by several fctors, including (1) DOE siting decisions, which are indeed based on sound information and data; (2) compliance with NRC procedures; (3) consistency with the National Environmental Policy Act (NEPA); and (4) public involvement.

Public involvement is considered to be of great importance by the DOE. At present, involved states have a role in reviewing the federal decision-making relevant to the site, design, and construction of high-level waste repositories. State groups include the State Planning Council (SPC), the State Working-Group on High-Level Nuclear Waste Management (SWG), the National Governor's Association (NGA), and the National Council of State Legislators (NCSL). All of these groups have contributed to the NWTS (see DOE, 1982).

In addition, consultation with state, local, and tribal governments is carried out by the DOE, and their consultation is sought on matters relevant to the NWTS plan.

Review of DOE plans and documents on radioactive waste is also assured by use of one or more of the following groups: Program Review Committee, Technical Advisory Committee, Earth Science Review Group, State Geologists Technical Review Group, Geological Exploration Review Group, Basalt Waste Isolation Plant (BWIP), Geology Overview Committee, BWIP Hydrology Overview Committee, BWIP Rock Mechanics Overview Committee, BWIP Hydrology Overview Committee, BWIP Intergovernmental Basalt Working Group, Nevada Nuclear Waste Storage Investigations (NNWSI), Geologic Investigations Peer Review Group, NNWSI Media Studies Experimental Planning Peer Review Group, and NNWSI Climax Spent Fuel Test Peer Review Group. These groups assure high-quality technical adequacy and achievement and a wide perspective on the NWTS program.

Any candidate site must meet NRC technical criteria (Federal Register, 1981) before the DOE can be licensed for the handling of nuclear waste. The DOE's plans, which are designed to meet the NRC criteria, are based on a division of the regulatory process into four parts: (1) characterization of the site, (2) authorization for construction, (3) licensing of the repository, and (4) decommission of the repository. Characterization of the site is the most important for siting considerations. Basically, the DOE will prepare a site Characterization Report (SCR) with the NRC. This report will discuss essentially all facets of the proposed site, including siting and screening criteria, quality assurance, status of other siting activities, and related issues.

By 10CFR60, the DOE is required to fully characterize a minimum of three sites representative of two different geologic media before the final site selection. One of these sites may be salt, but one specifically cannot be salt. Further, one EPA Implementation Plan for Siting High-Level Radioactive Waste Repositories has been prepared by the DOE. As part of this plan, an Environmental Impact Statement (EIS) will be prepared for the construction and operation of the test and evaluation facility. This document has been published (DOE, 1982), and is being used as the basis for environmental review.

The process of site selection is summarized in Table 2-1, where the important steps are shown. The DOE carries out the site selection process by a system of national survey, regional survey, area survey, and location survey. In this process, the following steps are carried out:

(1) Identification of factors and information important for the next screening decision (i.e., after each survey, do the data justify going to the next step and, if not, why not?)

(2) Gathering of required information with applicable consultation

Table 2-1. Planned siting document chain.[a]

Document	Content or Proposed Content
Commercial Waste FEIS	Provides assessment of potential impact of geologic disposal of HLW
National Siting Plan/EA	Describes siting strategy and its potential impact
Characterization/Integration Report[b]	Presents data and decision rationale used to screen larger land areas for potential sites
Site Characterization Report (SCR)/ EA or EIS[c]	Describes activities of site studies and their potential impacts
Test Evaluation Facility Recommendation Report/EIS	Describes T&E Facility site selection and potential impacts of the facility
Updated Site Characterization Report (SCR)/EA or EIS for land protection	Describes data on which site suitability judgment may be based, and assesses potential impact of land protection measures
Repository Site Recommendation Report/EIS	Provides data and rationale for site selection decision and potential impacts of the repository

Source: Modified from DOE (1982).

[a]This table lists the minimum variety of reports that will support the selection of a repository site. Each will appear (appeared) in draft for public comment before being finalized.

[b]May be a collection of region, area, and location reports or an integrated report containing equivalent information. Because this collection of reports describes results of various screening studies, they are sometimes referred to as "screening documents."

[c]Initial SCR will discuss design and test plans for an exploratory shaft. Other documents, such as Site Characterization Plans (SCPs), may be prepared where they facilitate state review of the DOE process.

(3) Identification of possible alternatives for the level of survey in progress

(4) Evaluation of alternatives

(5) Comparison of candidate alternatives with recommendation of that to be pursued further

(6) Review of screening decisions after consultation with the states involved.

By following these steps, the DOE and the states are assured of agreement based on in-depth study and exchange of scientific, socioeconomic, and environmental considerations.

After the location surveys are complete, an EIS or Environmental Assessment (EA) will be written to focus on a particular site, including

Fig. 2-2. Geohydrologic provinces as defined by ESTP subgroup 1. *Source:*
Modified from DOE (1982).

specific studies of the proposed site and the site for an exploratory shaft, and
necessary land-protection action. Site activities will include, in addition to
creating exploratory shaft, gathering environmental data, drilling for a variety
of tests, and digging trenches and test pits. The results of all these activities
are incorporated into the SCR, and the EIS or EA will contain the other
information required by NRC. The documents for each site are in sequence
as shown in Table 2-1.

Overseing all phases of the program for site selection is the internal DOE
organizational committee. The U.S. Geological Survey provides additional
expertise and advice as needed, in part through their Earth Science
Technical Plan (ESTP), which defines the technical efforts required for a
geologic waste repository. In Fig. 2-2 the 11 geohydrologic provinces as
defined by the ESTP subgroup are shown.

A summary of NWTS exploration efforts (as of September 1981) is
given in Table 2-2 and areas of potential interest for waste repositories are
shown in Fig. 2-3. Of the shaded areas shown in Fig. 2-3, active work is
progressing at the Hanford Site in Washington, the NTS, the WIPP in New
Mexico, the Paradox Basin region of Utah (and parts of surronding states),
and the salt domes in Texas, Louisiana, and Mississippi. These are discussed
in some detail in Chapters 6 and 7.

An assessment of environmental considerations, while of great importance, is outside the main thrust of this book. Therefore, the reader is referred
to DOE/NWTS-4 (1982) for detailed commentary on environmental and
related problems.

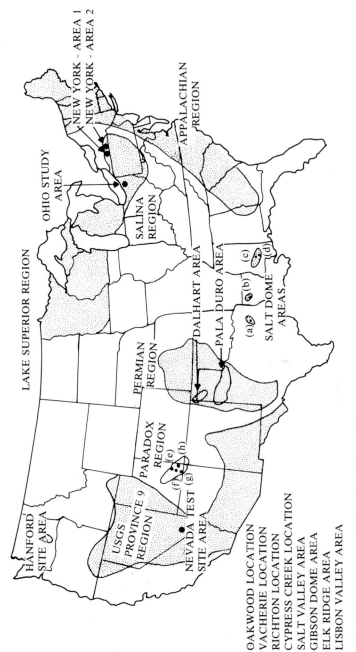

Fig. 2-3. Regions, areas, and locations of potential interest for terminal isolation of radioactive waste. *Source:* Modified from DOE (1982).

(a) OAKWOOD LOCATION
(b) VACHERIE LOCATION
(c) RICHTON LOCATION
(d) CYPRESS CREEK LOCATION
(e) SALT VALLEY AREA
(f) GIBSON DOME AREA
(g) ELK RIDGE AREA
(h) LISBON VALLEY AREA

Table 2-2. NWTS exploration efforts: Status as of September 1981.

National Survey Approach	Rock Types	Regions Identified	Areas Identified	Locations Identified
Geologic media	Bedded salt	Salina Region	Northeastern Ohio	—
			New York—area 1	—
			New York—area 2	—
		Paradox Region	Gibson Dome	—
			Elk Ridge	—
			Salt Valley	—
			Lisbon Valley	—
		Permian Region	Palo Duro area	—
			Dalhart area	—
	Domed salt	Gulf Coast Region	Texas Salt Domes	Oakwood Dome
			Mississippi Salt Domes	Richton, Cypress Creek Domes
			Louisiana Salt Domes	Vacherie Dome
	Crystalline rocks	Lake Superior Region	—	—
		Appalachian Region	—	—
	Argillaceous rocks	—	—	—
Land use (DOE Land)	Basalt	Not applicable	Hanford Site	Cold Creek Syncline
	Various (including tuff)	Not applicable	Nevada Test Site	—
Geohydrologic systems	Various	Basin and Range Province	—	—
National systems screenings	Various	—	—	—

Source: Modified from DOE (1982).

Table 2-3. Waste handling and isolation activities: Countries with major R&D programs.

	Transportation	Interim Storage	Shallow Land Burial	Geologic Isolation	Seabed Isolation	Airborne Waste Immobilization	Safety/Risk Analysis
Austria	X			X			X
Belgium				XX		X	X
Canada		XX		XX		X	X
Denmark				X			
France	XX	X	X	XX	XX	X	XX
Federal Republic of Germany	XX			XX		XX	XX
India				XX			XX
Ispra				X			X
Italy				XX			XX
Japan	XX			X	XX	X	X
Netherlands				X			X
Spain		X		X			
United Kingdom	X			X	XX	XX	XX
USSR	X		X	X			
Sweden		XX		XX			XX

Source: Modified from DOE (1982).

Key: X = R&D activities underway; XX = cooperative program with the United States.

International Programs

Nuclear wastes are an international problem. At present, there are 35
countries with one or more nuclear power plants either in operation or slated
for operational start-up by 1984, and all of these generate nuclear wastes.
International participation at meetings dealing with nuclear wastes in this
country reflect the keen interest and awareness of the efforts of the United
States to deal with nuclear wastes. The nuclear energy waste program is
heavily involved with increasing international cooperation among the nations
with nuclear capability. Some of these efforts are: (1) to minimize waste
management research and development expenditures of participating nations
by eliminating duplications of test facilities and experimental programs; (2)
to help establish an internationally accepted technological basis for waste
conditioning processes and waste disposal criteria and methods; (3) to
contribute to the base for systems evaluations; and (4) to improve plans for
U.S. research and development programs.

Many foreign waste management R&D programs duplicate or complement
research in the United States. Some of these efforts are summarized in Table
2-3. Studies of basalt, salt, granite, and especially shale in countries outside
the United States will help the United States program. Commercial waste is
not reprocessed in the United States, but defense wastes are routinely
processed to separate plutonium. Hence the HLW reprocessing programs in
Germany (FRG), France, Belgium, the United Kingdom, India, and the
U.S.S.R. are of great interest to the United States. Some of these efforts are
presented in Table 2-4. International study of waste forms is also going on, as
are programs dealing with airborne nuclear wastes, including tritium,
krypton, iodine, and carbon-14.

In general, however, many of the LLW are simply dumped into the oceans
or stored in mines, disposal methods not in favor in the United States at the
present time.

Significant studies, of use to United States work, has been conducted in the
United Kingdom and in Belgium. Combustible waste treatment studies in the
Federal Republic of Germany, Japan, Belgium, and the United Kingdom are
also of extreme value to the nuclear energy waste program.

Transportation studies in Canada, Japan, the United Kingdom, Sweden,
France, and the Federal Republic of Germany have been very helpful to
United States programs, and are in compliance with the basic standards and
model regulations of the International Atomic Energy Agency (IAEA).

Finally, subseabed disposal (see Chapter 8), if it is to be a viable nuclear
waste disposal alternative, will demand broad international agreement and
cooperation. Participation by the United States, Canada, France, Japan, the
United Kingdom, and The Netherlands is occurring at present under the
auspices of the Nuclear Energy Agency (NEA) of the Organization for
Economic Cooperation and Development (OECD). Some of the international programs and agreements are listed in Tables 2-5 and 2-6.

Table 2-4. Waste conditioning activities: Countries and agencies with major R&D programs.

	HLW Conditioning	HLW Form Characterization	Spend Fuel Packaging	LLRW and Intermediate Level Waste Conditioning	TRU Waste Conditioning
Australia	XX	XX			XX
Belgium	X	XX		XX	
Canada	X	XX	XX	X	
France	X	XX		X	X
Federal Republic of Germany	XX	XX		XX	XX
India	X			X	X
Ispra		XX			X
Italy	X	X		X	
Japan	X	X		X	X
USSR	X	X		X	
Sweden	X	XX	XX	X	
United Kingdom	X	XX	X	X	XX

Source: Modified from DOE (miscellaneous reports).
Key: X = R&D activities underway; XX = cooperative U.S. interest underway.

Table 2-5. Geologic waste isolation programs: Foreign countries.

Nation	Waste Type	Formations of Interest	Current Studies	Progress (Milestones)
Austria	Spent fuel, miscellaneous	Crystalline rock	Site evaluation, granite, safety	—
Belgium	HLW, non-HLW	Clay beds	Site evaluation, clay bed properties, *in situ* tunnel studies	Pilot repository at Mol for alpha and non-HLW
Canada	SURF, HLW, other	Bedded salt, crystalline rocks	Site evaluation, granite test site, drilling, safety assessment, repository design	Commercial repository: site—1982 demonstration—1990 start-up—2000
Denmark	Miscellaneous	Salt	Site characterization	—
France	Alpha wastes	Salt, crystalline rocks	Geologic survey, safety assessment, Oklo, repository design, site characterization	Pilot plant repository—1985
Germany (DRG)	Non-HLW	Salt	Bartheleben studies	—
Germany (FRG)	Spent HTGR fuel, LLW, ILW	Salt	Asse studies—safety assessment, repository design, *in situ* tests, engineering studies	Commercial repository at Gorlebed—late 1980s

Country	Waste type	Rock	Activities	Schedule
India	HLW, other	Igneous rocks, sediments	Site evaluation, rock properties	—
Ireland	Non-HLW	—	Geologic survey	—
Italy	HLW	Clay, salt	Geologic survey, *in situ* studies in clay, safety assessment, site studies	Pilot plant repository—mid-1980s
Japan	HLW	Several rocks	Geologic survey, safety assessment	—
The Netherlands	Non-HLW	Salt	Geologic survey, safety assessment, site study and design	—
Spain	HLW, non-HLW	Shale, salt	Old U mines, site evaluation	Pilot plant repository—mid-1980s
Sweden	SURF, HLW, non-HLW	Granite	Field tests(Stripa), safety assessment, design, *in situ* tests(Studsvik)	Pilot Plant repository—mid-1980s
Switzerland	HLW, non-HLW	Granite	Site evaluation, safety assessment, HLW test site	—
United Kingdom	Miscellaneous	Clay, crystalline rocks	Geologic survey, safety assessment, site evaluation	Commercial repository—2000
USSR	Non-HLW	Varied	Direct injection underground	—

Source: Modified from DOE (miscellaneous reports).

Table 2–6. Bilateral agreements with the United States.

Country	Date Concluded	Principal Subject
Sweden	July 1, 1977	Radioactive waste storage Stripa Mine test
Canada	Sept. 8, 1976	Radioactive waste management and system analysis of heavy water reactors
Federal Republic of Germany	Dec. 20, 1974	Radioactive waste management
United Kingdom	Sept. 20, 1976	Fast breeder reactors
Belgium	Pending	Radioactive waste management
Japan	Jan. 31, 1979	Fast breeder reactors
Australia	Pending	HLW immobilization, mine/mill tailings nuclide migration

Source: Modified from DOE (miscellaneous reports).

The member states in NEA are Australia, Austria, Belgium, Canada, Denmark, Finland, France, Federal Republic of Germany, Greece, Iceland, Ireland, Italy, Japan, Luxembourg, The Netherlands, New Zealand, Norway, Portugal, Spain, Sweden, Switzerland, Turkey, United Kingdom, United States, and Yugoslavia (note: New Zealand and Yugoslavia are special-status members).

The Commission of the European Communities (CEC) also conducts detailed studies of importance to nuclear waste disposal problems. Salt deposits are being studied by the Federal Republic of Germany and in The Netherlands, crystalline rocks in France and the United Kingdom, and argillaceous rocks in Italy and Belgium. The CED funds and operates the joint research centers at Ispra, Italy; Karlsruhe, Germany; Petten, The Netherlands; and Geel, Belgium.

The IAEA, an autonomous member of the United Nations, is charged with the mission of promoting peaceful uses of atomic energy. It is concerned with many aspects of nuclear power development, including information exchange, and the sponsoring of symposia covering many aspects of nuclear power, the overall fuel cycle, uranium deposits, and related topics. It is also charged with inventory of nuclear power plants of the signatory nations of the Treaty of the Non-Proliferation for Nuclear Weapons. They provide technical reports dealing with release of radionuclides into the environment as well as monitor actual releases from nuclear facilities. More recent is the formation of an Advisory Group on Radioactive Waste Disposal into Geologic Formations (formed in 1978). This group is developing guidelines for safe nuclear waste disposal underground, including the common options being investigated in many countries.

It is of interest to note how other nations plan to deal with the disposal of nuclear wastes. France has been investigating ways to store and treat radioactive wastes since the 1950s. From 1969 to 1973 the PIVER pilot plant was used to encapsulate HLW into borosilicate glass; this plant has been replaced by the AVM plant, in Marcoule, which started operation in 1978. France is considering several options to deal with the processed waste: surface storage of 150 years to allow the temperature to cool sufficiently to allow essentially "cold," compact burial, or a 30-year cooling period and burial of the waste packages farther apart (versus the closer-packed 150-year cooled packages), or even cooling the waste-containing glass *in situ* in a facility that would later become a repository facility. Rocks under investigation by the French include salt and crystalline rocks.

The Federal Republic of Germany has been examining the salt domes of the Gorleben District for several years. Specifically, the Aase Salt Mine was in use from 1967 to 1978 for pilot studies involving low-level and intermediate-level radioactive wastes. At present, interim storage plants are to be separate from a demonstration reprocessing plant, which in turn will be separate from a plutonium refabrication site, and all of these will be separate from the eventual disposal site. Salt is still the leading candidate in the Federal Republic of Germany. Further, there are plans to encapsulate the HLW into borosilicate glass after a 3- to 5-year cooling period.

Studies in Sweden, especially at the Stripa iron mine, have been carried out for several years. This old iron mine is sited in granite and has been used jointly by the Swedish Government and the United States, Finland, Japan, Canada, France, and Switzerland. At present, however, the Swedish Government apparently plans for their fuel reprocessing to be carried out in France. Any HLW material returned to Sweden will first be stored in air-cooled facilities and then buried deep in crystalline rocks.

In the United Kingdom a 50-year surface storage of HLW is presently planned (Feates and Richards, 1983). The liquid wastes, as acidic solutions, are stored in stainless-steel containers, primarily at Sellafield and Dounreay reprocessing plants. The United Kingdom plans to use the French AVM system for encapsulating the HLW. Ultimate disposal will either be on the ocean floor or in deep geologic repositories. Granitic plutons and crystalline metamorphic rocks are both under consideration for repository use.

Japan has a reprocessing plant that was started in 1977. Some liquid wastes are stored in stainless-steel containers, and some fuel reprocessing is being carried out in France. In addition to subseabed disposal, the Japanese Government is actively investigating granitic rocks, with some emphasis on zeolitized volcanic rocks. Despite the high seismicity in Japan, some of these rocks appear suitable for radwaste storage.

Canada is continuing its practice of surface and near-surface storage of irradiated fuel assemblies. Crystalline rocks in the Canadian Shield, mainly granites, are under investigation for the ultimate disposal of radioactive materials, presumably without reprocessing (note: at least for 75 years). Fyfe (1983) has recently discussed some of the important geologic work being done to evaluate granitic rocks in the Shield. New geophysical methods allow detection of shear zones at depth in some plutons, apparently formed in response to unloading after burial by glacial ice. Knowledge of these zones allows avoidance of rocks in which such shear zones can allow deep rock to be penetrated by meteoric water.

The Soviet Union has several plans underway. Both borosilicate glass and phosphates are being evaluated for their use in encapsulating HLW. Surface storage of liquid wastes in stainless-steel tanks is practiced at present, and a plant to vitrify many of these wastes is planned for the 1980s (Spitsyn et al., 1982). Some of the solidified waste may be stored in near-surface facilities, while several deep geologic disposal concepts continue to be investigated. Disposal of liquid wastes in deep in geologic formations (Spitsyn and Balukova, 1979) has been proposed, and subseabed disposal is also being considered as an alternative to land disposal.

Several countries in Europe, including Sweden, The Netherlands, Switzerland, and Belgium currently use fuel reprocessing in France (note: Belgium also stores liquid wastes in stainless-steel vats). Fuel from Finland and the USSR may also eventually be reprocessed in France. Since 1970, Italy has been reprocessing radioactive wastes in their EUREX pilot plant, although some wastes are stored as liquids. Italy is also investigating borosilicate glass and phosphatic glass as waste forms and impermeable rocks for waste isolation (Bresetti et al., 1980). In Finland, cyrstalline rocks are under consideration for radwaste storage, shales and other clay-rich rocks are being considered in Belgium, and evaporites are being considered in both The Netherlands and Switzerland.

Of the other 35 countires with nuclear power plants (including those to come on-line by the end of 1984), only India has active plans to deal with the wastes. India practices storage of acidic, liquid wastes in stainless-steel tanks at present, and, after encapsulating the waste in glass, disposal may be in deep igneous or impermeable sedimentary rocks. Subseabed disposal is also an alternative.

Concluding Remarks

There are obviously many plans under consideration by various countries for the safe handling and disposal of radioactive wastes. The reader is referred to the symposia on the Scientific Basis for Nuclear Waste management proceedings, which have been published since 1978 (see McCarthy, 1979b; Northrup, 1980; Moore, 1981; Topp, 1982; Lutze, 1982; and Brookins, 1983). In these references are the various overviews on nuclear waste management, details of waste form properties and processes, materials for canisters, overpack, and engineered backfill, and descriptions and evaluations of repository candidates.

CHAPTER 3,

Natural Radiation in the Geochemical Environment

Introduction

Natural radiation is primarily from U and Th and their radioactive daughters, and from K and Rb. In addition, there is minor radiation from ^{14}C and such species as ^{147}Sm, ^{176}Lu, and ^{187}Re. There is also a very small amount of radiation (less than 0.04 mrem/yr) from radioactive species formed by atmospheric testing of nuclear weapons. The radiation in rocks, however, is basically due to U, Th, actinide daughters of U and Th, and K and Rb. Only these sources will be discussed in this chapter.

Geochemical Environment

Natural radiation is everywhere. The principal radioisotopes that affect animals are ^{14}C, ^{40}K, ^{87}Rb, ^{210}Po, ^{222}Rn, ^{226}Ra, ^{228}Ra; these isotopes yield doses of approximately 25 mrem/yr to the whole body, with 47 mrem/yr to endosteal cells (bone) and 24 mrem/yr to bone marrow (Gera, 1975). The dose from minor radioisotopes, such as ^{3}H, and even ^{235}U, ^{238}U, and ^{232}Th, is only a few hundredths of a mrem/yr. The data for the principal radioisotopes are shown in Table 3-1. ^{40}K is the main contributor to the whole-body dose, while ^{210}Po is the main contributor of doses to endosteal cells.

The important radioactive species in the natural environment are ^{40}K, ^{235}U, ^{238}U, and ^{232}Th, as well as ^{226}Ra. The average U and Th contents in geologic materials are shown in Table 3-2. Some placers are greatly enriched in Th, for example, which is a reflection of the insolubility of Th-bearing phases, such as monazite, relative to more soluble U phases. Phosphatic rocks and carbonaceous and bituminous shales, on the other hand, are richer in U relative to Th due to the ease with which soluble U(VI) species, after

Table 3–1. Estimated average annual internal radiation doses from natural radioactivity in the United States.

Radionuclide	Annual Doses (mrem/person)		
	Whole body	Endosteal (bone)	Bone marrow
^{3}H	0.004	0.004	0.004
^{14}C	1.0	1.6	1.6
^{40}K	17	8	15
^{87}Rb	0.6	0.4	0.6
^{210}Po	3	21	3
^{222}Rn	3	3	3
^{226}Ra	—	6.1	0.3
^{228}Ra	—	7	0.3
Total	25	47	24

Source: From Gera (1975).
Note: the contribution from other radionuclides is so small that they are not listed.

separation from Th during weathering, are removed as a result of reduction to U(IV) in chemically favorble environments.

In many terrestrial materials U and Th are found in late-stage minerals and as minute U- and/or Th-bearing phases along grain boundaries, crystal defects, and other similar sites. When exposed to percolating groundwaters, the U(IV) is oxidized to soluble U(VI) and transported as UO_2^{2+}, UO_2 (CO_3), $UO_2 (CO_3)_2^{2-}$, or $UO_2 (CO_3)_3^{4-}$, whereas Th(IV) is unaffected by this oxidation and remains at or close to the site of uranium removal. Should Th be transported in a colloidal state or as particulate matter, exposure to water will result in precipitation (i.e., the solubility product constant for Th(OH)$_4$ is approximately 10^{-35}), which is so efficient that even ^{230}Th formed from decay from parent ^{238}U will be removed from solution under many conditions. The residence time for Th in seawater is only approximately 300–350 years.

Uranium, however, has a residence time in seawater of some 500,000 years. Precipitation of uranium can occur easily by reduction to insoluble U(IV). Thus environments in which carbonaceous and bituminous shales form are particularly favorable for U removal by reduction of U(VI) to U(IV). Lignites are also enriched in uranium, some as U(IV), as expected, and some as U(VI). U(VI) can easily be scavenged by coaly material without reduction, thus accounting for part of the high U content of such rocks. In the case of phosphatic rocks, coprecipitation of U^{4+} with Ca^{2+} is likely due to their very similar ionic radii. The exact mechanism for reduction of U(VI) to U(IV) is not known, however. The solubility of uranium is highest under very acidic and very basic pH; at pH roughly from 4 to 10 uranium is commonly insoluble due to U(IV) species. This is shown in more detail on the Eh–pH diagram for U species (Fig. 11-1).

Table 3–2. Uranium and thorium concentrations in rocks.

Type of Rock	Concentration (ppm)		
	U	Th	Th/U
Igneous rocks			
Acidic rocks (granites, etc.)			
North America	4	13	3.2
World	3.5	18	5.1
Intermediate Rocks			
North America	2.6	10	3.8
World	1.8	7	3.9
Basic Rocks			
North America	0.8	5	6.2
World	0.8	3	3.7
Average for all igneous rocks	3	10	3.3
Sedimentary rocks			
Placers (with U,Th minerals)	2	60	30
Sandstones	0.5–1	2–5	—
Shales (average)	3.7	12	3.2
Grey–green shales, USA	3.2	13.1	4.1
Bentonites, USA	5	24	4.8
Bauxites	9.3	53.1	5.7
Residual clays	1.8	13	7.2
Shales, USSR	4.1	11	2.6
C-bearing shales	50–80(+)	—	—
Limestones	1.3	1.1	0.9
Limestones, USA	2.2	1.1	0.5
Limestones, USSR	2.1	2.4	1.1
Phosphate rocks	2.1(+)	—	—
Lignites, coals	to 10,000	—	—

Source: From Gera (1975).

As expected, the Th/U ratio in nature varies widely. In the case of rocks from which U has been removed, high Th/U ratios result; similarly, in rocks precipitated under chemically reducing environments far from suspected source rocks, U is enriched over Th. Thus above-average Th/U ratios are observed in continental sediments, especially in laterites and other residual deposits. Low Th/U ratios are found in chemically precipitated marine sedimentary rocks, such as evaporites and limestones, and extremely low Th/U ratios are found in carbonaceous rocks.

Radium is most commonly present as ^{226}Ra, which is found where uranium occurs. The Ra/U ratio at radioactive equilibrium is about 3.6×10^{-7}. Ra is the most hazardous radionuclide in the natural environment, and is present in high-level waste some 50,000 years after storage. This is due to its extreme radioactivity of about 1 Ci/g. Radium and uranium may be enriched in natural waters depending on transfer of the elements into water, stability of soluble species, and effectiveness of separation from solution. Separation of Ra from U may occur in rocks and minerals as a result of incipient leaching. This is commonly thought to be due to the fact that Ra^{2+} is a large cation (1.4 Å) relative to U^{4+} (1.04 Å) and is metastable in the structure of such minerals as uraninite and coffinite. Loss of Ra probably occurs by diffusion in the original host mineral to the grain surface and by diffusion through the water layer adsorbed on the grain surface, and hence into solution. Ra^{2+} is moderately soluble in natural waters, although a high SO_4^{2-} content will favor its removal as $RaSO_4$ or mix crystals of $(Ba, Ra)SO_4$, or for high-CO_2 waters, removal as $(M, Ra)CO_3$. Ra in solution is not strongly dependent on anionic species.

Uranium is strongly influenced by the presence of anions, especially dissolved CO_2 species. In general, the higher the dissolved CO_2 content the greater the solubility of U(VI). This can be seen by inspection of Fig. 11-1a, in which the increase of fields of aqueous species with increasing dissolved CO_2 activity is shown. Both U and Ra are enriched in low- and high-pH waters, hence such waters are commonly more radioactive than waters at intermediate pH.

Removal of Ra and U can occur by hydrolysis, adsorption, breakdown of complex ions, reduction of U(VI) to U(IV), and formation of insoluble salts (see Gera, 1975). For U, formation of U(VI)-bearing minerals, such as cornotite, tyuyamunite, and uranophane, is an effective way of scavenging U without reduction. For Ra, coprecipitation with Ba salts is common, and some coprecipitation with Ca, Mg, and even Fe and Mn occurs.

Groundwaters are typically richer in both Ra and U relative to surface waters due to dilution of the latter from meteoric sources. The abundance of Ra and U in natural waters is shown in Table 3-3. Many natural waters are depleted in Ra due to its incorporation into sediments by parent ^{230}Th. The ^{230}Th is insoluble [as $Th(OH)_4$], and thus ^{226}Ra formed is trapped in the sediments. Yet some is leached and returned to seawater, accounting for the slightly low Ra/^{230}Th ratio commonly noted in seabottom sediments (Gera, 1975).

The Th content of natural waters, as expected, is markedly low. The average Th content of seawater falls between 2×10^{-8} and 5×10^{-8} ppm (Goldberg, 1965; Germanov et al., 1958). The U/Th ratio in seawater is about 100 while the Th/U ratio in seabottom sediment is roughly 4, essentially identical to the average value for continental rocks.

Some closed hydrologic basins contain very high concentrations of U (see Vogler and Brookins, 1982), which is the result of moderately high U source rocks coupled with a lack of an efficient mechanism for U removal in the basins. Ra content in the same basins is variable but usually high.

The amount of naturally occurring Np and Pu is extremely small; that which is present is a result of neutron-induced reactions in uranium minerals. The behavior of these elements in rocks cannot be investigated directly, but data from the Oklo natural reactor (see Chapter 11) indicate that both Np and Pu are much more immobile than U (see Figs. 11-5 and 11-6, for example).

The amount of alpha-emitters that can safely be ingested is moderately well agreed on (Gera, 1975). The limit to intake of ^{226}Ra and U is set at 26 pCi/day and 3800 μg/day, respectively, while the actual average intake is 1–2 pCi/day for ^{226}Ra and 1.3 μg/day for uranium in several areas of the United States. Radium, therefore, poses a far greater risk than uranium. Gera (1975) discusses the mechanisms by which Ra, U, and Pu may be increasingly concentrated by the soil–plant–animal chain. He notes that the potential hazard from heavy radionuclides is low due in part to: (1) strong adsorption on soil particles, (2) low solubility, (3) low uptake by plants, and (4) marked discrimination factors against transfer of these elements from one trophic level to another. Although Ra would be expected to be biologically more active than U (due to its similarity to Sr and Ba), no significant difference in soil–plant and plant–animal concentration factors are noted.

Aquatic ecosystems are more sensitive to the heavy metals, and Ra and U are commonly richer in aquatic life forms relative to water. Gera (1975) concludes that heavy-metal concentration in aquatic life is significant, although variable.

Nuclear atmospheric testing by several countries has led to contamination of the natural environment by many fission products and actinides and their daughters. This results in an opportunity to study such elements in the natural environment, although admittedly, the time available for chemical and other reactions is indeed very short. Studies of plutonium, americium, and ^{137}Cs have been carried out by numerous workers (see summary in Beasley et al., 1982). Basically, the behavior of these elements is consistent with theoretical predictions as to how they should be distributed in natural media, and no distribution pattern indicates some drastically unknown chemical behavior. This is fortunate, as it, in turn, indicates that models dealing with the fate of fission and actinide elements possibly released from a hypothetically breached radioactive waste canister can be better assessed based on the input of the Pu, Am, Cs, and related studies.

Airborne radioactive particles from natural sources are relatively few. Dispersion of ^{222}Rn is well known, and all other radioisotopes are not efficiently transported as airborne particles. Even ^{222}Rn is so quickly diluted from any point source that background values for atmospheric Rn are reached within tens of meters from the source. Interestingly, Rn gas is commonly observed in limestone caves, tunnels, and other underground sites, although especially in limestone. In fact, Wilkening and Watkins

Table 3–3. Content of uranium and radium in natural waters.

Type of Waters	Natural Conditions	Radium (g/L)		
		Minimum	Maximum	Mean
Surface waters	Oceans and seas	8.0×10^{-14}	4.5×10^{-11}	1.0×10^{-13}
	Lakes	1.0×10^{-13}	8.0×10^{-12}	1.0×10^{-12}
	Rivers	1.0×10^{-13}	4.0×10^{-12}	2.0×10^{-13}
Waters of sedimentary rocks	Zone of intensive water circulation	1.0×10^{-13}	6.0×10^{-12}	2.0×10^{-12}
	Zone of highly impeded water circulation	1.0×10^{-11}	1.0×10^{-8}	3.0×10^{-10}
Waters of magmatic acid rocks	Zone of intensive water circulation (waters of the weathering shell)	1.0×10^{-12}	7.0×10^{-12}	2.0×10^{-12}
	Zone of impeded water circulation (waters of deep tectonic fissures)	2.0×10^{-12}	9.0×10^{-12}	4.0×10^{-12}
Waters of uranium deposits	Zone of intensive water circulation (water of the oxidation zone)	8.0×10^{-12}	2.0×10^{-9}	8.0×10^{-11}
	Zone of impeded water circulation (waters of the reduction zone)	1.0×10^{-11}	8.0×10^{-10}	6.0×10^{-11}

Source: Modified from Tokarev and Scherbakov, (1956); AEC-tr-4100 (1960).

(1976) have shown that the accumulation of ^{222}Rn in the limestone of Carlsbad Caverns, New Mexico can be so severe that it poses a serious threat to full-time workers involved with the underground operations.

Background Radiation

Areas of high background radiation have been investigated with some regularity. Two areas in Brazil, one rich in monazite placer sands and the other an alkaline rock complex, yield extremely high Th and U. Monazite-

Uranium (g/L)			
Minimum	Maximum	Mean	Ra/U
3.6×10^{-8}	5.0×10^{-6}	3.0×10^{-6}	3×10^{-8}
2.0×10^{-7}	4.0×10^{-2}	8.0×10^{-6}	1×10^{-7}
2.0×10^{-8}	5.0×10^{-5}	6.0×10^{-7}	3×10^{-7}
2.0×10^{-7}	8.0×10^{-6}	5.0×10^{-6}	5×10^{-7}
2.0×10^{-8}	6.0×10^{-6}	2.0×10^{-7}	1×10^{-3}
2.0×10^{-7}	3.0×10^{-5}	7.0×10^{-6}	1×10^{-6}
2.0×10^{-7}	8.0×10^{-6}	4.0×10^{-6}	2×10^{-6}
5.0×10^{-5}	9.0×10^{-2}	6.0×10^{-4}	1×10^{-7}
2.0×10^{-6}	3.0×10^{-5}	8.0×10^{-6}	1×10^{-5}

bearing sands in the Kerala coastal district, India, are also well known for extremely high natural radiation. Monitoring of individuals in these areas shows few differences between high-dose and moderate-dose groups. In the United States, high radioactivity is noted at uranium mill tailings piles (see Chapter 10).

More recently, the People's Republic of China (1980) has reported on an in-depth study of two groups, each of about 70,000 people, in Yanjiang County, Guangdong Province. One group was from an area underlain by monazite-bearing granite in which the background radioactivity was three or more times as great at the other group. The group from the high-radioactivity

Table 3–4. Annual estimated external gamma whole-body doses from natural terrestrial radioactivity (mrem/person).

Political Unit	Average Annual Dose	Political Unit	Average Annual Dose
Alabama	70	New Jersey	60
Alaska	60[a]	New Mexico	70
Arizona	60[a]	New York	65
Arkansas	75	North Carolina	75
California	50	North Dakota	60[a]
Colorado	105	Ohio	65
Connecticut	60	Oklahoma	60
Delaware	60[a]	Oregon	60[a]
Florida	60[a]	Pennsylvania	55
Georgia	60[a]	Rhode Island	65
Hawaii	60[a]	South Carolina	70
Idaho	60[a]	South Dakota	115
Illinois	65	Tennessee	70
Indiana	55	Texas	30
Iowa	60	Utah	40
Kansas	60[a]	Vermont	45
Kentucky	60[a]	Virginia	55
Louisiana	40	Washington	60[a]
Maine	75	West Virginia	60[a]
Maryland	55	Wisconsin	55
Massachusetts	75	Wyoming	90
Michigan	60[a]	Canal Zone	60[a]
Minnesota	70	Guam	60[a]
Mississippi	65	Puerto Rico	60[a]
Missouri	60[a]	Samoa	60[a]
Montana	60[a]	Virgin Islands	60[a]
Nebraska	55	District of Columbia	55
Nevada	40	Others	60[a]
New Hampshire	65		
		Total United States average annual dose	60

Source: Modified from DOE (1980).

[a]Assumed to be equal to the United States average.

area received about 25 mrem/month and the control group only about 9 mrem/month. The areas were chosen because, in part, the inhabitants could trace their roots through three to five generations. There was no significant difference between the health of the two groups; in fact, the group exposed to above normal radiation was more healthy.

The natural background gamma radiation in the United States is quite variable. In Tables 3-4 and 3-5 the estimated annual external gamma

Table 3–5. Estimated annual cosmic-ray whole-body doses (mrem/person).

Political Unit	Average Annual Dose	Political Unit	Average Annual Dose
Alabama	40	New Jersey	40
Alaska	45	New Mexico	105
Arizona	60	New York	45
Arkansas	40	North Carolina	45
California	40	North Dakota	60
Colorado	120	Ohio	50
Connecticut	40	Oklahoma	50
Delaware	40	Oregon	50
Florida	35	Pennsylvania	45
Georgia	40	Rhode Island	40
Hawaii	30	South Carolina	40
Idaho	85	South Dakota	70
Illinois	45	Tennessee	45
Indiana	45	Texas	45
Iowa	50	Utah	115
Kansas	50	Vermont	50
Kentucky	45	Virginia	45
Louisiana	35	Washington	50
Maine	50	West Virginia	50
Maryland	40	Wisconsin	50
Massachusetts	40	Wyoming	130
Michigan	50	Canal Zone	30
Minnesota	55	Guam	35
Mississippi	40	Puerto Rico	30
Missouri	45	Samoa	30
Montana	90	Virgin Islands	30
Nebraska	75	District of Columbia	40
Nevada	85		
New Hampshire	45	Total United States average annual dose	45

Source: Modified from DOE (1980).

radiation whole-body dose due to natural terrestrial and cosmic ray sources are shown. In Colorado, for example, the combined annual terrestrial and cosmic ray dose is about 225 mrem/person, while in Louisiana the combined total is only 75 mrem/person. The highest terrestrial radioactivity is noted for Colorado and South Dakota (Table 3-4). The U–Th-bearing granitic rocks account for the elevated figure in Colorado, while U-bearing coaly rocks account for the high values in South Dakota. The average value from terrestrial sources in the United States is only 60 mrem/person. States with high mean elevation, Colorado, Wyoming, New Mexico, Utah, receive the highest cosmic ray dose (Table 3-5), and values for these four states range from 105 (New Mexico) to 130 (Wyoming), far above the average value of 45 mrem/person for the United States.

There is no evidence for any obvious health effect due to the high natural radioactivity in the states mentioned above. As pointed out by Sauer and Brand (1971), the highest incidence of malignant neoplasms is in metropolitan centers nationwide and lowest in nonmetropolitan areas. Further, the highest incidence of malignant neoplasms by area is found in the Louisiana coastal area, the Southeast coastal area, and the northeast Atlantic coast from New Jersey to Boston. The lowest incidence of malignant neoplasms is reported for the Great Plains and Rocky Mountain States. Thus Wyoming, Colorado, Utah, and much of New Mexico and South Dakota are in the lowest octile for malignant-neoplasm deaths, although these states receive very high radiation from natural sources (see Tables 3-4 and 3-5). These observations are consistent with the in-depth report on the effect of low-level ionizing radiation by the People's Republic of China (see discussion elsewhere in this chapter), where no adverse health and disease effects due to low-level radiation has been found.

Rubidium occurs in nature as Rb^+ ion (radius $= 1.47$ Å); it is always camouflaged by K^+ and forms no Rb minerals. The ratio of K/Rb in igneous rocks is commonly close to 200, although syenitic rocks commonly exhibit much higher ratios (e.g., 500–650). Thus K-enriched minerals, such as biotite, muscovite, and K-feldspars, will contain most of a rock's Rb, with only minor amounts in K-poor minerals such as plagioclase, most amphiboles, pyroxenes, olivine, and magnetite. The importance of Rb distribution in rocks for geochronologic purposes is discussed in Chapter 5. Radiation from ^{87}Rb is very slight. The very long half-life (48.8×10^9 yr) for the decay results in essentially negligible radiation from any ^{87}Rb source. In sedimentary rocks Rb behavior is predictable. In shales, argillaceous sandstones, and other nonevaporitic sedimentary rocks, the K/Rb ratio will vary as a function of detrital phases and amount of authigenic phases. In some clay minerals, for example, Rb is preferentially enriched relative to K due to ion exchange and other processes such that low K/Rb ratios result. In evaporitic rocks, K-rich phases (e.g., sylvite, carnallite, polyhalite, leonite, langbeinite) will also contain virtually all the Rb present, and such minerals may be well suited for Rb–Sr geochronologic work (see Chapter 5).

Potassium occurs in nature as K^+ and makes up 2.8 wt.% of the earth's crust. It is widely distributed in nature, and enriched in the common rock forming micas, feldspars, and other minerals. Consequently, the amount of total radioactivity due to ^{40}K decay is large for the entire crust, and, along with U and Th, simple measuring devices allow one to quantitatively measure the K–U–Th abundance in rocks by simple gamma spectrometry.

Yet in the evaporitic rocks the K/Rb ratio is much higher (near 3500 to 5000) because the K/Rb ratio in seawater is only 5000. This results from the fact that Rb^+ is more effectively scavenged by clay minerals and other media in river and deltaic environments relative to K^+.

Table 3–6. Distribution of U, Th, K, and Rb in the crust of the earth (data in ppm except for K).

Element	Crust	Basalt	Granite	Shale	Seawater
Uranium	2.7	0.5	5.0	3.5	0.0032
Thorium	8.5	1.5	20.0	12.0	10^{-5}
Potassium	2.1%	0.8%	3.2%	2.5%	380 ppm
Rubidium	90	30	150	140	0.12

The distribution of U, Th, K, and Rb for major rocks of the eath's crust is given in Table 3-6.

Some Comments

The natural radiation from U, Th, Ra, K, Rb, and other species in rocks and soils can be quite high, and, if combined with radiation from cosmic rays, results in an area of high background radiation (e.g., Colorado). Basically, however, even in areas of very high background radiation, such as the study areas in China, adverse health effects may not be present.

The studies of natural radiation conducted over the last two decades are important for many reasons. One obvious product of this research is the identification of various areas as "high" or "low" background for radioactive species, and this, in turn, is of importance for siting studies not only for possible repositories but for other facilities as well. Of even greater importance is that the reasons for the high or low background radiation are now known. In short, combined geologic and geochemical characterization studies of any particular area easily explain its natural radiation, and this ability to understand various media make the job of selection of repository sites that much easier.

CHAPTER 4

Types of Radioactive Wastes and the Multibarrier System

Introduction

In this chapter the various types of radioactive wastes are briefly described. Specifically, high-level waste (HLW), spent uranium rod fuel (SURF), cladding wastes (CW), transuranic wastes (TRU) and low-level wastes (LLW) are introduced in terms of their radioactive components, sites of generation, and some other features.

The concept of multibarriers is also introduced in this chapter, primarily because some brief discussion is needed before the geochronologic and geologic discussions that follow later in the book. Details of the system are found in later chapters (see Chapters 7, 10, 11, 12, and 13).

Radioactive Wastes

Spent unreprocessed fuel (SURF) consists of reactor fuel elements removed from the core after a set time for the generation of energy and the buildup of fission products. SURF contains both fertile and fissile actinides, and fission products. High heat is generated by the decay and radioactivity. Present plans call for storage of SURF at temporary reactor sites for 10 years prior to shipping to repositories.

Some of the characteristics of spent fuel include: (1) preset form and geometry; (2) greater actinide content than equivalent high-level waste (HLW); (3) a long-lived thermal power potential; (4) quantities of fissile actinides great enough to cause concern about criticality and safeguards; (5) fission product gases free in the fuel assembly and bound in the UO_2 matrix; (6) resistance to leaching and dissolution; (7) a resource value contained in the package materials and, for the future, actinides (see Carr *et al.*, 1979). These are discussed below.

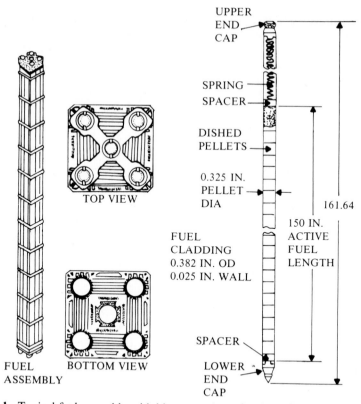

Fig. 4-1. Typical fuel assembly with blown-up view of a single fuel rod. *Source*: DOE.

Most of the spent-fuel assemblies will come from boiling-water reactors and pressurized-water reactors, and will consist of uranium dioxide pellets (UO_2) in sealed metal (Zircaloy, stainless-steel, inconel) tubes in a geometric grid array with end plates (see Fig. 4-1). Typical lengths are 12–15 ft with varying cross sections dependent on reactor model and core type. Important to discussion of the spent fuel is knowledge of its history in the reactor as well as on-site, post-reactor handling operations prior to shipment to a repository. During this time the fuel pellets can possibly be damaged structurally such that cracking may occur, providing pathways for leaching and surface alteration. Breaches between the fuel pellet and cladding may also occur. All of these raise the possibility for fission gas (krypton, xenon, etc.) release within the fuel assembly and increase the chance for possible exchange with the external surroundings of the fuel assembly. As a consequence of these possible effects, in addressing elemental migration from SURF, elements such as those in fission gases (possibly including iodine) must be treated as though they had already moved from within a fuel pellet to the surface. Thus, for example, if the diffusion of any such element to the surface of a pellet is on the order of 10^4 years, then to

Table 4–1. Biologically significant radionuclides in high-level waste (HLW).

Radionuclide	Half-Life	Activity (Ci/canister) (Time out of Reactor)			
		10 yr	10^2 yr	10^3 yr	10^4 yr
^{89}Sr	50.5 days	0	0	0	0
^{90}Sr	29 years	1.4×10^5	1.5×10^4	3.5×10^{-6}	0
^{90}Y	64 hours	1.4×10^5	1.5×10^4	3.5×10^{-6}	0
^{91}Y	58.6 days	0	0	0	0
^{93}Zr	9.5×10^5 years	4.3	4.3	4.3	4.3
^{93m}Nb	12 years	1.8	4.3	4.3	4.3
^{95}Zr	65.5 days	0	0	0	0
^{95}Nb	35.1 days	4.3×10^{-5}	0	0	0
^{99}Tc	2.13×10^5 years	3.2×10^1	3.2×10^1	3.2×10^1	3.2×10^1
^{106}Ru	369 days	1.3×10^3	0	0	0
^{106}Rh	2.18 hours	1.3×10^3	0	0	0
^{125}Sb	2.73 years	1.6×10^3	0	0	0
^{126}Sn	10^5 years	1.3	1.3	1.3	1.3
^{129}I	1.59×10^7 years	1.8×10^{-4}	1.8×10^{-4}	1.8×10^{-4}	1.8×10^{-4}
^{134}Cs	2.06 years	1.9×10^4	0	0	0
^{137}Cs	30.1 years	2.0×10^5	2.4×10^4	2.3×10^{-5}	0
^{144}Ce	284.4 days	3.5×10^2	0	0	0
^{147}Pm	2.62 years	1.8×10^4	8.1×10^{-7}	0	0
^{154}Eu	8.6 years	1.0×10^4	2.1×10^2	0	0
^{210}Pb	22.3 years	0	1.7×10^{-6}	1.6×10^{-4}	6.5×10^{-3}
^{210}Po	138.4 days	0	1.7×10^{-6}	1.6×10^{-4}	6.5×10^{-3}
^{226}Ra	1600 years	2.5×10^{-7}	2.6×10^{-6}	1.6×10^{-4}	6.5×10^{-3}
^{229}Th	7340 years	9.6×10^{-8}	1.7×10^{-6}	1.6×10^{-4}	1.3×10^{-2}
^{230}Th	7.7×10^4 years	4.9×10^{-5}	7.4×10^{-5}	8.8×10^{-4}	8.1×10^{-3}
^{231}Pa	3.25×10^4 years	5.7×10^{-5}	5.7×10^{-5}	6.0×10^{-5}	8.8×10^{-5}
^{233}U	1.58×10^5 years	3.4×10^{-5}	3.4×10^{-5}	3.5×10^{-3}	3.6×10^{-2}
^{237}Np	2.14×10^6 years	8.1×10^{-1}	8.1×10^{-1}	8.8×10^{-1}	8.8×10^{-1}
^{238}Pu	87.8 years	2.4×10^2	1.2×10^2	2.8×10^{-1}	0
^{239}Pu	2.439×10^4 years	3.8	3.8	4.7	9.6
^{240}Pu	6540 years	1.0×10^1	2.0×10^1	1.8×10^1	7.3
^{241}Pu	15 years	7.4×10^2	1.1×10^1	7.2×10^{-1}	3.4×10^{-1}
^{241}Am	433 years	4.1×10^2	3.8×10^2	8.1×10^1	3.4×10^{-1}
^{243}Am	7370 years	4.0×10^1	4.0×10^1	3.7×10^1	1.7×10^1
^{242}Cm	163 days	1.6×10^1	1.1×10^1	1.8×10^{-1}	0
^{244}Cm	17.9 years	4.0×10^3	1.3×10^2	0	0

Source: Data from NUREG (1979), Table A-1.

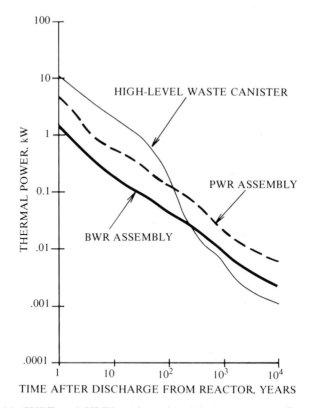

Fig. 4-2.(a) SURF and HLW canister thermal power curves. *Source:* Modified from Storch and Prince (1979).

be safe, it must be treated as though it had already diffused to the surface at or prior to the time of storage in a repository. Alternately, one can estimate the amount of any such element likely to be concentrated at the pellet surface relative to that left in the pellet and use this information as the basis of predicting elemental retention or migration.

The actinide content of spent fuel is very high. Carr *et al.* (1979) estimate that one-third of PWR and one-fourth of BWR fuel assemblies are taken from reactors each year and classified as spent fuel containing about 3% fission products as well as transuranic elements. The thermal output, radioactivity, and potential for criticality of the spent fuel are affected by the amount of fission products and transuranic elements present. In addition, the large amounts of actinides, many of which have half-lives ranging from millions to billions of years (see Table 4-1), necessitate repositories in which isolation will be maintained for much longer time than necessary for HLW. This results from the fact that the HLW contains mostly relatively short-lived isotopes or only small amounts of very long-lived radionuclides (i.e., ^{230}Th, ^{235}U, ^{238}U) such that the total radioactivity and heat output of the HLW will have decreased dramatically in a few hundred years, while the

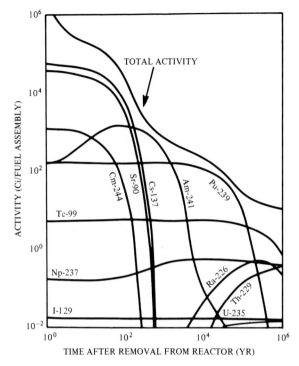

Fig. 4-2.(b) Radioactive decay of spent fuel from power water reactor. *Source:* Modified from NUREG (1979).

radioactivity of the SURF will remain high for hundreds of thousands of years (Fig. 4-2a, b).

The possibility of movement of fission gases arises from structural damage to fuel pellets or cladding. Gases could leak into defective sites along cracks or partly altered pellet surfaces or cracks. Thus only the cladding will prevent escape of krypton and xenon, and possibly some volatile iodine.

A view of emplacement of canistered spent fuel in a sleeved hole is shown in Fig. 4-3.

Cladding wastes (CW) result from the initial phases of reprocessing spent fuel. After the spent fuel has been stripped from the total fuel assembly and dissolved in acidic solution, the cladding material remains. The cladding is radioactive due to ^{60}Co and other radionuclides formed during the use of the fuel assembly in the reactor. The thermal power generated by CW is about 1–2% of the heat generated by reprocessed HLW after removal of uranium and plutonium (see Claibourne, 1976). The cladding varies in composition, but can be considered as a mixture of zircaloy, stainless steel, and inconel, plus some small amounts of actinides and fission products mixed in from the spent fuel. In Table 4-2 are shown the major elements present in BWR and PWR cladding materials.

Table 4–2. Composition of charged BWR and PWR fuel assembly materials (in g/metric ton heavy metal).

Element	PWR Assembly[a]	BWR Assembly[a]
H	3.05	
B	0.21	
C	69.61	9.2
N	9.27	
O	224.31	
Al	82.69	67.0
Si	440.23	161.9
P	286.65	5.2
S	21.51	3.5
Ti	107.35	120.6
V	4.69	
Cr	10,872.74	4,731.0
Mn	857.13	276.9
Fe	31,327.15	10,040.0
Co	3.4040	157.0
Ni	12,433.02	8,185.0
Cu	17.47	40.2
Zr	230,090.85	281,414.5
Nb	710.83	696.8
Mo	384.10	428.8
Sn	3,760.10	4,285.5
Hf	18.33	

Source: Modified from Storch and Prince (1979).
[a]BWR assembly from zircaloy 4, stainless steel 302, 304, inconel 718, and nicrobraze 50. PWR assembly from zircaloy 4, stainless steel 304, and inconel 718.

High-level waste (HLW) is defined by the Federal Government (U.S. Code of Federal Regulations, Title 10, Part 50, Appendix F) as " . . . aqueous wastes resulting from the operation of the first cycle solvent extraction system, or equivalent, in a facility for reprocessing irradiated reactor fuel." HLW consists of fission products as well as some actinides and radiogenic actinide-daughter elements. In Table 4-1 is shown many of the biologically significant radionuclides in HLW, their half-lives, and their activities (in Ci/L) 10, 10^2, 10^3, and 10^4 years out of the reactor. HLW possesses very high heat, high, penetrating radioactivity, and long periods for radioactive decay. At present, Federal regulations demand that the HLW be solidifed within 5 years after production, followed by shipment to a repository within 10 years. At present, it has not been decided whether a glass or ceramic waste form will be used for the HLW (see discussion in Chapter 12).

Transuranic wastes (TRU) are those radioactive wastes with long-lived

Table 4–3. Important Radionuclides of transuranic wastes (TRU).

Radionuclide	Half-Life	Activity (Ci/l of waste)
^{90}Sr[a]	29 years	3.52 (10^{-1})
^{60}Co	5.27 years	2.20 (10^{-3})
^{106}Ru	369 days	3.08 (10^{-3})
^{132}Cs	4.2 hours	1.76 (10^{-3})
^{152}Eu	13 years	4.40 (10^{-3})
^{154}Eu	8.6 years	1.76 (10^{-3})
^{232}Th	1.4×10^{10} years	1.03 (10^{-6})
^{234}U	2.44×10^{5} years	8.06 (10^{-9})
^{235}U	7.04×10^{8} years	3.44 (10^{-7})
^{238}Pu	4.47×10^{9} years	7.44 (10^{-6})
^{28}Pu	87.8 years	9.12 (10^{-5})
^{239}Pu	2.439×10^{4} years	1.06 (10^{-3})
^{240}Pu	6540 years	2.54 (10^{-4})
^{241}Pu	15 years	6.46 (10^{-3})
^{241}Am	433 years	1.75 (10^{-5})
^{244}Cm	17.9 years	4.40 (10^{-3})

Source: Modified from DOE (1982).

[a]This is the "reference waste" for transuranic waste generated at Rocky Flats, Colorado. It is assumed to be representative of all TRU waste.

alpha activity greater than 10 nCi/g [10 nCi/g is the upper range of activity of some uranium ore deposits according to Storch and Prince (1979)] due to the presence of actinides and transuranics. These radionuclides have long half-lives and possess such high radiotoxicity that their isolation is required. Data for some TRU are given in Table 4-3. For convenience, TRU is divided into low-level and intermediate-level subcategories.

Intermediate-level transuranic wastes (IL-TRU) include those wastes in solid form that contain long-lived alpha-emitting radionuclides with an activity of more than 10 nCi/g as well has combined beta and gamma emission so high as to require remote handling (i.e., about 100–1000 mrem/hr). These wastes result from the reprocessing of spent fuel and thus contain small amounts of actinides and fission products as well.

Low-level transuranic wastes (LL-TRU) are those wastes that contain plutonium and other long-lived alpha emitters in amounts greater than 10 nCi/g but with radiation levels low enough that they can be handled without remote controls and without shielding. The surface dose rate of LL-TRU is about 10 mrem/hr. LL-TRU results from fuel fabrication and reprocessing facilities and consists largely of salts and sludges arising from treatment of liquid waste streams as well as contamined paper, plastic, ceramics, etc., from the handling of materials. It has been proposed that LL-TRU wastes be shipped to repositories 5 years after their generation.

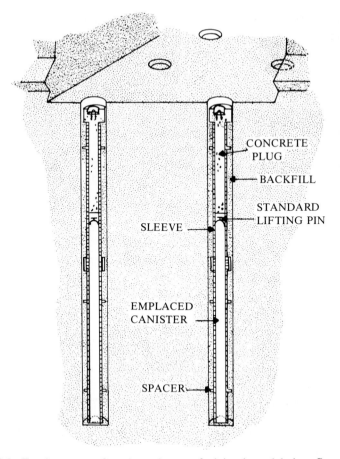

Fig. 4-3. Emplacement of canistered spent fuel in sleeved holes. *Source*: From DOE.

Low-level radioactive wastes (LLW) are those wastes that, by NRC definition, are (1) not high-level waste, and (2) contain less than 10 nCi/g (i.e., less activity than TRU wastes) of transuranic and other actinide elements. LLW do not, however, include wastes from the mining and milling of uranium. Commercial sources of LLW include nuclear power reactors, research and medical institutions, and industrial facilities. LLW from nuclear power reactors include sludges, resins, used gloves and clothing, plastic, paper, and tools. LLW from research and medical institutions include the above wastes plus organic wastes (i.e., chemicals and animal remains contamined with slight amounts of radioactivity), while industrial LLW result from the manufacturing of smoke detectors and pharmaceuticals, luminous dials, emergency exit signs, as well as various measuring and testing devices. In short, LLW includes just about everything that contains, or is suspected to contain, very low amounts of radioactivity. Volume-wise, LLW is far greater than HLW, TRU, or SURF, yet its radioactivity is very slight. The LLW generated in 1979 and 1980 amounted to 80,000 m^3 and 92,000 m^3, respectively; these wastes were disposed of by shallow land burial (see Chapter 9) at three sites in the United States: Barnwell, South Carolina, Beatty, Nevada and Richland, Washington. About 50% of LLW comes from commercial nuclear power reactor facilities; another 9% comes from government or military operations; research and medical institutions provide 19%; and other industrial sources provide the remaining 22%. The amount of LLW generated by each state is shown in Fig 9-1. It is apparent (DOE, 1980) that the volume of LLW in the year 2000 will be approximately 224,000 m^3. Thus, by sheer volume alone, the LLW are indeed of major importance in discussion of all radioactive wastes. Fortunately, however, the handling and disposal technologies and U.S. Government policies for LLW have been addressed with great care. Further, LLW have been stored at various facilities for close to 30 years, hence we have practical as well as theoretical and experimental experience to help guide us in addressing the problem of LLW disposal.

The Multibarrier Concept

The multibarrier concept for radioactive waste repository confinement has been developed to guarantee the safe isolation of radioactive wastes from the biosphere. In this concept, it is believed that the only solvent that might eventually reach radioactive waste forms in stored canisters is water. Hence the multibarrier concept has been developed to guard the waste form from water and, moreover, should water reach the waste form, then minimize the amount of radioactive species that might be locally dissolved and transported by water. The types of barriers proposed for HLW (and TRU, for that matter) are shown in Fig. 4-4. The multibarrier system starts with the waste form itself. As discussed elsewhere, the waste form for HLW will no doubt be either glass or ceramic. In the case of SURF or CW the actual fuel assembly or parts thereof will constitute the waste form. For HLW, the waste form itself will be surrounded by container walls consisting of Ti-bearing stainless steel or an equivalent. In the case of SURF and CW, air or helium under pressure will act as filler for the voids in the fuel assembly as well as surround the fuel asseembly within the canister wall (see Fig. 4-4). The canister with the contained fuel assembly is held in another cylinder, which constitutes the package, and the package boundary constitutes the overpack. The waste package is inserted into a preset sleeved hole (see Fig. 4-4) with the hole sleeve and sleeve coating serving as two additional barriers. Surrounding the hole sleeve is material, commonly a form of clay minerals $\pm$ reductants such as magnetite or other Fe^{2+} -bearing minerals $\pm$ other minerals, designated as engineered backfill. The purpose of the engineered backfill is to minimize or totally stop any radionuclides that might escape from a breached radioactive waste canister. In short, the engineered

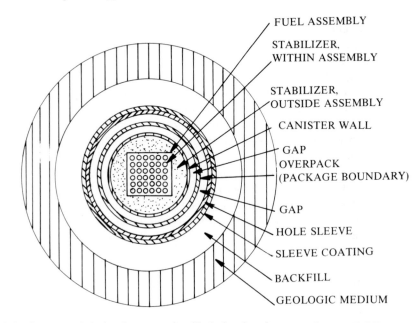

FUEL ASSEMBLY

STABILIZER,
WITHIN ASSEMBLY

STABILIZER,
OUTSIDE ASSEMBLY

CANISTER WALL

GAP

OVERPACK
(PACKAGE BOUNDARY)

GAP

HOLE SLEEVE

SLEEVE COATING

BACKFILL

GEOLOGIC MEDIUM

Fig. 4-4. Conceptual design incorporating likely barrier elements. *Source*: DOE.

backfill will absorb or repel water and sorb radionuclides from the waste form. The suitability of different materials for engineered backfill will be discussed in Chapter 13. Surrounding the engineered backfill is the rock in which the repository is located. This is the ultimate barrier. The rock site (see Chapter 6) is of considerable depth in a geologic site in which the rock permeabilities are low and the sorption potentials are high. The site is chosen so that water would have great difficult in reaching the repository and, if it did and breaching occurred, even more difficulty in transporting any of the radioactive materials to the biosphere. The time for such an unlikely event to occur is so long that, in theory, all (or almost all) of the radioactive species would have decayed.

In the multibarrier system, each barrier may actually act somewhat independently of other barriers. This points out the importance of single-barrier reactions, coupled reactions, and multireactions. Models of these reactions are divided by near-field and far-field conditions.

Near-field reactions are those in which the entire waste package, from waste form to repository rock, may be affected. To study near-field reactions, both theoretical and experimental approaches must be used. For far-field reactions, i.e., those involving hypothetically transported waste products from the canister site to some distance into the repository, theoretical considerations are best examined by natural analogs, because the duration of experiments under far-field conditions is too long for most experiments to be carried out. The near-field reactions will be controlled by forces arising from the waste package through engineered backfill, while the far-field reactions will be controlled by forces from the overall repository rocks.

Interaction between the near field and far field may occur, for instance, if fracturing occurred from the canister site into the repository rocks. If a source of water is present (we will not worry about how this would happen for the moment), then water-soluble species of radionuclides may be transported into the far field. Among the many parameters to be considered concerning the effect of fracturing of the near-field setting are (1) possibly short times for water to reach the waste package, (2) increased permeability of the repository rocks, thus allowing for more rapid water transport, and (3) flow of water along fractures rather than through the rocks. Although this is possible for a repository, it must be emphasized that sites are chosen with these facts in mind: (1) likelihood of fracturing is small, (2) certain rocks (i.e., halite) will self-seal if fracturing occurs, (3) fractures are usually very rapidly filled with secondary minerals, thus choking the water flow, and (4) the multibarriers have been designed for the unlikely event of water intrusion and should, short of total side flooding, be adequate to withstand such an event.

Barriers are of two types; natural and man made. The natural barriers will result from a combination of such circumstances as long flow-paths for groundwater, slow rates for water flow, high sorptive capacity of the rocks. Each of these would favor isolation of radionuclides. Man-made barriers in the overall system will also lead to radionuclide isolation; these will include a low-solubility waste form (or chemically inert fillers, for SURF or CW), corrosion-resistant canister and sleeving, and the sorptive capacity of the engineered backfill. Although each of these barriers is small relative to the repository rocks, each has been chosen to effectively work against water intrusion or, if necessary, against radionuclide flow away from the hypothetically breached waste package. Admittedly one cannot unequivocally say that a repository waste package is fail-safe, yet the various man-made barriers, especially if considered in the overall repository framework, will be adequate to hold up during the lengthy time of the maximum thermal periods of the rock repository.

Uncertainties and Their Resolution

There are several uncertainties for radioactive waste repositories that must be resolved. Some of these uncertainties can be resolved by existing knowledge or by further observation and research, while other uncertainties can be addressed by considering limiting conditions. At the present time, our state of the art knowledge has advanced to the point where the sites can be selected and their characteristics worked out. Further, by studying the sites, knowing their geology, geochemistry, and geohydrology, and by laboratory studies, theoretical considerations, and application of information from natural analogs, plausible scenarios as to how the entire waste package behaves in a repository environment can be worked out. Although complex, the results (Chapter 7) clearly indicate that it is likely that radioactive wastes can indeed be safely stored in the earth's crust.

Before addressing the nature of some of the uncertainties, it is necessary to address future accidental intrusion by man into a repository site. With the technology for site protection against such accidental entry presently available, this should not be an important possibility for further consideration. The protective measures to be taken will include encouraging public awareness and education, installing permanent markers, enforcing institutional controls, addition of tracers to the waste, the rock, and engineered barriers, and, most important, choosing the repository site so as to prevent any releases into the biosphere (or else designing it so that any releases to the biosphere will be so small as to not pose any biological problems).

For a considerable time only one rock type, bedded salt deposits, received any detailed study for use as a repository for radioactive waste. It is now our practice to consider several rock types and the engineered barriers that will be used in the overall system. Thus the entire geochemical and geohydrologic environment and the engineered part of the system are considered.

Criticality

Allen (1978) has compiled data for hypothetically buried SURF canisters in which is shown that enough actinides are present to allow criticality to be achieved. In order for criticality to be achieved, however, the necessary amounts of the actinides must be concentrated at a point source and segregated from each other and from neutron poisons. Possible criticality scenarios include within a package, between several waste packages, and by buildup of actinides by very selective leaching, transportation, and deposition. Criticality in a single package is unlikely, as the geometry of the assembly coupled with package fillers and neutron poisons are more than adequate to prevent criticality. In addition, package-to-package criticality is equally unlikely for the same reasons. The possibility of actinide concentration by selective leaching, transport, and deposition has been shown by Brookins (1978e) and Behrenz and Hannerz (1978) to be highly improbable for generic studies in the United States and in Sweden, respectively. Brookins (1980, unpublished) has also evaluated the possibility of criticality occurring at the WIPP site (see discussion in Chapter 7) and found it to be highly improbable. Further, even if the remote possibility of criticality were a reality, the reaction would result in a small, very localized release of thermal energy—far less important in terms of radionuclide transport than normal geochemical processes (see Chapter 7).

Summary

The radioactive products of fission, the radioactive parent actinides, and radioactive and nonradioactive daughters are known for the various types of radioactive wastes, as are those of greater potential biological hazard versus nonhazardous species. The handling and disposal of the various radioactive wastes is then a straightforward consequence of designing facilities to isolate the various species described.

The multibarrier concept, introduced here, is based on the assumption that the radioactive wastes, encapsulated into a waste form, should be isolated as thoroughly as possible from interaction with surrounding media or with fluid. Thus the waste forms considered must pass rigorous testing (see Chapter 12 for details), as must the surrounding canister (Chapter 13), the engineered backfill (Chapter 13), and the repository rocks proper (Chapters 6 and 7). A multiplicity of safety features would thus be present. The various aspects of the multibarrier waste system will be explored in the remainder of the book.

CHAPTER 5

Geochronology and Radwaste

Introduction

Absolute age determination based on radioactive decay of parent isotopes to stable daughter isotopes is an accepted tool for geologic research. Rocks as old as the age of the earth (i.e., meteorites and some lunar samples) have been successfully dated, and the time scale for the earth has been rigorously worked out by a combination of several geochronologic methods.

It is important to point out that the standard geochronologic methods involve an understanding of the geochemistry of U, Th, the actinide daughters, and Rb, Sr, K, Ar, Sm, Nd, Re, Os, Lu, Hf, and others, and that isotopes of many of these elements are either products of fission in man-made reactors or are some of the heavy metals encountered in radwaste. The importance of this relationship will be addressed in this chapter. As examples, repository rocks can be dated in order to test their chemical and isotopic integrity, and their radiometric ages are valuable tools for investigation of natural analogs as well.

The Importance of Radiometric Age Determinations

Geologists and other scientists have been using radiometric age determinations for most of the 20th century. The main objective of many of these determinations has been the quantitative assessment of the relative time scale given by paleontologic and other evidence, although there are many applications of the radiometric age data to other problems such as rock petrogenesis (see Faure, 1977).

Several methods are widely used today. Historically, the radioactivity of uranium (at first it was not recognized that there were two isotopes of uranium) and its decay to daughter lead were used for geochronologic purposes in the very early 1900s (Boltwood, 1907). By the 1930s theory and experimental testing of theory had led to the documentation of the potassium–argon method, and in the 1940s the rubidium–strontium method was tested successfully. By the 1960s these major methods were used widely for the dating of many rocks and rock materials. The relevant age methods are briefly discussed below.

Uranium–Lead

^{235}U decays to ^{207}Pb with a half-life of 0.713×10^9 years. The decay scheme for ^{235}U is shown in Fig. 5-1. Similarly, ^{238}U decays to ^{206}Pb with a half-life of 4.468×10^9 years (Fig. 5-2). The age of a U-bearing mineral is then calculated from the following generic equation:

if

$$D^* = P_0 - P = P(e^{\lambda t} - 1) \qquad (5\text{-}1)$$

then

$$t = 1/\lambda \ \ln\left[(D^*/P) + 1\right], \qquad (5\text{-}2)$$

where $D^* =$ number of atoms of daughter product, $P =$ number of parent atoms remaining, $P_0 =$ initial number of parent atoms, and $\lambda =$ the decay constant for the parent isotope. Usually, there is some amount of the daughter isotope (D_0) already present in the rock or mineral, such that the total amount of the daughter isotope, D, which is measured, actually contains D^* and D_0. In order to successfully determine the radiometric age of a rock or mineral, both D and D_0 must be known in order to calculate D^*. In addition, the following criteria must be known for a successful age determination:

(1) The rock or mineral must be a closed system. After a radioactive isotope starts to decay, no gain or loss of any P, D, or D^* can take place from the rock or mineral.

(2) The amount of D_0 must be known.

(3) The rate of decay must be known.

On a practical note, there must be measurable amounts of P, D^*, or D present, and the rock or mineral must have formed over a short period of time relative to its absolute age.

Now the equations for uranium–lead age determinations can be stated rigorously:

$$t_{207} = 1/\lambda_2 \ \ln\left[(^{207}Pb^*/^{235}U) + 1\right]. \qquad (5\text{-}3)$$

But this is not useful unless we know the precise number of atoms of ^{207}Pb* and ^{235}U present. A convenient way to relate the mass spectrometric isotopic ratios to the above equation is to ratio each $D*$, P isotope to a stable, nonradiogenic D isotope (i.e., of different mass than $D*$). Thus the Pb and U data are ratioed to ^{204}Pb and, when $D - D_0$ rather than $D*$ is used, then the above equation becomes:

$$t_{207} = 1/\lambda_2 \ln \left[\frac{(207/204)_m - (207/204)_0}{(235/204)} + 1 \right]. \qquad (5\text{-}4)$$

Similarly:

$$t_{206} = 1/\lambda_1 \ln \left[\frac{(206/204)_m - (206/204)_0}{(238/204)_m} + 1 \right] \qquad (5\text{-}5)$$

where the numbers in parentheses refer to the number of atoms of ^{207}Pb, ^{204}Pb, ^{235}U, etc., and, by dividing Eq. (5-4) by (5-5) we obtain:

$$\frac{^{207}\text{Pb*}}{^{206}\text{Pb}} = \frac{^{235}\text{U}}{^{238}\text{U}} \frac{e^{\lambda_2 t} - 1}{e^{\lambda_1 t} - 1}, \qquad (5\text{-}6)$$

which eliminates the D_0 and total uranium. The equation is useful for across check of the independently determined ^{235}U and ^{238}U ages. Realistically, most U–Pb ages are discordant to some degree [i.e., the ages from Eqs. (5-4) and (5-5) disagree, with the ^{235}U–^{207}Pb age usually somewhat older than the ^{238}U–^{206}Pb age, although both ages are commonly too young]. This results from the fact that the structural sites for uranium as U^{4+} (ionic radius 1.04 Å) are not favorable for daughter Pb^{2+} (ionic radius 1.4 Å). Hence Pb^{2+} is metastable in such a site and may diffuse (i.e., some $D*$ is lost). Further, in the decay of ^{238}U there are several possible places when intermediate, radioactive decay daughters may escape from the chain. When ^{234}U is formed, it wil be as U^{6+}, which is more easily lost from rocks by solution than U^{4+}. Next, ^{230}Th may be chemically different from U and be removed. Next, ^{226}Ra may be lost from the U-bearing mineral, and there may be loss of ^{222}Rn, which is well known in nature. Loss of intermediate, radioactive decay daughters from the ^{235}U chain is less likely to occur (see Fig. 5-1) because their half-lives are much shorter than their counterparts in the ^{238}U chain (see Fig. 5-2).

Thorium–Lead

In nature, thorium is about four times as abundant as uranium, and many rocks possess Th/U ratios of 4. In most rocks, uranium occurs as U^{4+} with

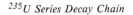

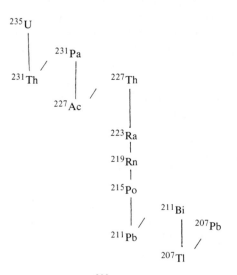

Fig. 5-1. ^{235}U series decay chain.

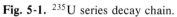

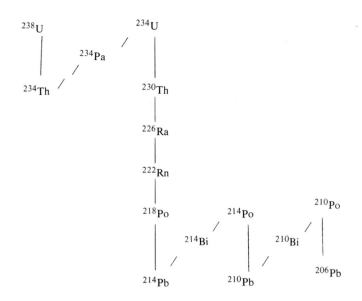

Fig. 5-2. ^{238}U series decay chain.

an ionic radius of 1.08 Å, although oxidation can cause formation of U^{6+}, which is very soluble relative to U^{4+}. Thorium, on the other hand, only occurs in natural substances as Th^{4+} with an ionic radius of 1.12 Å. From Goldschmidt's Rules, then, U is slightly more enriched in Zr^{4+} (ionic radius = 0.8 Å) than Th and the ratio of Th/U in Zr-bearing (or equivalent Ca-bearing minerals) is commonly much less than 4, while the matrix to these minerals will result in Th/U considerably greater than 4. Th is commonly enriched in minerals such as monazite; Dating of a Th-bearing mineral or rock is done by the formula:

$$t_{208} = 1/\lambda_3 \ln \left[\frac{(^{208}Pb/^{204}Pb)_m - (^{208}Pb/^{204}Pb)_0}{(^{232}Th/^{204}Pb)_m} + 1 \right], \qquad (5\text{-}7)$$

where λ_3 is the decay constant for $^{232}Th = 4.95 \times 10^{-11}$/y. If the U, Th dates are obtained on the same mineral, then the two U–Pb ages and the Th–Pb date allow a triple check on the age to be made. If the three ages are in close agreement, then there is high degree of confidence in the age. The decay scheme for ^{232}Th is shown in Fig. 5-3.

One drawback to the U–Pb and Th–Pb age determinations is that the D_0 estimate is commonly made from a separate mineral, such as pyrite (FeS_2, but which contains Pb), instead of Pb_0 from a major rock-forming silicate or oxide. Ideally, since Pb^{2+} substitutes for K^+ because of their similar ionic radii, Pb_0 values for K-feldspar would be more suitable than values for pyrite for age determinations on zircons, sphenes, and other silicates and oxides. The errors introduced cannot be too large, however, or the ages will be very erratic, whereas, in most cases, the agreement is quite good.

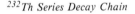

^{232}Th Series Decay Chain

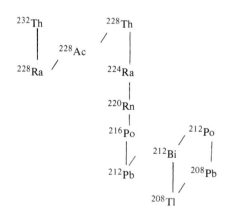

Fig. 5-3. ^{232}Th series decay chain.

Potassium–Argon

The decay of ^{40}K is complex as it results in both ^{40}Ca (89%) and ^{40}Ar (11%) being formed, but most radiometric ages are calculated by the potassium–argon method rather than by the potassium–calcium method. This is due to the fact that ^{40}Ca is the most abundant isotope of total Ca, plus the fact that calcium is more abundant than potassium. Hence when ^{40}K decays in most common rock-forming minerals, then, since ^{40}K only makes up 0.012% of total K, too little ^{40}Ca* is formed to be accurately measured. But these same minerals can be dated by the buildup of ^{40}Ar*, as the amount of ^{40}Ar$_0$ in most igneous rock and minerals is below measurement (i.e., $D_0 \sim$ O). The age equation is:

$$t = \frac{1}{\lambda} \ln \left[\left(\frac{^{40}Ar^*}{^{40}K} \right) \left(\frac{\lambda}{\lambda_e} \right) + 1 \right] , \qquad (5\text{-}8)$$

where λ = total decay constant of ^{40}K = 5.48×10^{-10}/yr and λ_e = decay constant for ^{40}K$\rightarrow^{40}$AR = 0.566×10^{-10}/yr; ^{40}Ar* and ^{40}K need not be ratioed to any other Ar$_0$ isotope because D_0 for ^{40}Ar $\sim$ O.

The potassium–argon method works very well for many unmetamorphosed igneous minerals and some whole rocks. In the event of one or more metamorphisms or some heating event, ^{40}Ar, because it is an inert gas and not bonded electrically to any other elements in the mineral structure, will be lost by diffusion. Hence the age of such a mineral will usually be the age of the last thermal event affecting the mineral, and not the age of mineral formation. Thus most K–Ar ages should be interpreted as minimum ages. It must be re-emphasized, however, that for many minerals useful K–Ar age information can be obtained that could not be obtained by any of the other radiometric age methods. The importance of K–Ar dates for specific repository sites will be commented on later.

Rubidium–Strontium

The decay scheme of ^{87}Rb is by beta-minus emission to ^{87}Sr. In nature, Rb^{1+} is camouflaged by K^{1+} and always occurs in K-bearing minerals, although it never forms a pure Rb (K-free)-mineral. Strontium occurs as Sr^{2+} and is more common in Ca-bearing minerals, although some enters K^{1+} sites. The age equation is:

$$t = \frac{1}{\lambda} \ln \left[\frac{(^{87}Sr/^{86}Sr) - (^{87}Sr/^{86}Sr)_0}{(^{87}Rb/^{86}Sr)} + 1 \right] , \qquad (5\text{-}9)$$

where λ = the decay constant for ^{87}Rb = 1.42×10^{-11}/yr and both ^{87}Sr

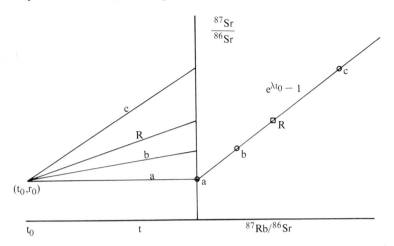

Fig. 5-4. Rb–Sr growth curves (a) and isochron (b) for rock with no postformational metamorphism.

(measured) ^{87}Rb are ratioed to ^{86}Sr for convenience (i.e., in many terrestrial materials the abundance of ^{86}Sr and ^{87}Sr are close).

An advantage of the Rb–Sr method is that Rb follows K, a common rock-forming element, and Sr follows Ca, another common rock-forming element. Hence if a rock or mineral is not too K poor, then it can be dated by this method (i.e., limestone, with very high Ca/K, cannot be dated by this method, but most other common rocks, with proper caution, can be dated). Further, unlike ^{40}Ar*, the ^{87}Sr* formed by decay from ^{87}Rb is compatible with the structure in which it is formed. Thus loss of ^{87}Sr* by a metamorphic (or equivalent) event may move ^{87}Sr* from its original site but not from the rock in which was first formed. Thus, as shown in Fig. 5-4, so-called whole rocks (i.e., those rock samples with an approximate rock/largest mineral diameter ratio of 10 or greater) are suitable for dating. If no metamorphism has occurred, the individual mineral as well as whole rock points will give the age of formation, T_f, of the rock (Fig. 5-4) and will fall on a line of slope ($e^{\lambda t}$ − 1), the isochron. If a metamorphism occurs, then ideally, ^{87}Sr* may be lost from a K-rich mineral (e.g., biotite, K-feldspar) but incorporated into a Ca-rich mineral (e.g., plagioclase) such that the reset minerals will fall on a younger, mineral isochron (Fig. 5-5), which gives the age of the last thermal event. Thus one can determine both the age of formation, t_f, and the age of the last thermal event, t_m, for such a rock.

A possible disadvantage of the Rb–Sr methods, however, is that age determination may be affected by weathering. When in contact with surface waters, the rocks affected may encounter HCO_3^- ions (resulting from CO_2 from the atmosphere being dissolved in the surface waters and forming

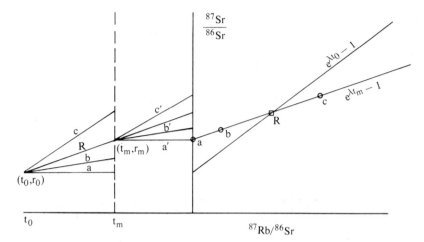

Fig. 5-5. Growth curves (a) and isochrons (b) for a rock with constituent minerals that has undergone a metamorphism at t_m.

H_2CO_3, which is then ionized under the pH ranges 6.4–10.3 to HCO_3^-). Since Ca^{2+} is commonly released from rocks being weathered, there is a strong possibility for the precipitation of $CaCO_3$ in rocks affected by such groundwaters. If this weathering has occurred at any significant time after formation, t_f, then it is, by definition, not a closed system and the amount of D will be too high. If samples are properly taken, however, this weathering should not be a critical factor, which is attested to by the truly impressive numbers of Rb–Sr ages of many minerals and rocks of ages from less than 10 million years to the age of the earth (4.5 billion years).

Of the utmost importance in discussing Rb–Sr ages in conjunction with radioactive waste disposal studies is the behavior of Rb, Sr, and Cs in rocks and minerals. First, if a successful age is obtained by the Rb–Sr method, then that rock or mineral has remained closed to gain or loss of Rb and Sr since the age (i.e., formational or last thermal event). Second, Cs^{1+} (ionic radius = 1.5 Å) behaves much like Rb^{1+} and is commonly camouflaged in K, Rb-bearing minerals. Third, when the alkali metals are released by weathering or other processes, Cs is more highly retained than either Rb or K in most sorptive media, and Rb more so than K. Thus is a rock or mineral is closed to Rb and Sr, it is also closed to Cs. In terms of HLW disposal, if a rock or mineral has exhibited closed-system behavior for Rb and Sr, then ^{90}Sr and ^{137}Cs may also be immobile in such rocks.

Uranium–Ruthenium Age Determination

The radiometric age methods described above rely on emission of beta (+ or −) and/or alpha particles and electron capture. Spontaneous fission, especially of ^{235}U and ^{238}U, also occurs. Recently, Maeck *et al.* (1978a), based on their earlier work on fissiogenic Ru isotopes at Oklo, investigated U–Ru dating of uranium ores from several occurrences. The relevant age equation is:

$$t = \frac{1}{\lambda_\alpha} \ln \left[\frac{\lambda_\alpha N_{SF,i}}{\lambda_{SF} N_{238} \gamma_{SF,i}} + 1 \right], \qquad (5\text{-}10)$$

where λ_α is the alpha-decay constant for $^{238}U = 1.551 \times 10^{-10}$/yr, λ_{SF} is the spontaneous fission decay constant of $^{238}U = 8.46 \times 10^{-17}$/yr, $N_{238} =$ measured number of ^{238}U atoms in the ore sample, $N_{SF,i} =$ no. atoms of fission element, and $\lambda_{SF,i} =$ the ruthenium spontaneous fission yields in ^{238}U taken as ^{99}Ru: 6.0%, ^{101}Ru: 7.25%, ^{102}Ru: 8.0%, ^{104}Ru: 4.2%.

Maeck et al. (1978b) report U–Ru ages for 13 U-ore occurrences of Precambrian age. Their results are in good agreement with reported U–Pb ages for the same deposits, although systematically slightly higher. This is not not surprising. From crystal chemical arguments, Ru^{4+} should be much more compatible in U^{4+} sites than Pb^{2+}, and loss of daughter Pb may occur while fission daughter Ru is retained. More important, though, is that the Ru produced is a product of fission and is retained in the samples analyzed, thus showing that nature has provided firm evidence that, under proper conditions (low Eh, compatible structural sites) fissiogenic Ru can be easily retained in natural geomedia.

Importance of Combined U–Pb, Th–Pb, K–Ar, and Rb–Sr Ages

The importance of the various major methods for radiometric age determinations lies in the diversity of the parent, intermediate, radioactive daughters, and stable daughters involved. If a rock has been dated by all methods, then one knows a great deal about the daughters with moderate half-lives (^{234}U, ^{230}Th, ^{226}Ra, ^{222}Rn, ^{210}Pb, ^{210}Po, etc.), nonradioactive, mutagenic isotopes (Pb), alkali elements (Rb, K), alkaline earths (Sr), and an inert gas (Ar). Further, advances in studies of samarium–neodymium dating (i.e., ^{147}Sm–^{143}Nd) show that both Sm and Nd are retained in rocks datable by other major methods, hence indirect inferences can be made concerning lanthanide element retention in a rock dated by the major methods.

In examining the list of elements and their isotopes present in HLW and TRU (Tables 4-2, 4-4) we note that we are also concerned with actinides (including transuranics), intermediate, radioactive daughters (primarily Ra), some Pb, alkali elements (^{137}Cs), alkaline earths (^{90}Sr, Ba), lanthanides, and inert gases (Kr, Xe). In radioactive waste, we are also concerned with several other elements (I, Br, chalcophile and lithophile elements), but, if it can be demonstrated that the Rb, Sr, K, U, Th, Pb, and Ar (and Sm and Nd) of a rock have not migrated, then this suite is diverse enough to strongly suggest that other elements in the rock should not migrate as well. This idea will be explored for several of the proposed major rock repositories (Chapter 7).

The radiometric age determinations are a great asset in assessing the suitability of a rock for radioactive waste storage consideration. If the radiometric age data indicate a rock has in relatively recent geologic time lost variable and large amounts of either parent or daughter isotopes, then this rock may not be a good candidate for retaining hypothetically encased radioactive waste. If, on the other hand, the rock yields concordant radiometric ages, then it could be a good candidate for the retention of any radioactive waste that might escape from a canister.

Some Examples

The following are some examples where radiometric age determinations can or have been used to address various aspects of disposal of radioactive wastes in rocks. Some of the brief examples given here will be discussed more fully with specific rock types in Chapter 6.

Bedded salt deposits can be studied by several of the geochronologic methods, but have actually not received much study. I have investigated the Rb–Sr and K–Ar systematics of the bedded evaporites of southeastern New Mexico (the WIPP site and surrounding area) and find that those minerals suitable for age study yield values in excellent agreement with the known age of sedimentation. Hence these rocks have remained closed systems since their formation some 210–230 million years before the present (MYBP).

The basaltic rocks at the BWIP project in the state of Washington have been studied very little by conventional geochronologic means. The isotopic composition of Sr in the basalt has been determined (Michel, oral communication), but the secondary minerals have not been studied, although they certainly are suitable for age work. A key question: Did the secondary minerals form at about the end of the volcanism when the basalts were formed or, alternately, did they form in relatively recent times by infiltration of the rock system by groundwater? For some minerals containing K, ages of near 12 MYBP would indicate an early formation history; ages from about 0 to perhaps 1 to 2 MYBP would make meteoric infiltration of waters a strong possibility. In the latter case, then, one would expect to find "fingerprints" of certain elements in the secondary minerals. In the case of strontium isotopes, meteoric waters usually possess significantly higher $^{87}Sr/^{86}Sr$ ratios (0.708–0.709) relative to the basalt (0.705–0.706) (these ratios can be measured with very high precision—better than $\pm$ 0.00005—hence a variation of $\pm$ 0.002–0.003 is very large indeed). Secondary minerals formed early in the history of the rock, or formed late but with Sr from the basalt, should yield Sr isotopic signatures near 0.705–0.706, while any meteoric material should yield values nearer 0.708–0.709. Unfortunately, simple as these measurements are to make, they have not yet been determined.

Granitic and tuffaceous rocks from the NTS have also not yet been studied by geochronologic methods, yet another granitic and tuffaceous–rhyolitic rocks have. Since some of the studied rocks are similar to the NTS rocks, the radiometric data can be applied to the NTS rocks. I have studied the Eldora–Bryan stock (granitic rock) and also the Alamosa River Stock (silicic rock intrusive into tuffaceous rocks and others), both in Colorado. At the Eldora site (see Brookins et al., 1981a,b, 1982a,b) the evidence is strong that no, or very little, transfer of elements occurred between the high heat sink Eldora–Bryan magma and the intruded rocks were reset to near the age of the intrusion. Cooling was by conduction at these sites, and the results (discussed in more detail in Chapter 11) show the favorability of both the stock and intruded crystalline rocks for possible radioactive waste storage. At the Alamosa River Stock, cooling was by convection and conduction, hence this area is more interesting in attempting to see if elements from either the stock or intruded rocks were moved about considerably due to the high-temperature intrusion. Yet no such widespread migration has been noted, and, like the Eldora–Bryan stock, elemental transfer between the stock and intruded rocks appears to have been very local (i.e., on the order of 1–2, if that). These and other studies like them are quite important. Several years ago Hart and his co-workers (see Hart et al., 1968) clearly demonstrated that the effect of intrusions on the surrounding rocks was to cause age lowering by loss of $^{40}Ar^*$ and remobilization of $^{87}Sr^*$ in the intruded rocks as a function of distance from the contact; more recently Brookins et al. (1981 a,b) have shown that the whole rocks of the intruded rocks still yield the age of formation even where the most pronounced mineral loss or resetting had occurred, thus demonstrating that these rocks had remained closed systems to Rb and Sr despite the high-temperature intrusion.

In the case of dome salts, no radiometric age studies have been carried out in the United States. In Germany studies have shown that the dome salts can indeed be studied by the Rb–Sr and K–Ar methods (see Lippolt and Raczek, 1979a). The data of these investigators indicate stability and chemical and isotopic integrity of the dome salts for over 100 million years.

Argillaceous rocks have received some consideration for radioactive waste studies, both as possible repository rocks and as backfill material. Unfortunately, most of these are untested. I have examined the radiometric ages and other geochemical data for many such rocks, and have demonstrated that not only can these rocks be successfully dated by the Rb–Sr method, but that the data can be applied to problems of radioactive waste element behavior in such rocks. If one is careful to work with the authigenic clay mineral fraction of either shales or argillaceous sandstones, then these rocks can be dated by the Rb–Sr methods. Further, the dates indicate closed-system behavior to Rb and Sr and, by inference, to Cs. Rocks dated include the Westwater Canyon and Brushy Basin members of the Morrison Formation, the Pierre Shale, the Dakota Sandstone, the Mancos Shale, the Madera Limestone (by dating the clay-rich parts), and others. I have also applied the results of his studies of argillacous rocks to backfill problems (see Brookins, 1980a, 1982a), and shown that the combined radiometric and other geochemical data show the suitability of several of these rocks for use as backfill because any hypothetically released Rb, Sr, or Cs would be sorbed by the rocks. Further, since experimental studies show that actinides and many other heavy elements are sorbed more effectively than Sr and Cs, the Rb–Sr studies imply that the rocks would serve as effective getters for many of the HLW and TRU components.

I have also examined the geochemistry of uranium deposits in conjunction with radioactive waste studies. In this approach, it is recognized that most trace elements associated with uranium deposits are found in minerals associated with, but not mixed in solid solution with, uranium minerals. Thus, to a first approximation, the uranium ore can be considered as analog to a spent fuel rod (or HLW) in which the nonactinide elements have already hypothetically moved to the surface of the waste (i.e., the lengthy time for movement of any radionuclide to the waste form surface then need not be considered). As such, migration or retention of the nonactinide elements may be taken as a worst-case scenario of leaching from a HLW form. Yet studies show (Brookins, 1979b, 1980a) remarkable retentivity of the HLW analog elements, especially since they are confirmed by Rb–Sr dates identical to U–Pb dates on the ore. In turn this implies that the clay minerals and other minerals mixed with the uranium ore are effective as getters during the uranium mineralization and add to its stability for subsequent events. The data from these and other studies also indicate that even if some elements should escape from such uranium ore they would be quickly sorbed by nearby clay-rich layers. Finally, it should be pointed out that some uranium ore has been remobilized due to severe oxidizing conditions, yet the Rb–Sr and other geochemical studies show that a new generation of clay minerals forms during the reconcentration–reprecipitation of the uranium, and the formation of this new generation of clay minerals helped stabilize and retain the uranium ore in its new site. Even under the extreme oxidizing conditions the uranium was not remobilized too far from its original site (see discussion in Chapter 11).

It is appropriate to here digress to the topic of experimental sorption studies. Often these are carried out using argillaceous rocks or clay minerals, and the ability of these materials to sorb elements like Cs, Sr, etc., reported as "K_ds" (in ml/g or m^3/kg). Very often wide variation in K_d values for different samples of any particular rock are noted, and a convenient explanation is that the sorptivity of any species in question is also widely variable. I propose, however, that the explanation for this variation may arise from sample-to-sample internal mineral heterogeneity. Studies of many argillaceous rocks show that they contain both detrital and authigenic minerals; the former will commonly yield ages of provenance and the latter age of sedimentation. If the two are not separated from each other, then a mixed age somewhere between the two ages will result. Similarly, the sorptive capacity of the detrital and authigenic minerals are commonly different, hence if one sample possesses more detrital than authigenic minerals, or vice versa, then one would expect to get different K_d values. This can be monitored by studying the Rb–Sr systematics of the samples prior to the sorption studies and attempt to relate the apparently different sorption characteristics to Rb and Sr behavior. This has, surprisingly, not been done.

Dating of tuffaceous rocks from many places has been successfully carried out by combined K–Ar, Rb–Sr, and fission track techniques. Briefly, K–Ar and Rb–Sr dates on cogenetic biotites and sanidines and fission track dates on zircons yield results that are in remarkable agreement, and attest to the ability of these rocks to prevent movement of internally distributed U, Rb, Sr, K, and Ar. By inference, the rocks should also be closed systems for Th, Pb, and some other elements, but often the rocks are too young for U, Th–Pb techniques to be used.

Not much information is available for ocean floor materials, yet K–Ar dates of basalts show the same "mirror image" trends noted for magnetic reversals away from spreading ridges, thus demonstrating that the rocks are closed to K and Ar. Further, studies of aeolian clay minerals dredged from the ocean floors show that these minerals have often maintained their chemical and isotopic integrity despite their direct contact with ocean water and ooze and other bottom materials. While the residence time of these samples is not precisely known, estimates of several million years to perhaps in excess of some 100 million years indicate that they have not been severely affected by the high-salt waters even under great hydrostatic and total pressure.

HLW–TRU Radionuclide Behavior in Rock Systems: Evidence from Oklo and Other Sources

Oklo, discussed more fully in Chapter 11, is the only known "fossil" natural nuclear reactor. This fission reactions occurred relatively soon after uranium mineralization, which followed sedimentation relatively soon; all three dates somewhere in the vicinity of 1.95–2.05 BYBP. That criticality was achieved is in itself remarkable; that the 2-billion-year-old event can be studied today is even more remarkable. There have been, in the past decade or so, some very detailed studies of Oklo materials, many of which are in French (see IAEA, 1976, 1978). The importance of the Oklo studies to aspects of radioactive waste disposal in rocks can be summarized as follows: The host rocks to the uranium ore at Oklo consist of folded, faulted, and jointed sanstones, shales, and conglomerates with some carbonate. These rocks may have exhibited a wide range in porosity and permeability at the time when the reactor was operative. Certainly the rocks at the Oklo uranium mine would not receive high consideration for disposal of radioactive waste today; yet these rocks did retain many fission products and radiogenic actinides and transuranices and mutagenic lead and bismuth not only during the reactors operative lifetime but for at least 25–30 million years to, in some cases, 2 billion years after shutdown. The behavior of the fissiogenic and other elements at Oklo has been the subject of much of my research over the last few years and is summarized in Chapter 11.

Some Conclusions

Successful geochronologic age determinations include proper formational ages, as well as ages of some postformational event that can be identified by the methods (i.e., Rb–Sr mineral date on metamorphic mineral). Preformational information (provenance, magma assimilation, etc.) can also be addressed. The extensive geochronologic literature demonstrates just how sophisticated the age studies have become. More important, the different methods allow investigation of individual chemical systems of interest to radwaste disposal, such as behavior of U, Th, ^{230}Th, ^{226}Ra, Pb, K, Ca, Ar, Rb, Sr, Sm, Nd, and others. Further, the chemical and isotopic integrity of candidate repository rocks can be evaluated (see Chapter 7), and, in one system, evidence has been obtained for preservation of fissiogenic Ru for in excess of a billion years.

The message from the geochronologic studies is clear. Rocks and minerals often do behave as closed systems, and, when they do not, the geochronologic methods allow a quantitative assessment of the reasons for open system behavior. All this is critical to our understanding of how elements will behave under repository conditions.

CHAPTER 6

Geologic Sites: Generic Cases

Geologic Disposal of Radioactive Wastes

In this chapter we shall explore the geologic criteria for selection of possible repository sites in different settings. Specifically, mined repositories will be explored here. Geologic considerations for deep-well injections and other nonmined repository alternatives will be explored later.

Basically, radioactive wastes placed in a mined repository must be isolated from the biosphere for several thousand years, or so contained that the amount of radioactive material that might escape from a waste container and reach the biosphere would be insufficient to pose a health hazard. Any geologic site considered as a repository will contain a set of site-specific inherent geologic conditions that will not apply to other sites, consequently each proposed site must be very carefully studied. A repository site can then be either approved or rejected based on these geologic, chemical, and related studies. To date, geologic rock types under consideration for repositories include bedded and dome salt, granite, basalt, argillaceous rocks, and tuffaceous rocks. Other rock types are not necessarily ruled out, it is just that the leading candidate sites have been proposed for the rock types mentioned.

First, however, it is important to mention some general information concerning any possible rock repository. The depth of proposed storage areas must be great enough so that surface processes and natural phenomena, including wind action, fluvial erosion, glaciation, at or near-surface water flow, meteorite impact, and weathering, will have no impact on the buried waste. Since lithostatic pressure increases with depth, the burial depth must be such that the lithostatic pressure will not pose a threat to repository openings (i.e., if the lithostatic pressure exceeds to rock repository strength then mine openings would be in danger of collapse, and thus retrievability of the waste would be jeopardized). Since the lithostatic pressures of rocks vary widely from point to point, the pressure–depth relationships must be worked out for each prospective site, and any facilities to be built will need to be very site specific.

Since a major objective of waste disposal in rocks is to isolate the waste, and it is recognized that water is the most likely solvent to come in contact with buried waste, then great care must be taken to select sites in which water infiltration will be unlikely or so slow as not to pose a problem. Thus rocks with very low water content, porosity, and permeability are better candidates than rocks through which water flows at high rates due to high supply and high porosity and permeability. By selecting rocks with low permeability, water flow is minimized in the repository host rocks. The combined low porosity and the very low water content of the host rocks, will also assure the isolation of the waste, even if the near field is affected by the heat from the canister (see discussion on near-field effects in Chapter 4). Further, the host rocks' chemistry, including radioactive species, should be predetermined so that any hypothetical reaction between any part of the waste package and the surroundings will not be effective in causing radionuclide migration. Finally, the host rock dimensions around the buried waste should be large enough so that if any radionuclides are unavoidably released from the waste package, their travel time to the biosphere will be so long that either nonhazardous amounts reach the biosphere or the radionuclides are destroyed by decay.

Fundamental to the study of any site is its hydrology and, if possible, its fossil hydrology. By knowing the present and past conditions controlling water flow in rocks, it is possible to locate a repository so that the chances of water reaching the repository are minimal. Further, in the unlikely event that water reaches the site, the site can be so situated that the water can, effectively, be imprisoned at the site. Finally, no site should be located close to, or with access to, possible discharge routes to aquifers, where contact with the biosphere is possible.

Geologic data for the tectonic stability, including faulting and seismological data, must be carefully assessed. Ideally, a site should be located in the most tectonically stable and seismically inactive rocks possible. Alternately, sites in more tectonically active zones can be considered if it can be demonstrated that the effects of such tectonism do not affect the candidate repository rocks and the waste buried therein. Faults should be avoided, as should other rock fractures, and volcanic activity should have been absent in the repository rocks for the last few million years. It is anticipated that a particular site may possess some structural features that indicate some type of tectonic activity and, if such features are present, it must be demonstrated that no recent tectonism is responsible for, or has affected, these features.

Natural resources must not be drastically affected by waste repositories. By their nature, the repository demands isolation of not only the buried waste but much of the surrounding host rocks as well, and none of this entire system can be excavated for any reason. Hence ascertaining the natural resources present at any site is of much importance, and if present, the repository site must be chosen to minimize the effect on these natural resources as it will concern future generations.

As mentioned throughout this book, the use of natural barriers to isolate any lost radionuclides, or to impede their flow so that only nonhazardous amounts may reach the biosphere, is an important consideration. Some rocks contain more natural multiple barriers than others. These natural barriers will be addressed for specific rock types. Generic parameters include the abovementioned depth, low porosity–permeability–water flow, the geomedia above and below a mined repository site, tectonic stability, and natural resources. Of the many rocks available on the earth's surface, the geologic conditions are best met by bedded salt, dome salt, granites, basalt, argillaceous rocks, and tuffaceous rocks.

Bedded Salt and Dome Salt Deposits

Historically, bedded salts deposits have received a great deal of consideration as potential waste repository sites. The basic arguments for the favorable consideration of these rocks include extremely low water content (if the water content were high, the rocks would have dissolved), extremely low porosity and permeability, high thermal conductivity (to minimize thermal near-field effects), and the ability of salt to anneal after fracturing (i.e., should faulting occur, the fault would self-seal, thus preventing radionuclide escape) among others.

Bedded salts under consideration are properly classified as marine evaporites. These rocks formed by slow evaporation of seawater in large tectonic basins, several of which are shown in Fig. 6-1. The most common rocks in a marine evaporite sequence are the result of the precipitation of minerals at different times caused by decreases in original water volume. As shown by Usiglio (1849), the sequence of mineral precipitation as a function of water loss by evaporation is given as follows: at about 30% original volume (POV), calcite precipitates; at 19.5 POV, calcium sulfate (gypsum) precipitates; at 10 POV, halite precipitates; and at 2–4 POV, K- and Mg-salt precipitate. Thus the borders of these marine basins will consist of alternating layers of calcite mixed with some gypsum (or anhydrite, which forms from it as a function of burial). The alternating layers are primarily a result of fluctuations in seawater level due to flooding or other influx, which causes the volume remaining to oscillate between slightly above and below POV. Similarly, these local variations will occur throughout the deposition so that any zone very rich in a particular mineral will commonly show more alternating layers both near the top and bottom of the layer. Calcium carbonate precipitates in abundance because of the relatively large POV for precipitation coupled with addition of dissolved CO_2 from the atmosphere, which facilitates $CaCO_3$ formation.

Fig. 6-1. Bedded salt deposits and salt domes in the contiguous United States. *Source:* Modified from DOE (1979).

Toward the center of the marine basins calcium sulfates become more abundant, and they become mixed with layers of halite, NaCl, near the center. This is the most common marine evaporite sequence: a thick series of alternating layers of anhydrite and halite. The amount of water influx can commonly vary a few percent at any time, hence conditions for precipitation as opposed to dissolution can vary near POV values of 30, 19.5, and 10 (i.e., points where calcite, gypsum, and halite, respectively, precipitate). By the time evaporation has reached 2–4%, the influx can remain about the same, but the effect on the residual brine is more drastic. Commonly the residual brine is so dispersed that buildup of K- and Mg-salts is difficult, if not impossible. Inspection of Fig. 6-2 will reveal that only infrequently do the bedded salt deposits contain widespread accumulations of K- and Mg-salts. When such salts occur, however, it implies not only relative quiescence for salt formation, but essentially complete isolation of these areas from all but insignificant amounts of water in its postsedimentation history. This is an important point because the presence of these K- and Mg-salts means the overall rocks have kept the rock sequence isolated from water infiltration, thus attesting not only to its dry nature but to its inert chemical and isotopic integrity as well.

At random intervals during the evaporation of seawater, aeolian materials, usually clays, may be deposited by storm-like conditions. This action results

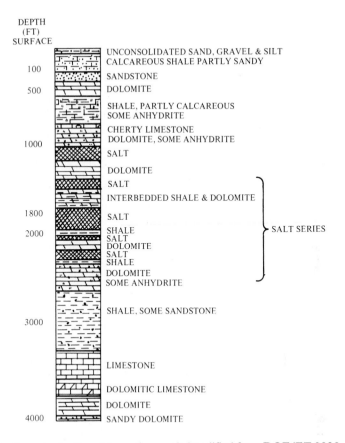

DEPTH
(FT)
SURFACE

UNCONSOLIDATED SAND, GRAVEL & SILT
CALCAREOUS SHALE PARTLY SANDY
100
SANDSTONE
500   DOLOMITE

SHALE, PARTLY CALCAREOUS
SOME ANHYDRITE
CHERTY LIMESTONE
DOLOMITE, SOME ANHYDRITE
1000
SALT

DOLOMITE

SALT
INTERBEDDED SHALE & DOLOMITE

1800   SALT

2000   SHALE
SALT
DOLOMITE   SALT SERIES
SALT
SHALE
DOLOMITE
SOME ANHYDRITE

SHALE, SOME SANDSTONE
3000

LIMESTONE

DOLOMITIC LIMESTONE

DOLOMITE
4000   SANDY DOLOMITE

Fig. 6-2. Generic stratigraphic section—salt (modified from DOE/ET 0028, 1979).

in interbeds of clay minerals with other oxides–silicates mixed with the clay. These interbeds are usually thin, but cover very wide lateral extents. These interbeds may be of great importance in the multibarrier concept, as the clay minerals have been demonstrated to be effective sorption agents for many radionuclides.

Nonmarine evaporites form by the evaporation of relatively large bodies of water not in direct communication with seawater. Thus their chemistry is controlled by the elements fed into these basins by weathering and other processes and not on the composition of normal seawater. Nonmarine evaporites will typically contain thick sequences of sodium carbonates and sulfates, calcite, and dolomite, but virtually no chloride minerals, as dissolved chlorine content in these shallow seas is, unlike seawater, very low. Figure 6-1 also contains the larger nonmarine evaporites as well as the marine evaporites and dome salts.

The distribution of different types of major salt accumulations in the United States is shown in Fig. 6-1. In Fig. 6-1 are shown occurrences of bedded salt basins, dome salts, and smaller, nonmarine salt basins as well. The bedded salt deposits of the Michigan Basin, Appalachian Basin, and Permian Basin have received the most attention for repository considerations. The Delaware Basin, a subdivision of the Permian Basin, is the site of the WIPP project (see Chapter 7). A brief description of some of the salt basins shown in Fig. 6-1 follows. The Michigan and Appalachian Basins contain bedded evaporites of the Salina Group, which were deposited in the Late Silurian. The salt deposits cover an area (largely in the subsurface) of over 260,000 km^2 in Michigan, Pennsylvania, New York, Ohio, and West Virginia, and part of Canada as well. The Salina Group consists of basal red and green shales (Vernon Formation), interbedded salt, anhydrite, dolomite, shale (Salina salt), and upper layers of shale, dolomite, gypsum (Camillus shale). Several major bodies of salt, commonly separated by anhydrite or thin shale layers, are common in the Salina Salt. The salt series is more than 400 m thick in places. Depths of the salt vary from near surface in western New York to more than 1800 m in Ohio. The Michigan Basin contains both salts of the Salina Group and the Lucas Formation (Devonian). The stratigraphy of the Salina Group is essentially the same as that of the Appalachian Basin, and that of the Lucas Formation consists of salt interbedded with layers of shale and anhydrite. The salt layers of the Salina Group occur at depths of more than 2000 m in Michigan, and are thinner than the salt units in the Appalachian Basin.

The Permian Basin contains extensive salt deposits of Permian age. The Delaware Basin, which is the site of the WIPP project, is a subbasin that contains large potash reserves (Chapter 7). The WIPP site infringes only on subeconomic parts of the potash-bearing rocks. The salt beds occur over 300,000 km^2. The salts range in age from Early Permian in the Anadarko subbasin of Kansas and Oklahoma to Late Permian in the Delaware subbasin in southeastern New Mexico. The bedded salts are often more than 75 m thick.

The Williston Basin, which extends into Canada, contains at least 11 salt beds recognized in the subsurface. The salts range in age from Jurassic (Dunham Salt) to Devonian (Prairie Formation), and all occur at depths greater than 600 m.

Dome salts are most common in the Gulf Basins, as is shown in Fig. 6-1. They are formed by lateral and upward diapiric movement of massive bodies of salt. Domes result from penetration of the overlying rocks. If upwelling without penetration of the overlying rocks occurs, then an anticlinal structure results. The depths of the salts vary from about 900 m, in the Gulf Interior Basin, to more than 3000 m, in the Gulf Coastal Basin. The low density of the salt compared to overlying, more dense, materials causes it to move upward as the pressure from overlying sedimentary rocks and sediments increases, commonly forming diapiric structures. In order for this upward flow to occur, the salt must be fairly thick, be at a sufficient depth, and have sufficient mass and volume to allow the diapiric movement to take place. The evaporites consist of Jurassic or Permian (Louann salt) and Permian (Buckner Formation) age formations.

The other salt occurrences are described in the readings listed at the end of this chapter.

A generic site description for bedded salt is now in order. U.S. Government guidelines for repositories in bedded salt require a salt bed to be at least 76 m thick at a depth of less than 600 m. These parameters are used to meet the requirements of long-term mine openings and repository waste containment. In actual practice, depths of a possible repository may be somewhat greater (see discussion of the WIPP site, Chapter 7). A simplified generic section is shown in Fig. 6-2, in which the salt series is shown as alternating layers of salt and shale and/or dolomite. This generic section was compiled by using data from existing basins, rejecting those in which the depths of the salt beds were too shallow or too great, and taking the composite geologic columns from the remaining basins into consideration. It is intended only as a working guide, but it is interesting to note that the model shown in Fig. 6-2 is essentially that of a typical marine evaporite. The salts considered for the generic model assume the near-zero dip of bedded salts in large, essentially underformed sedimentary basins. The regional groundwater flow is assumed low, due to topographically near-identical recharge and discharge areas. The assumed porosity and permeability data for the rocks shown in Fig. 6-2 are given in Table 6-1.

It is important to briefly discuss the physical properties of bedded salts now. First, however, let us remember that a repository involves a very large mass of rock. This is important to remember when discussing a material like salt, as data from laboratory experiments on a very small amount of material

Table 6–1. Hydrologic parameters for Fig. 6–2.

Rocks	Hydraulic Conductivity (cm/sec)	Range	Porosity (%)
Till	7×10^{-6}	$10^{-7}-10^{-3}$	20
Calcareous shale	1.2×10^{-5}	$10^{-10}-10^{-1}$	13
Sandstone and dolomite	8×10^{-5}	$10^{-10}-10^{-2}$	20
Calcareous shale with anhydrite	1.2×10^{-5}	$10^{-10}-10^{-3}$	13
Cherty limestone, dolomite with anhydrite	6×10^{-5}	$10^{-7.5}-10^{-3}$	20
Dolomite	$10^{-4.3}$	$10^{-7.5}-10^{-2}$	30
Salt	Nil	$10^{-21}-10^{-8}$	0.5
Dolomite–shale	$10^{-5.3}$	$10^{-10}-10^{-3}$	20
Shale	$10^{-6.3}$	$10^{-10}-10^{-3}$	16

Source: Modified from DOE (1979).

may not directly be applicable to a large rock body. Movement of fluid inclusions in salt is a good example. Laboratory measurements on small halite specimens clearly show that the fluid will migrate toward a heat source to which it is exposed. This has been interpreted as an argument in support of the idea that a relatively high heat source, such as radioactive waste, will cause migration of fluids from surrounding rocks toward the waste, thus increasing the changes for fluid–waste interaction. Although heating experiments are currently underway at the WIPP site in southeastern New Mexico to test this idea, we can also discuss it in terms of an experiment kindly conducted by nature some 33 million years ago. At that time the bedded evaporite sequence some 10 km from the WIPP site was intruded by an igneous dike (see discussion in Brookins, 1981c). This dike was emplaced at a temperature of 800–900°C, well above the temperatures expected for buried HLW. Further, no barriers were present to stop any transfer of elements or isotopes between the dike and the evaporites. While the length of time for crystallization is unknown, the fact that the dike is crystalline, not glassy, indicates that a fairly long time elapsed before crystallization occured. The host rocks to this dike have been studied by Loehr (1979) and by Brookins (1981c). Loehr's work shows that while there is some recrystallization of clay minerals in the evaporites near the contact with the dike, the normal clay mineral assemblage is reached within about 2 m of the contact. Further, fluid inclusions give temperatures that are reset by the dike within 1.5 m of the dike contact. The writer has examined the uranium, thorium, and lanthanide content of the dike and host rocks and finds no evidence for transfer of elements from the dike into the evaporites, which strongly suggests that the salt acts as an effective barrier against elemental transport. More important, Loehr's work shows that a high-temperature source such as the dike is sufficient to perturb fluid inclusions only within a meter or so of the contact, and even this fluid source was not so remobilized as to cause formation of large pockets of high-temperature, high-salinity brine. What this means is simply that the laboratory studies of in-grain diffusion cannot safely be extrapolated to massive rock conditions. Were a direct extrapolation possible, large amounts of fluid should have been generated near the dike, but this was not the case. Hence caution must be used in evaluating the physical properties of such a medium as salt.

It is well known that salt will, if subjected to a large lithostatic load, creep and readjust until lithostatic equilibrium is attained. The ability of salt to creep has both favorable and potentially unfavorable implications for radioactive waste repository conditions. The fact that faults in salt anneal readily due to creep prevents water from flowing through the salt, hence its extremely low porosity and nil permeability. Yet keeping mined openings stable is also a major problem. At depths of 800 m the creep may be such that openings may be difficult to maintain, hence depths of less than 800 or

600 m are advocated, and even at the shallower depths contingency plans will have to be made to ensure availability or retrievability at the storage areas.

Temperature has a pronounced effect on the physical behavior of salt. The rate of creep is increased with increasing temperature, and it has been shown (see Lomenick and Bradshaw, 1969) that the strain rate is increased by a factor of seven for a temperature increase from 23°C to 100°C, and the compressive strength drops by 10% for a temperature increase from 23°C to 200°C. Their data also show that the heat capacity increases and the thermal conductivity decreases with increasing temperature. However, we must keep temperature, buried waste, and rocks in perspective. Unlike the case of a very large heat sink emplaced in salt, the radioactive waste canisters are extremely small, the dimensions about one foot in diameter by 12 feet in length. The canisters are placed so that there will not be a sympathetic overall increase of heat by more than one canister for a given volume of salt. Hence the temperature effects on the salt will be due principally to the heat from only one canister. Due to above-ground storage prior to burial, the temperature at the canister wall will be no greater than about 100°C, and ambient temperature at a depth of 500 to 800 m will be approximately 60°C. This means that the overall heating effects of the waste package on the salt will be very small.

Pressure effects in salt are difficult to assess, but it is generally agreed that the creep rate of salt will increase with an increase in stress.

The water content of bedded salt is moderately well known. Some is present in fluid inclusions, some along grain boundaries, in crystal defects, and in isolated brine pockets. The brine pockets contain much less by volume than the structurally bonded water, but are conspicuous because of their locally important volume. If the brine is heated, it can force release with considerable energy. For this reason areas of high brine density are avoided. It should also be pointed out, though, that these brines are usually saturated with the major ions of the salt deposits, hence they do not pose a major problem to the repository, as they are unable, by themselves, to cause salt dissolution. The laboratory experiments conducted on salt from the Kansas salt mine occurrences near Lyons and Hutchinson showed that fracturing occurred at temperatures from 260°C to 320°C. For massive salt, and canisters placed so that the near-field temperatures are well under 150°C, no fracturing will occur. Further, even if local, temperature-induced fracturing were to occur, the salt would self seal and the overall effect on the repository would be negligible.

The irradiation effects on salt have been well studied (Bradshaw and McLain, 1971). It is known that irradiation exposure doses of some 10^8 to 10^9 rems produces some changes in structural properties of salt, but these changes are negligible when considering the mass of salt repository.

Fig. 6-3. Granitic rock in the contiguous United States. *Source:* Modified from DOE (1979).

Granites

Large bodies of granites are common in the United States. These rocks, many of which are shown in Fig. 6-3, consist of varying amounts of plagioclase and K-feldspar with quartz, with accessory biotite and, less commonly, hornblende. Other accessory minerals include zircon, apatite, magnetite, pyrite, sphene, beryl, and monazite. The formulas and other chemical data for granites are shown in Table 6-2. Granites typically occur as plutons, fairly large masses formed at depth and moved upward by tectonic forces. Many are fairly homogeneous and fine to medium grained. Batholiths consist of many plutons in close association. Granite forms well below the earth's surface, hence its porosity is low and it contains very little natural moisture. The permeability of such rocks is also low, especially for unweathered granite. Weathered granite may be more permeable due to chemical attack, which is more pronounced along grain boundaries, and it is known that weathered granites may lose elements easily affected by oxidation. Unweathered granites, however, usually retain their chemical and isotopic integrity (see Brookins *et al.*, 1981a,b). The strength of granites is high, and it is for this reason, as well as its resistance to normal weathering conditions, that it can be successfully used as an ornamental and building stone. Its component minerals are hard, thus adding to its strength, and it is remarkably durable. Granite is also rigid and does not deform under most stresses at shallow depth (i.e., at repository depth the stress increase will be negligible). It is also less susceptible to heat effects than bedded salts; its resistance to heat lies in the fact that the minerals formed at several hundred

Table 6–2. Chemistry and minerology of granite.

Mineralogy
  Essential Rock Forming Minerals
  Quartz SiO_2 K-Feldspar $KAlSi_3O_8$
  Plagioclase $(Na,Ca)Al_{1-2}Si_{3-2}O_8$
  Common Accessory Minerals
  Biotite $K(Mg,Fe)_3Al_2Si_3O_{10}(OH)_2$
  Hornblende $Na_{0-1}(Mg,Ca,Fe,Al)\ (Al,Si)_4O_{11}(OH)_2$
  Zircon $ZrSiO_4$ Apatite $Ca_5(PO_4)_3(OH,F)$
  Magnetite $Fe''Fe_2''O_4$ Pyrite FeS_2
  Sphene $CaTiSiO_5$ Beryl $Be_3Al_2Si_6O_{18}$
Chemistry (Average Value)
  Major Oxides (%)
  SiO_2—77.0 Al_2O_3—12.0 Fe_2O_3—0.8   FeO—0.9
  CaO—0.8 Na_2O—3.2 K_2O—4.9 H_2O—0.3
  TiO_2—0.1   MgO—0.08   others—0.1–0.2

Source: Modified from DOE/ET-0028 (1979).

degrees centigrade (greater than 570°C). The minerals are not highly reactive at moderately low temperatures such as would be reached at radioactive waste canister walls (100–150°C).

The chemistry and mineralogy of granites is of interest when addressing their potential to hold radionuclides that might escape from buried radioactive waste. When a granite forms, several radioactive species are distributed throughout. Potassium, followed by rubidium, is held in K-feldspar and in lesser amounts in biotite, initial strontium is held in calcium sites in plagioclase, apatite, sphene and K-feldspar and in small amounts in biotite. Thorium and uranium compete for the same sites, and, due to its slightly smaller ionic radius, U^{4+} is preferentially incorporated into zircon, sphene, and apatite. Some uranium and thorium are concentrated in the last magma and form minor phases along grain edges throughout the rock. Thorium is found in zircon, apatite, and sphene also, as well as in monazite, allanite, and in the widely distributed minor phases with uranium.

Uranium is often leached from weathered granite but is preserved in unweathered granite. For unweathered granite, radiometric ages indicate the successful retention of radioactive K, Rb, U, and Th and radiogenic Pb, Sr, and Ar. The ages often show good agreement and attest to suitability of the rock for retention of these radioactive and radiogenic nuclides. Even during a strong metamorphism, in which radiogenic ^{87}Sr is lost from Rb-rich phases and is incorporated into Ca-rich, Rb-poor phases, this redistribution takes place on the scale of fractions of millimeters. Thus whole-rock isochron dating (see Chapter 5) still yields the age of original crystallization. Hence the whole rock behaves as a closed system despite the metamorphism(s) for Rb and Sr. Only radiogenic ^{40}Ar is effectively lost from K-bearing minerals during metamorphic events. How effectively ^{40}Ar is lost can be gauged by noting that the dominant isotope of Ar in the earth's atmosphere is ^{40}Ar, and not ^{38}Ar or ^{36}Ar, and that the ^{40}Ar has been derived from the earth.

Areas containing large bodies of granite are shown in Fig. 6-3. Granitic rocks occur throughout the areas shown, but are mixed with and surrounded by many other rocks. For ease of presentation, only areas of major occurrences of granitic rocks are shown, and no attempt is made to separate granite from quartz monzonite, granodiorite, or other "granitic" rocks. Granites are not bedded like sedimentary rocks, and none of the channel ways common along layers in sedimentary rocks are found in them, although directions of flow banding are common and minerals are often somewhat layered due to late magmatic movements. Granite bodies are typically fractured, usually as well-defined joint systems. These joints are usually filled with secondary minerals, thus preventing extensive fluid flow along the joints. If the vein- or joint-filling minerals formed at many different times during the history of the rock, then it can be argued that these fractures may make the rock susceptible to infiltration by fluids at virtually any time, and thus, the rocks might not be suitable for consideration as a repository site. Alternately, if it can be demonstrated that the fracture-filling material was formed early in the rock history, or else that the material found in the fractures was derived from depth and not from sources in contact with meteoric water, then the rocks are presumably closed to contact with surface or near-surface sources. Not all granites have been investigated to determine the history of fracture-filling materials. Useful information has been obtained from the granitic rocks being studied as part of the Hot Dry Rock (HDR) Program for geothermal energy at the Los Alamos National Laboratory. In this program, deep holes (about 3 km) are drilled into granitic rocks in areas of high heat flow, and the rocks are fractured at depth. Water is then injected into one hole, and is heated by heat flow at 3 km, producing steam from an interconnecting recovery hole. In order for the HDR plumbing system to work, the fractures must not be filled with newly formed minerals. Further, existing fractures must be studied to determine if the minerals found therein formed early or late in the history of the rocks. Study of the HDR site rocks at Fenton Hill, New Mexico has proved very interesting. In this area the Precambrian granitic rocks are overlain by Paleozoic limestones and other sedimentary rocks and, in turn, by younger sedimentary rocks and very young volcanics. The use of strontium isotopes as a tracer for the fracture-filling calcite allows one to determine if waters dissolved material from the limestone and transported it to depth. The ^{87}Sr/^{86}Sr average ratio in limestone is approximately 0.708 ($\pm$0.0006), while the ^{87}Sr/^{86}Sr ratio in many of the granites from Fenton Hill is on the order of 0.75 or greater. If the calcite in fractures is found to contain a ^{87}Sr/^{86}Sr ratio of 0.708 or so, then it is indistinguishable from the limestone value from which it may have been derived (note that some parts of even granitic rocks possess low ^{87}Sr/^{86}Sr ratios, hence a low ratio does not, without supporting data, prove either rock as the source). A ratio significantly higher than 0.708, however, would indicate a source other than the limestone. In fact, the calcite from the fractured granitic rocks at the HDR site possesses $^{87}Sr/^{86}Sr$ ratios that vary from 0.73 to 0.735, thus showing that they have not been derived from the overlying limestone sequence. Further, leaching studies of the granite yield leachates with Sr ratios nearly identical to those found in the fracture-filling calcites. The results of these studies prove that the granitic rock plumbing system was favorable for HDR development, yet the data can also be used to show that these rocks have been isolated from the overlying meteoric sources or Paleozoic section. Very simply, if these rocks were being considered for a radioactive waste repository, the Sr isotopic data would indicate the rocks to be suitable for such consideration.

Selecting a "generic" granite poses many difficulties. The assumption is made (see DOE/ET-0028, 1979) that a typical site will consist of a continuous granite body overlain by a fairly thin cover of weathered material. Jointing and other fractures, many of which may be filled with secondary minerals are expected to be found in such a granite. It is further assumed that the topography is low over the area where the granite is found, and that groundwater will follow this topography. For the sake of a generic site, it is assumed that groundwater flow is through joints and other fractures and persists to depths of about 600 m, the depth of a possible repository. Studies at the HDR site at Fenton Hill, New Mexico suggest that the actual communication between surface waters and depth may be on the order of no more than about 200 m, and much of this due to flow in sedimentary and other covering rocks. The actual flow in the granitic rock section may be less than 50 m, although more followup work is needed to verify this. Hydraulic conductivities will range from about 10^{-9} to 10^{-8} cm/sec, with lower values expected at depth because of closing of fissures due to pressure and/or more tightly sealed fracture-filling minerals at depth. The frequency and size of the joints will control the porosity, a reasonable value being 0.005 (0.5%). Stress in the granite can be divided into a vertical component, equal to the overburden, and a horizontal component, expressed as a fraction of the vertical stress, and varies between values of 0.5–2.0.

Many granites contain zones of pronounced shearing that must be avoided as they are zones of potentially high fluid flow and, in general, zones of weakness.

Experiments of long duration on granite are not as plentiful as those on salt, but are adequate to comment on some environmental aspects. Creep in granites, for example, is much less than that in salt or shale, and is not considered to be significant in most cases. Temperature effects are more produced, for example, a 200°C increase will result in lower Young's modulus, Poisson's ratio values, and stresses. Pressure effects in granite are more difficult to assess due to the different behavior of highly jointed versus relatively fracture-free parts of a granite, but, in general the tendency is for confining pressure to increase the strength as well as the Young's modulus

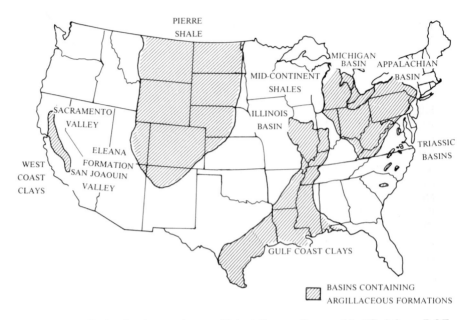

Fig. 6-4. Shales in the contiguous United States. *Source:* Modified from DOE (1979).

value. Poisson's ratio would show an initial increase followed by a decrease in a dry granite. Moisture effects in granite are low, but will vary widely during experimental investigation (Obert and Duvall, 1966). Jenks (1975) has studied irradiation effects in granite and found them to be insignificant.

Shale

The areas of the contiguous United States underlain by shale are shown in Fig. 6-4. In these areas shales are usually mixed with other rock types, such as salt, carbonates, and some arenaceous units. Shale forms from compacted muds and contains a widely variable amount of detrital minerals, such as quartz, feldspars, clay minerals, rock fragments, and authigenic minerals, such as different types of clay minerals, quartz, calcite, and occasional other minerals. The degree of compaction in shales is variable, with younger, poorly sorted shales less compacted than older, indurated varieties. The important minerals and chemistry for some shales is shown in Table 6-3. Shales are usually fissile, although they may be so finely grained that breakage occurs across as well as parallel to bedding. Shales are characteristically weak due to the soft and friable minerals and often-weak cementing agents. The natural moisture content and porosity are high due to the abundance of clay minerals, but the permeability can be quite low because many of the zones of high porosity are not interconnected. In addition, many

Table 6–3. Chemistry and mineralogy of shales.

	Chemical Composition of Average Shales (%)		Average Mineralogy (%)	
	I[a]	II[b]		
SiO_2	58.1	61.9	Clay minerals[c]	58
Al_2O_3	15.4	16.9	Quartz	28
Fe_2O_3	4.02	4.20	Feldspar	6
FeO	2.45	3.0	Carbonates	5
MgO	2.22	2.40	Iron oxide	2
CaO	3.11	1.49		
Na_2O	1.30	1.07		
K_2O	3.24	3.70		
CO_2	2.63	1.54		
C	0.80	—		
H_2O	5.00	3.90		

[a]From Pettijohn (1975).
[b]From Mackenzie and Garrels (1971).
[c]Mainly montmorillonite, illite, kaolinite, and chlorite.

shales can, when subjected to tectonism, have a potential for plastic flow. Clay minerals are known to be effective for exchanging cations from solutions with which they come in contact, and slaking can be caused by alternating wetting and drying. The chemical reactivity of shales is, in many instances, unknown. Yet this is merely due to a lack of measurement, not unresolvable properties of shale. For example, it is known that Cs and Sr can be sorbed and exchanged readily in shale, based on laboratory experiment, but extrapolations to naturally occurring clay minerals in shales is problematic. A more basic question is: Have the clay minerals of shales remained closed systems since formation or not? If they have, then the rock has retained its chemical and isotopic integrity; if they have not, then it has remained at least a partially open system. This question can be answered by simple geochronologic studies. When the authigenic fraction (usually -2μ) is studied for its Rb–Sr systematics, a radiometric age equal to the age of sedimentation is commonly found, thus attesting to closed-system conditions. In these cases, since Cs has a greater retentivity than Rb (Kharaka and Berry, 1973), the rock is also closed to Cs as well as Rb and Sr. Hence any fissiogenic Sr (i.e., ^{90}Sr) or Cs (^{137}Cs), as well as Rb and Ba (i.e., retentivity of Ba is greater than Sr), will be retained in the shale. Further, should any radium be present, it is more immobile than either Sr or Ba, hence it, too, will be retained. The Rb–Sr method can also be used to study the parts of the shale that are more highly oxidized near or at the surface. When this is done, it is commonly noted that the Rb–Sr systematics break down, usually due to a combination of Sr addition by admixed, secondary carbonate minerals and Rb-fixation following K-fixation. Scattered about the Rb-Sr isochron and an overall anomalous date are the results. Such rocks will need further study to assess their suitability for repository consideration. One possible application of Rb–Sr work on shales could be to test the different K_d values for different samples of shale (or other rocks for that matter) from the same occurrence. It is probable, for example, that a sample of slightly to moderately oxidized shale will yield a different K_d than unoxidized shale, but without detailed petrography and Rb–Sr work, chemical clues to explain K_d variation may be nothing more than speculation. The Rb–Sr method offers the investigator a quantitative tool to assess and evaluate this problem.

Shales of the midcontinent range in age from Ordovidian to much younger. Many of these shales are thicker than 50 m and are found at depths of 300 to 450 m. Carbonate rocks commonly over- and underlie the shales, and are mixed with them in many places. Many of the shales are carbonaceous. The chemistry of carbonaceous shales is not well understood. For radioactive waste considerations, an assessment of "pro" versus "con" aspects of the organics present in the shales is important. It has been argued that organometallic transport of many radionuclides may make carbonaceous shales unattractive for radioactive waste repository consideration, yet it has also been argued that the highly sorptive, reducing environment of the carbonaceous shales should be well suited for retention of radionuclides and fission products. Surprisingly there are not many data that address this problem. Brookins' study of shales in many parts of the United States (and from foreign occurrences, too) shows that most unoxidized carbonaceouse shales have retained their Rb and Sr chemical and isotopic integrity since sedimentation, often for hundreds of millions of years. Further, these shales are closed to Ba and Cs and, by inference, to Ra. Still further, their actinide (U, Th) abundances and distributions are normal, thus suggesting no movement. Where lanthanide and other trace element data are available (Brookins, 1979b), these elements appear to not have migrated as well. Investigations of the reducing characteristics of shales suggest that their reduced carbon and iron contents are sufficient to cause reduction and retention of any $M^{5+,6+}$ actinides that might escape from hypothetically breached radioactive waste packages (Brookins, 1980a; Bondietti and Francis, 1979). The shales usually contain pyrite (FeS_2), which will also help to cause reduction of any U^{6+} present and the oxidized sulfur will trap any radioactive radium present. The theoretical and experimental data for shales (Brookins, 1980a) show that they are well suited for backfill purposes in all repositories and are suitable for consideration as a repository site. One distinct advantage of shales is the fact that, by definition, an infinite number of barriers are present for the effective retention of fissiogenic and radioactive species.

The Triassic basins ranging from South Carolina to Virginia, contain local thick accumulations of shale, many of which are coal bearing.

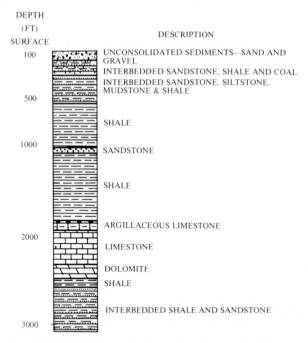

DEPTH
(FT)
SURFACE

DESCRIPTION

100 UNCONSOLIDATED SEDIMENTS—SAND AND
     GRAVEL
     INTERBEDDED SANDSTONE. SHALE AND COAL
     INTERBEDDED SANDSTONE. SILTSTONE.
     MUDSTONE & SHALE
500

     SHALE

1000 SANDSTONE

     SHALE

     ARGILLACEOUS LIMESTONE
2000
     LIMESTONE

     DOLOMITE
     SHALE

     INTERBEDDED SHALE AND SANDSTONE
3000

Fig. 6-5. Stratigraphic section for generic shale. *Source:* Modified from DOE (1979).

The Pierre Shale (Cretaceous) covers an extremely wide area (960,000 km^2), encompassing parts of 10 states (Fig. 6-4). It is basically a bentonitic mudstone mixed with clays and other materials. It varies in thickness from 150 m to more than 1500 m. The Pierre Shale may be organic rich, silt rich, calcareous, or marly.

The Eleana Formation (Mississippian) of Nevada is of interest because it occurs at the NTS. It consists of argillite, siltstone, quartzite, and some limestone. It is lithologically very inhomogeneous, and it is usually underlain by dolomite and overlain by volcanics, Permo–Pennsylvania limestone, or pediment.

A generic shale is, by definition, somewhat of a misnomer, due to the widely varying properties of shales from site to site. Yet the generic shale proposed (DOE/ET-028, 1979) is a good estimate of a typical shale (Fig. 6-5). It is assumed that a repository will be sited no more than about 450 m deep. The generic section shown is fairly typical of many shale sections measured throughout the United States. The assumed repository shale is a marine variety mixed with alternating layers of argillaceous limestone and carbonates with minor sandstone layers. It is fairly strong, more permeable than weaker shales, and is presumed to be calcareous, clayey, or carbonaceous. Further assumptions include siting at the edges of tecronically stable, undeformed geologic basins with low regional dip, and flat-lying strata. The potentiometric surface is equal to the land surface and the

Fig. 6-6. Potential repository basalts in the United States. *Source:* Modified from DOE (1979).

hydraulic gradient is low. Vertical stress in such rock is about equal to overburden stress, and horizontal stress is about 0.5 to 1.0 times that of the vertical stress.

Creep in shales is pronounced in some shales, yet for deeply buried, tight shales the creep is small. The strength of shale should decrease with increasing temperature, and it is noted that shales are commonly broken where heated near intrusives. Swelling in montmorillonitic shales is common, and may cause buckling of steel structures footed in shale or collapse of tunnel walls. Such effects are far less common in nonmontmorillonitic shales. Irradiation effects in shale are not well studied, but are presumed to be small from studies peripheral to salt studies.

Basalt

Basalts occur in many parts of the United States, and in great abundance in the western states. More or less continuous flows are found in the Keeweenawan Peninsula of Michigan and in parts of the Triassic basins in New England. The largest series of flows are shown in Fig. 6-6. Basalts are volcanic rocks that commonly consist of fine-grained crystals of plagioclase and pyroxene with many other minerals. Their average chemistry and mineralogy are given in Table 6-4. Basalts are very dense and possess high strength, thus their porosity and permeability are low and their water content is negligible. Their strength is not affected by mild heating but they will expand. Columnar or platy jointing is common, and many of the joints are filled with secondary minerals such as calcite, chalcedony, hydromica, and

Table 6–4. Chemistry and mineralogy of basalt.[a]

Chemistry		
SiO_2	47.65	
Al_2O_3	15.28	
Fe_2O_3	3.57	
FeO	7.54	
MgO	7.52	
CaO	9.91	
Na_2O	2.98	
K_2O	1.23	
TiO_2	2.14	
P_2O_5	0.44	
H_2O	1.51	
Mineralogy[b]		
Plagioclase	$(Ca,Na)(Al)_{2-1}$ $(Si)_{2-3}O_8$	
Pyroxene (augite)	$Ca(Mg,Fe,Ti,Al)(Si,Al)_2O_6$	
Olivine	$(Mg,Fe)_2SiO_4$	
± Hypersthene	$(Mg,Fe)SiO_3$	
Accessory Minerals		
Apatite	$Ca_5(PO_4)_3$ (OH,F)	
Ti-magnetite	$(Fe,Ti)Fe_2O_4$	
Pyrite	FeS_2	
Ilmenite	$FeTiO_3$	
Zircon or baddeleyite	$ZrSiO_4$ or ZrO_2	
Nepheline, quartz, others		

[a]From more than 4200 basalts; Chayes (1972).
[b]The mineralogy is extremely variable; pyroxene and plagioclase commonly make up more than 80% of basalts).

clay minerals. The development of these fractures may have occurred early (i.e., essentially deuteric) or very late (i.e., due to meteoric infiltration) in the rock history. It is important to know the ages of fracture filling but this has not yet been established for most basalts under consideration for radioactive waste repositories. The extensive flood basalts of the Pacific Northwest average about 100 m in thickness and cover about 200,000 km². Locally, they may be over 3000 m thick. The basalt flows range from 17 to 6 MYBP in age.

Unlike the generic salt and shale, the generic basalt is specifically intended at the basalts of the Pacific Northwest. This is justified because these basalts represent the most pronounced series of flows known in the United States and are, in turn, the only basalts being considered for repositories (see Fig. 6-7).

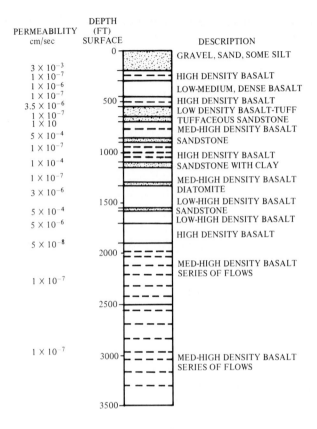

Fig. 6-7. Generic stratigraphic section—basalt. *Source:* Modified from DOE (1979).

The Columbia River Basalts may transmit considerable water because of brecciated or jointed surfaces; some flow occurs vertically due to the jointing. Confined flow is restricted to the interflow zones. The permeability of basalts is low, varying between 10^{-6} and 10^{-8} cm/sec, with the lower values occurring for thick, dense flows with decreased numbers of discontinuities (i.e., due to the higher *in situ* stresses). Porosity in basalt is related to the size and frequency of fractures, joints, and other openings such as vesicles. The effective porosity is less than one percent. The vertical stress in basalt is equal to the overburden pressure, and the horizontal stress varies from 0.5 to 2 times the vertical stress.

Creep in basalts is essentially nonexistent. Temperature effects in basalt are not pronounced, although lower stresses and the Young's modulus ratio are noted with increasing temperature. Pressure effects include an increase in strength and the Young's modulus ratio. Natural moisture content of basalt is very low, but its effects are not well understood. Radiation damage (see Jenks, 1975) is slight, but not well characterized.

Tuffaceous Rocks

There is no "generic tuff" currently proposed by the DOE, yet tuffaceous rocks are being investigated in great detail at the NTS, hence a brief discussion of such rocks will be given here.

Tuffs are composed mainly of pyroclastic material. When siliceous material is ejected by volcanic processes, the high viscosity of such material under low-pressure conditions usually results in explosion of the siliceous material in such a fashion that glassy products form. The tuffaceous rocks are usually rhyolitic to dacitic in composition, and locally make large volumes of volcanic terrain, sometimes reaching thicknesses in excess of several thousand meters. Primary materials, such as phenocrysts, are not abundant in tuffaceous rocks, commonly making up less than 10–20% of the rocks, and lithic fragments usually do not exceed 10% of the rock. Hence some 70–80% of the rock is composed of glass. Because the conditions accompanying the volcanism vary, as do conditions under which the tuffs are deposited, there is ample opportunity for reworking of the tuffaceous material. The pyroclastic pumice, perlite, glass shards, and ashy material are deposited downwind from the volcano vent. If undisturbed, the material is called air-fall or ash-fall. In the vent area, accumulation of the highly viscous, pyroclastic material may cause collapse over the magma chamber, in turn causing calderas to form, among other features, and resulting in a thick accumulation of pyroclastic rocks. Here rocks often form a single unit, commonly with a core of welded material. This welded core may devitrify to form feldspar and quartz with some cristobalite (Smyth and Sykes, 1980). The dense, glassy material at the base of the welded zone is called vitrophyre. Volcaniclastic sediments may occur at any location in which the earlier pyroclastic material has been reworked by surface waters or entrapped waters. Alteration in tuffaceous rocks is abundant, especially since glass is thermodynamically metastable in the surface environment (although the metastability may be on the orders of tens of millions of years; see Chapter 12). When the glass alters, zeolites are commonly produced. The rate at which zeolitization proceeds is a function of both porosity and water content, with more rapid alteration going hand in hand with increased porosity and water content. Vitrophyres may exist undisturbed for very long periods of time if the porosity and water content are low. Quartz and feldspar do not alter to zeolites (Sykes and Smyth, 1980), although the feldspars may alter to clay minerals. In Fig. 6-8 (from Smyth and Sykes, 1980) is shown an idealized ash-flow tuff cross section. Variation of the bulk density and grain density are shown as a function of the rocks, including alteration products present, in the cooling unit.

A series of experiments, continuing the present time, have been conducted

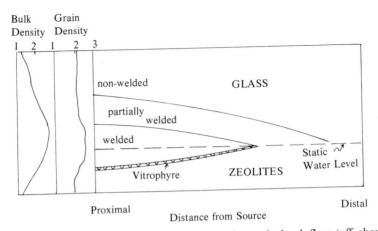

Fig. 6-8. Generalized schematic cross section of a typical ash-flow tuff showing typical grain and bulk densities. From Smyth and Sykes (1980).

on the tuffaceous rocks at the NTS (see Lappin *et al.*, 1981; and Johnstone and Wolfsberg, 1980). The findings of these investigations are numerous, and only a few will be elaborated on here. Mineralogical and petrological studies reveal that the dominant alteration process is devitrification, in which glass crystallizes to feldspar plus quartz or cristobalite, especially in densely welded parts of the tuff. In addition zeolitization occurs, in which glass reacts with groundwaters, especially in zones of high porosity. Clinoptilolite is the dominant zeolite mineral in the upper parts of the tuffaceous rocks, although analcime may be dominant in the lower parts.

Geochemical studies have been compiled by Wolfsberg (1980), who reports that, basically, tuffaceous rocks can effectively sorb many of the radionuclides that may be released from a breached radwaste canister. Some of the data are given in Table 6-5. The data are reported in terms of a sorption ratio, R_d, defined as:

$$R_d = \frac{\text{Activity in solid phase per unit mass of solid}}{\text{Activity in solution per unit volume of solution}};$$

R_d is the same as the distribution coefficient, K_d, except that no approach to equilibrium is assumed. As pointed out by Wolfsberg (1980), in an equilibrium situation identical values for R_d should be obtained for all conditions intermediate between all activity initially on the solid (desorption) and in solution (sorption). Yet wide variations are noted in the examination of the data in Table 6-5, due in part to the fact that the sorptive properties of the tuff vary with the abundance and composition of the constituent minerals, glass, and lithic fragments. Yet although the approach to equilibrium is not reached, the R_d values are in general agreement whether the R_d value is

Table 6–5. Ranges of sorption ratios (mL/g).

Element	Devitrified Tuff	Zeolitized Tuff
Cs, Sorption	150–870	8600–29,000
Cs, Desorption	310–630	13,000–33,000
Sr, Sorption	53–190	1800–20,000
Sr, Desorption	56–200	2700–20,000
Ba, Sorption	430–1500	15,000–130,000
Ba, Desorption	440–1300	34,000–190,000
Ce, Sorption	80–15,000	550–2000
Ce, Desorption	400–15,000	1200–13,000
Eu, Sorption	90–7500	1200–2500
Eu, Desorption	800–7300	2100–8700
Am, Sorption	130	180
Am, Desorption	2200	1100
Pu, Sorption	110	120
Pu, Desorption	1100	340
Tc (air), Sorption	0.3	0.2
Tc (air), Desorption	1.2	2.0
Tc (N_2), Sorption	8–26	13
Tc (N_2), Desorption	18–79	118
U (air), Sorption	1.6–2.2	2.3–5.1
U (air), Desorption	6–13	15
U (N_2), Sorption	0.5–1.5	15
U (N_2), Desorption	2–14	57
I, Sorption	0	0

Source: From Wolfsberg (1980).

obtained by sorption or desorption, and this information in turn indicates that the tuffs are effective scavengers of the radionuclides shown in Table 6-5.

Zeolite-rich tuffaceous rocks generally show the highest sorption. This is especially true for Sr, Cs, and Ba. Wolfsberg (1980) suggests that cation exchange processes may account for this, for the devitrified, but non-zeolitized tuffs, yield lower, although still high, sorption values. Wolfsberg (1980) proposes that this type of tuffaceous rock may be best suited for repository study because of favorable mechanical and thermal properties, and the fact that such rocks are commonly surrounded by the zeolite-bearing rocks (thus providing an additional barrier to radionuclide migration). Of the radionuclides listed in Table 6-5, Cs, Sr, and Ba are strongly sorbed by the zeolitized tuff as well as by the devitrified tuff. The sorption of Ce and Eu appears to be independent of zeolite content, being high in both tuff types. Am and Pu are also strongly sorbed by both tuffs, and the species responsible for transport of these elements in ionic form is not well understood. Technetium is moderately well sorbed even under air-saturated conditions, and more strongly sorbed under a nitrogen atmosphere condition maintained during the experiment. This must in part be the result of reduction of Tc(VII) to Tc(IV), as the higher-valence species is known to be more mobile (see Brookins, 1978a). Uranium is also moderately to very well sorbed under both air- and nitrogen-saturated conditions by both tuff types, in part due to reduction (U(VI) to U(IV)) and in part to the ready sorption of U(VI) by tuffaceous materials. Iodine is not sorbed by the tuffs, yet additives to the tuffs will effectively sorb this element. Further, ^{129}I may not be a component of HLW.

Of further interest, Wolfsberg's studies reveal that sorption is, in general, greater at 70°C than at 22°C, although the difference is slight. This implies that, for a buried radwaste package, when the temperature is somewhere near or below 100°C, sorption will effectively retain or retard radionuclide movement.

Sorption is further complicated by changes in groundwater composition, which in turn may affect speciation of elements under study. Further, concentration of minor constituents may not be problematic. Wolfsberg (1980) reports no changes in sorption measurements for technetium between 10^{-3} and 10^{-12} mol, and none for Pu between 10^{-6} and 10^{-13} molar. Variations in composition of other elements is currently under study. Microautoradiographic studies (Wolfsberg, 1980) show that minerals, such as zeolites, clay minerals, and alteration rims on some minerals are better sorbing media than quartz or feldspar.

Studies on the thermal conductivity of NTS tuffs were conducted by Lappin et al. (1982). Data are given for numerous samples from drill hole USW-G1 (see Fig. 7-8 for location) for examination of grain density, porosity, and thermal conductivity (W/m°C) at different temperatures. The nonzeolitized tuffaceous rocks show a higher thermal conductivity than the nonzeolitized samples (Lappin et al., 1981).

Some of the mechanical properties of NTS tuffs have been compiled by Olsson and Teufel (1980), although many of the earlier data are from samples well removed from Yucca Mountain. Nevertheless, they are of interest. Of more interest, however, are the recent data from UE25a-1 and G-tunnel, which show that the higher the degree of welding in the tuff, the higher the strength and the higher Young's moduli, and that higher confining pressures cause greater strength. Of interest to the modeling effort is the fact that the welded tuff is stiffest normal to bedding while the nonwelded tuff is stiffest parallel to bedding. Knowledge of this, however, will make repository design more clear by allowing for this difference as different rock units are examined. Further, results from G-tunnel (Olsson and Teufel, 1980) yield the following data: average values for tensile strength, 2.75 MPa; Young's modulus, 7.43 GPa; and Poisson's ratio, 0.17. These authors demonstrate that there is a regular, predictable relationship between strength and porosity, as well as an effect due to water content. For example, dry tuff is 25% greater in compressive strength than its saturated counterpart. Also, temperature effects show that at 200°C there is a 30% decrease in strength from that at room temperature.

Radiation effects in NTS tuff are reported by Johnstone (1980), who notes that any radiation effects are restricted to the very near field.

Geohydrologic Considerations

The geohydrology of any site under consideration for radioactive waste disposal is of utmost importance. Parameters that influence the geohydrology are: intergranular or fracture hydraulic conductivity, hydraulic gradient, porosity (or tortuosity), leach rate of radioactive species from the waste form, hydraulic dispersion, radioactive decay, ion exchange, and distance to the biosphere.

The velocity of groundwater flow depends on hydraulic gradient, hydraulic conductivity (or permeability), and porosity of the geomedia. These parameters are related as follows (Darcy's Law):

$$V_D = Q/A = kH \qquad (6\text{-}1)$$

or

$$V_t = V_D/P = kH/P, \qquad (6\text{-}2)$$

where

V_D = Darcy velocity, length (d)/time (t),
V_t = average pore water velocity = d/t,
Q/A = flow of water per unit area (specific discharge) = $d^3 t^{-1}/d^2$,
H = hydraulic gradient, d_{H_2O}/d,
k = hydraulic conductivity, d/t,
P = porosity (void volume fraction).

The Darcy velocity (flux) yields the flow of water per unit area, although in the case of flow through a medium such as rock, the average flow is greater than the flux velocity by the factor of the reciprocal of the medium's porosity (Eq. 6-2)). Bailey and Marine (1980) have tabulated many sets of velocities for different conditions. Equations (6-1) and (6-2) are, however, directly applicable only for intergranular flow. For flow in a fracture, analogy with a narrow crack is made (Bailey and Marine, 1980) so that:

$$Q = \frac{2}{3} \left[\frac{(P_0 - P_d)\, d_B^3\, d_w}{v d_L} \right],$$ (6-3)

where

Q = flow, $d^3 t^{-1}$,
P_0 = pressure at inlet of crack, $Md^{-1}t^{-2}$,
P_d = pressure at outlet of crack, $Md^{-1}t^{-2}$,
d_B = half-width of crack,
d_w = length of crack perpendicular to direction of flow,
v = viscosity of liquid, $Md^{-1}t^{-1}$ (0.0089 g/cm/sec for water at 25°C),
d_L = length of crack from inlet to outlet,
$(P_0 - P_d)/d_L$ = hydraulic gradient,

and the velocity of liquid through the crack, as function of distance from crack centerline is given by:

$$V_x \left[\frac{(P_0 - P_d)}{2 v d_L} \right] [1 - (d_x^2/d_B)],$$ (6-4)

where

d_x = distance from centerline of crack.

The average velocity in the crack is two-thirds the centerline velocity (Bailey and Marine, 1980).

The *leach rate* (an alternate way of discussing leach rate is given in Chapter 12) of a solid is commonly expressed in units of grams of solid leached per cm^2 of surface area per day (g/cm^2/day). The total amount of solid leached per day is obtained by consideration of the weight fraction of the radionuclide in the solid, viz:

$$M_I = W_I S l,$$ (6-5)

where

M_I = mass of radionuclide I leached per day, g/day,
S = surface area of waste form, cm^2,
W_I = width fraction of radionuclide in solid,
l = leach rate of solid waste form, g/cm^2/day.

The leaching of the waste form is proportional to the surface area of the waste form. If the leach rate is expressed in units of the fraction of amount of waste form leached per unit time then we obtain:

$$F = \frac{S l}{M_S} = (S/V)\,(1/\rho),$$ (6-6)

where

F = fraction of waste form leached per day,
M_S = mass of solid waste form, in g,
V = volume of waste form, cm^3,
ρ = density of waste form, g/cm^3.

Thus the fraction of the waste form leached is proportional to the surface/mass ratio of the waste form and to the leach rate. This ratio is a function of the density of the waste form as well. The S/V ratio [Eq. (6-6)] is dependent on the size and shape of the waste form. Data for spheres and cylinders have been tabulated by Bailey and Marine (1980). For cylindrical waste form storage of 30- to 60-cm diameter and about 250-cm length, the S/V ratio ranges from 0.072 cm^{-1} to 0.138 cm^{-1} for the 60- and 30-cm-diameter cylinders, respectively. The actual S/V ratios may be higher, however, due to internal cracking of materials such as glass (see Bailey and Marine, 1980). For a spherical waste form, $S/V = 6/D$ and $S/M = 6/D$, where D is the diameter of the sphere.

Interestingly, Bailey and Marine (1980) have shown that changes in the ratio S/V does not drastically affect the leach rate until at least 50% of the waste form has been removed by leaching, at which point the leach rate will increase moderately.

The effect of dispersion on migrating radionuclides for one-dimensional flow is given by:

$$\frac{\partial C}{\partial t} = \left(\frac{D}{R_d} \right) \frac{\partial^2 C}{\partial X^2} - \left(\frac{V}{R_d} \right) \frac{\partial C}{\partial X} - \lambda C, \qquad (6\text{-}7)$$

where

C = concentration of radionuclide in water, M/d^3,
D = hydrodynamic dispersion coefficient, d^2/t,
V = velocity of groundwater, d/t,
λ = radionuclide decay constant $= 0.693/t_{1/2}$, ($t_{1/2}$ = half-life),
$R_d = 1/(1 + (\rho/P)k_d$,

and

ρ = bulk density of the geologic material, M/d^3,
p = porosity,
k_d = ion exchange coefficient, (volume water)/(mass of rock) $= d^3 M^{-1}$.

Both mechanical and molecular dispersion are included in the parameter D. If the release is by pulse rather than by steady state, dispersion will vary (see Bailey and Marine, 1980).

In the case of continuous release, which is, logically, more likely to be the case than by pulse release, the total quantity of a radionuclide that will pass a given point is of interest. First, however, it is necessary to comment on the amount of radionuclide, A_T, released from a point source:

$$A_T = \int_{t_s}^{t_e} q(t)dt, \qquad (6\text{-}8)$$

where

$q(t) =$ radionuclide release rate, amount released/unit time,
$t_s \;\;=$ beginning time of release,
$t_e \;\;=$ ending time of release.

For continuous release with a constant release rate, the total amount released (Bailey and Marine, 1980) is:

$$A_T = \int_{t_s}^{t_e} q_c dt = a_c(t_e - t_s) = q_c \Delta t, \qquad (6\text{-}9)$$

where

$q_c \;=$ constant release rate,
$\Delta_t =$ duration of the release.

Now, for continuous release with the release rate decaying with half-life equal to the radionuclide half-life:

$$A_T = \int_{t_s}^{t_e} q_0 e^{-\lambda(t-t_s)}dt = \frac{-q_0}{\lambda}(e^{-\lambda(t_e - t_s)}), \qquad (6\text{-}10)$$

where

$q(t) = q_0 e^{-(t-t_s)}$,
$q_0 \;=$ initial radionuclides release rate,
$\lambda \;\;=$ decay constant of radionuclides $= 0.693/t_{1/2}$, t^{-1},

and, if the release is assumed to continue indefinitely (i.e., $t_e \to \infty$) then:

$$A_T = \frac{q_0}{\lambda}(e^{-\infty} - 1) = \frac{q_0}{\lambda} = 1.4427 \cdot t_{1/2} \cdot q_0. \qquad (6\text{-}11)$$

Bailey and Marine (1980) show that the quantity of the radionuclide released per unit time decreases, and most of the radionuclide release occurs in a time equal to five half-lives after the release begins. This is of interest, as mention is commonly made of ten half-lives as a safe limit for which radioactive wastes must be stored, although five half-lives may be adequate. When A_T has been determined, then the amount passing a given point can be determined by multiplying the quantity released by the fraction of the radionuclide that will reach the point before decaying.

Removal of radionuclides by ion exchange (here used to include sorption as well) can be expressed by a distribution coefficient, K_d. The K_d can be expressed as equal to (ml water/g of geologic material). The velocity of a cation in terms of the velocity of the groundwater flow is given as:

$$V_c = \left[\frac{1}{1 + (\rho/P)k_d} \right] V_w = \left[\frac{1}{1 + \alpha k_d} \right] V_w, \qquad (6\text{-}12)$$

where

V_c = velocity of cation, dt^{-1},
V_w = velocity of groundwater, dt^{-1},
ρ = bulk density, Md^{-3},
p = porosity,
K_d = distribution coefficient,

$$\alpha = \frac{\text{mass of geologic material}}{\text{volume of water}} = \rho/p, \ Md^{-3}.$$

As shown by Bailey and Marine (1980), for K_d near zero, little retardation of cations occurs, but retardation is marked above zero.

The effect of distance to the biosphere is generally considered to be critical. It is more important for radioactive elements than for stable elements (i.e., Pb, As, Hg, Se, etc., which do not decay with time). Because different rock types are encountered above a repository horizon, Eqs. (6-1)–(6-12) must be dealt with independently for each and the results then added to each other.

Bailey and Marine (1980) have presented an example of the use of the preceeding equations and their parameters for a hypothetical repository. In their example, the hydraulic conductivity, porosity, and dispersivity values are typical of impermeable crystalline rock, and reasonable values for other parameters are used. The distance to the biosphere was taken at 0.2 km, and a value of 10 ml/g was chosen as K_d. Two values for radionuclide half-life, 30 and 25,000 yrs, were chosen because these are near the half-lives of ^{90}Sr and [137]Cs (each 30 yrs) and [239]Pu (24,400 yrs). The parameters used are listed in Table 6-6.

The fraction of the radionuclides remaining as a function of time are shown in Fig. 6-9. For the 25,000 half-life radionuclide, 1/1000 the original amount is left after 250,000 yrs and $1/10^6$ after 500,000 yrs.

Bailey and Marine (1980) show that for the groundwater to flow to a point 200 m downstream (i.e., optimum leaching direction), 500,000 years would

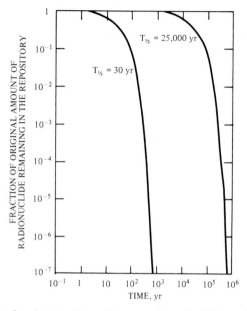

Fig. 6-9. Amount of residual radionuclides remaining (half-lives of 30 and 25,000 years. *Source*: From Bailey and Marine (1980).

be required. Thus the 30-yr half-life radionuclide has virtually disappeared (by a factor of 10^{-5}) and the 25,000-yr half-life radionuclide is only 9.3×10^{-7}. Although some of the remaining radionuclide would reach the point in the next 100,000 years, most of it will have been removed by radioactive decay. Further, the radionuclide will actually travel at a much lower rate than the groundwater due to removal by ion exchange. For the conditions given (Table 6.6) the average groundwater velocity is calculated at 5×10^{-5} (i.e., groundwater velocity without exchange removal is calculated at 4×10^{-4} m/yr). The average radionuclide velocity is thus reduced to approximately 2×10^{-8} m/yr (Bailey and Marine, 1980). Then it requires 50 million years for a radionuclide to move one meter, or 2000 half-lives for the 25,000 year half-life radionuclide.

The conclusion reached by Bailey and Marine (1980) is that virtually none of the radionuclides will move more than a very small fraction of a meter from the repository before decaying to vanishingly small amounts. The reader is reminded, however, that this is a fairly ideal case with specifically given parameters and, as such, widely different results may be obtained for actual repository conditions. However, in the absence of catastrophic events, this "generic" study reinforces the view that normal groundwater flow in and about a repository is not adequate to leach radioactive waste forms so that the radionuclides released will ever reach the biosphere. In addition, were bedded salt to be chosen instead of crystalline rock, the results would be very

Table 6–6. Parameters that Characterize a hypothetical repository site and a hypothetical waste form.

Site Characteristics	
Hydraulic conductivity	10^{-6} m/day
Porosity	10^{-3}
Density of dry, uncompacted geologic medium	2.5 g/cm^3
Hydraulic gradient	1 m_{H_2O}/km $= 10^{-3}$ m_{H_2O}/m
Characteristic axial dispersion length	10 m
Distance to biosphere	0.2 km $=$ 200 m
Distribution (ion-exchange) coefficient	10 g/ml
Waste Form Characteristics	
Waste form size	Cylinder D $=$ 0.328 m (1 ft); L $=$ 2.62 m (8 ft)
Waste form density	3 g/cm^3
Leach rate	10^{-6} g/cm^2/day
Radionuclide half-lives	30 yr; 25,000 yr
Repository Characteristics	
Dimensions	400 m $\times$ 200 m $\times$ 20 m
Free volume	10^6 m^3
Orientation of long axis to direction of groundwater flow	30°
Average inleakage rate during filling with water	1 m^3/day

Source: From Bailey and Marine (1980).

close, again in support of the conclusions reached by Bailey and Marine (1980).

Concluding Statement

There are large parts of the contiguous United States in which rocks potentially favorable for repository consideration occur. Some of these outcrop; others occur at favorable depths. At present, study is focused on bedded and dome salts, basalts, tuffs, granitic rocks, and shales. Generic studies of all but tuffs have been reported, and herein are some useful information on tuffs. Each rock type discussed herein, with the exception of shales, will be discussed in detail in Chapter 7.

CHAPTER 7

Geologic Repositories: Specific Sites

Introduction

Several potential sites for radioactive waste repositories are under investigation in the contiguous United States. The rocks under study include bedded and dome salts, basalts, tuffaceous rocks, and granitic rocks. Bedded evaporites of the Waste Isolation Pilot Plant (WIPP) site in southeastern New Mexico are being considered for disposal of defense TRU wastes; all other sites are being considered for HLW disposal. In this chapter, specific sites to be considered are the WIPP site, the Basalt Waste Isolation Project (BWIP), the tuffs and granite of the Nevada Test Site (NTS), and the bedded salts of the Paradox Basin and dome salts of coastal Louisiana and Texas. No shales are under consideration for HLW radwaste disposal; the reader is referred to the discussion of generic shale given in the last chapter.

The Waste Isolation Pilot Plant Site

The Waste Isolation Pilot Plant (WIPP) site in southeastern New Mexico is under investigation for the storage of defense-generated TRU waste. Realistically, the site could well be considered for all HLW and TRU, should that option ever be exercised.

The location of the WIPP site is shown in Fig. 7-1. The geology of the site rocks has been presented in SAND (1978), and some of the material to follow is summarized from this report and from references cited therein.

The repository site is planned for a depth of 700 m in the 500-m thick Salado Formation. This formation consists of massive halite with minor anhydrite, sylvite, clay seams, and other minor rocks. The waste, together with other evaporites, will be placed near the middle of a 1200-m-thick

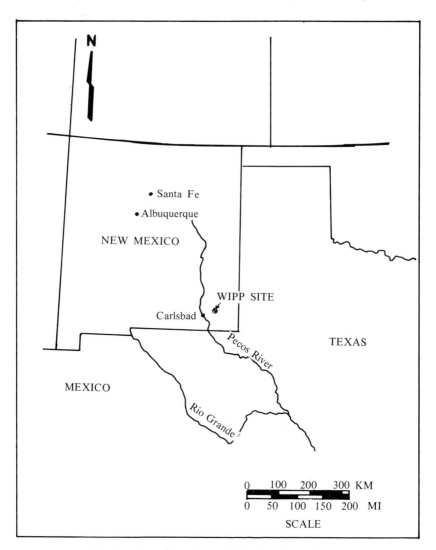

Fig. 7-1. Location of the WIPP site, New Mexico.

sequence of evaporites, where they will be isolated from any waters in the WIPP area.

The evaporites of the area are Late Permian. The total thickness of the geologic section above the Precambrian basement is on the order of 6000 m. The evaporites extend for at least 160 km north and south of the WIPP site; the area of the Salado Formation is about 64,750 km^2. The stratigraphic column for the WIPP site rocks is shown in Fig. 7-2.

Above the Salado Formation are found the calcium sulfate-rich (gypsum and anhydrite) rocks of the Rustler Formation, which are interbedded with

SYSTEM	SERIES	FORMATION	DESCRIPTION	THICKNESS (ft)
Recent		Surficial Sand	—caliche and sandstone	0–100
Quaternary	Pleistocene	Mescalero, Gatuna	Non-marine sandstone	0–35
Triassic	Trias	Santa Rosa Sandstone	Marine mudstone—siltstone, some impure sandstone	0–250
		Dewey Lake Redbeds		100–550
PERMIAN	Ochoan	Rustler	Gypsiferous anhydrite with minor sandstone and carbonates. Contains Culebra and Magenta dolomites	275–425
		Salado	Mainly halite, plus minor anhydrite, K-minerals, clay	1750–2000
		Castile	Massive anhydrite-calcite; halite	1250(+)
	Guadalupian	Bell Canyon	Sandstone, some siltstone, mudstone, limestone	1000(+)
		Cherry Canyon	Sandstone with interbeds of shale, carbonates	1000(+)
		Brushy Canyon	Sandstone with minor shale, carbonates	1800(+)
	Leonardian			3400
	Wolfcampian			1400

(older units, pre-Permian, below; about 4,000 feet to Precambrian basement)

Fig. 7-2. Modified stratigraphic column, WIPP site area. *Source:* Modified from SAND (1978).

siltstone toward the top and silty sandstone toward the bottom. Halite content increases toward the bottom. This formation contains two well-known dolomite marker beds, the upper Magenta dolomite and the lower Culebra dolomite. The importance of these units is discussed below.

Above the Rustler formation are found the Dewey Lake Redbeds, which consist of silty sandstones and siltstones and mudstones, all reddish-brown in color. The sandstones are less abundant than the other types. This unit marks the uppermost limit of the Ochoan Series of the Permian System (Hiss, 1975).

Triassic rocks occur above the Dewey Lake Redbeds. The Santa Rosa sandstone is a red-to-grey, nonmarine sandstone that varies in thickness from about 10 ft to near zero across the site.

Overlying the Santa Rosa Sandstones is the Gatuna Formation, consisting of friable, fine-grained sandstone capped by a siliceous limestone (caliche) crust, the Mescalero caliche. The uppermost 10 ft of the site consist of surficial blanket and dune sands plus some alluvium.

None of the rocks above the Salado Formation are as halite rich or as rich in K-bearing minerals such as sylvite, carnallite, polyhalite, or langbeinite. Further, although the Magenta and Culebra dolomites and the sandstones are aquifers, exchange of water from these units to the underlying formations is negligible in the site area (Dosch, 1979; SAND, 1978). Were this not the case, then recrystallized, water-bearing minerals, such as polyhalite, would document the influx of water; yet K–Ar radiometric ages of polyhalites from the Salado (Brookins, 1980d, 1981b) are commonly in excess of 200 MYBP. Register and Brookins (1980) have also reported a 214 ± 15 MYBP date for primary evaporite minerals by the Rb–Sr method, which confirms that these rocks have remained closed systems since approximately 200–225 MYBP.

Rocks below the Salado Formation include the lowermost member of the Ochoan Series, the Castile Formation, which consists of finely laminated anhydrite–calcite interbedded with massive halite containing thin interbeds of halite. The uppermost anhydrite layers do not contain calcite inter-layers.

Below the Castile, Permian rocks of the Guadalupian Series are found. These consist of sandstones of the Bell Canyon, Cherry Canyon, and Brushy Canyon Formations, each distinguished by different amounts of interbedded siltstones, limestones, calcareous shales, and dolomite. For this report, the rocks of the Guadalupian Series and those below it are not considered as important in addressing the hydrologic and chemical effects on the Salado Formation.

Important facts about the Salado Formation include the following:

(1) Extremely low porosity; interconnecting pores are absent except on extremely local scales (i.e., less than one cm). Hydraulic conductivity of the beds is therefore near zero. The small amounts of gas and brine along halite grains noted by Jones (1973) are considered to be of insignificant

importance as migration paths. Similarly, brine pockets are rare (SAND, 1978) and are only of local importance. Further, fluid inclusions (discussed below) are too sporadically distributed to be of importance to migration of fluids.

(2) Halite, and the less abundant K-bearing units, are extremely soluble. If water had penetrated the area, these rocks would have dissolved. Indeed, the rocks are partially dissolved west of the site where the Salado rises to the surface. Where this dissolution is noted, the halite has dissolved, leaving a residuum of oxide–silicate–clay minerals. Register *et al.* (1980) and Brookins *et al.* (1980) have examined these clay–oxide layers in the evaporite sections from drill core and mine samples, and find that they represent original detrital material that has only partly been altered by interaction with the evaporitic brines. This is documented by Rb–Sr clay mineral dates of 390 ± 77 MYBP, well in excess of 225–235 MYBP (i.e., the possible range of the age of formation for the Salado Formation K-bearing minerals). Since these clay–oxide seams are rich in Mg due to interaction with the brine, the reactions have not been severe enough to reset the Rb–Sr isotopic systematics of the original clay minerals. Stated another way, these minerals have at least in part been closed systems even in the presence of brines.

(3) A dissolution zone is present between the Salado and Rustler Formations in the Nash Draw area (Mercer and Orr, 1977). Water, which penetrates overlying rocks by movement along fractures and due to dissolution, moves to the south over the top salt and is discharged into the Pecos River. The brine at the Pecos River contains 300,000 ppm (+) dissolved solids. The brine discharge into the river is about 200 gpm (Theis and Sayre, 1942); Hale *et al.* (1954) have calculated a transmissivity of 8000 ft^2/day based on aquifer tests. Robinson and Lang (1938) have proposed that downward leaching by this brine is prevented because of a low hydraulic conductivity residuum at the base, hence dissolution has occurred following the Nash Draw structural control.

Potential sources of fluids in the Salado Formation include the above-mentioned hydration water of minerals such as polyhalite and carnallite, the water absorbed along grain boundaries, the local brine pockets, and fluid inclusions. Jockwer (1981), in his study of the Asse salt mine, Germany, has concluded that hydration water and grain boundary water are the most important for fluid migration. Roedder and Belkin (1979, 1980) have argued that fluid inclusions may migrate, and therefore may be important in the presence of a thermal field. More recently, Loehr (1979) and Brookins (1980b) have examined the contact zone of a lamprophyre dike that cuts the evaporite sequence some 10 km from the WIPP site. They found that despite recrystallization of clay minerals and evaporite minerals near the dike, fluid inclusion temperatures are ambient at 10 m from the dike contact (Loehr, 1979; citing Roedder and Belkin, 1979). Temperatures at the evaporite–dike contact are anomalously low at 150°C (Loehr, 1979), thus indicating disequilibrium of the fluid inclusion–rock system. More significant, though, is the fact that no widespread evidence for fluid migration toward the dike is noted. For example, extremely soluble minerals, including sylvite and carnallite, are noted in the zone next to the dike, along with polyhalite; these minerals are cogenetic with 31–33-million-year-old polyhalite.

Bodine (1978) has speculated that polyhalites may have formed more or less continuously throughout the Mesozoic and Cenozoic. If this is true, then water from unspecified sources would have to be available for formation of polyhalite from anhydrous mineral assemblages. This in turn would seem to make an assemblage such as sylvite–halite–polyhalite unlikely, as sylvite would be metastable to polyhalite under these conditions. Brookins *et al.* (1980) and Brookins (1980d; and in press) have shown that more pure polyhalites yield K–Ar dates near or in excess of 200 MYBP. Cases where ages below 200 MYBP are noted are explained as being due to the presence of sylvite and/or K-bearing halite, which easily lose radiogenic ^{40}Ar (see discussion in Brookins, 1980d). All the polyhalites, with the exception of that formed in the dike contact zone (mentioned above), appear to be penecontemporaneous with the sylvite–langbeinite–halite ($\pm$ anhydrite). Further, Lambert (personal commun., 1980) has indicated that at least some polyhalite once thought to be postcarnallite is, based on petrography, actually precarnallite. This is consistent with my new thermodynamic calculations. Thus the data and preliminary calculations argue for hydration water having been incorporated into minerals near 200–225 MYBP and not continuously throughout the Mesozoic and Cenozoic, as proposed by Bodine (1978). This also indicates that even grain boundary water must be of very local importance, as polyhalite (or other hydrous K-bearing phases) are not noted on grain edges. Finally, fluid inclusions apparently have not migrated, although near the dike contact the inclusions have been disturbed. My conclusions are therefore similar to Jockwer's (1981) studies, except that, due to the greater stability of the Ochoan rocks in New Mexico (i.e., flat lying) compared to the folded and metamorphosed sequences at Asse, no intrinsic or extrinsic waters have affected the Salado Formation at the WIPP site while hydration and grain boundary waters have affected the Asse rocks.

Geochronologic Studies

Radiometric age determinations at the WIPP site have proven to be extremely successful, as have such studies at the Asse Salt Deposit in Germany. The WIPP site evaporites were deposited in the Late Permian, roughly 225–235 MYBP. Before discussing the recent geochronologic studies at the WIPP site, however, it is first necessary to briefly discuss which minerals are suited for geochronologic study and which are not. Sylvite, KCl, is, at first glance, a mineral that should be favorable for K–Ar and Rb–Sr dating. Yet K–Ar dates on sylvite usually yield ages well below the age of sedimentation due to loss of ^{*40}Ar from the sylvite structure, a fact that has been known since the late 1940s. Thus K–Ar dates for the sylvites from the WIPP area have yielded values from 17 to 160 MYBP, all well below 225–235 MYBP. This apparently low range in ages has led some investigators to argue for extensive salt dissolution and reprecipitation throughout the postsedimentation history of the evaporites. If this is true, it would imply a new source of water in order for H_2O-bearing minerals like polyhalite and carnallite to form. This, in turn, would cast serious doubts on the chemical and isotopic integrity of the WIPP site rocks. For several years my students and I have been studying the geochemistry and geochronology of the WIPP site evaporites, and some of our results are shown in Table 7-1. Since it is well known that sylvites are not suitable for K–Ar age work, other minerals or methods must be applied. Studies by the Rb–Sr method on sylvites have been very successful. The primary evaporite minerals, sylvite, anhydrite, and polyhalite, yield a 32-data isochron that defines at 214 ± 15 MYBP age, well in the range of the age of sedimentation. More recently, I have investigated polyhalite by the K–Ar method, with a resultant age of 210–226 MYBP, also well in agreement with the age of sedimentation. The polyhalite ages are of interest because, when mixed with small amounts of sylvite and halite, the ages at first appear to be too low, ranging from about 154 to 220 MYBP, with many under 200 MYBP. When the data are plotted on coordinates of age versus Na_2O (or Cl) content, the data are remarkably linear and extrapolate to ages of 210–226 MYBP. This indicates that the ages can be corrected by an empirical factor dependent on Na_2O (or Cl) content, and all ages thus corrected yield 220 ± 10 MYBP. For polyhalites with approximately no admixed halides, the ages range from 210–220 MYBP. Further, model Rb–Sr dates by Tremba (1969) and K–Ar dates on langbeinite by Schilling (1973) range from 225–240 MYBP, thus showing that the age of approximate rock formation is determined on a variety of minerals and by different methods. These dates show that the primary evaporite minerals have remained closed systems to Rb, Sr, and probably K since formed approximately 220 MYBP, and that only sylvite has lost ^{*40}Ar.

Geochronologic studies are not only of use in testing the ages of primary minerals and early secondary minerals. Some 10 km from the WIPP site an igneous dike has intruded the evaporite sequence. The dike has been dated at 33 MYBP by the K–Ar method. The contact zone evaporite (discussed earlier) contains a new generation of mixed sylvite–polyhalite. The polyhalite–sylvite sample yields an apparently young date of 21 MYBP, which, when corrected for loss of ^{*40}Ar due to the sylvite present, yields an age of 31 MYBP, within the limits of error (± 1.3 MY) for the dike. This shows that the contact zone evaporite minerals affected by the intrusion were not later affected by possible water flowing along the wall of the dike (i.e., if the age were really much less than say 15 MYBP, then a new, postdike

Table 7-1. Summary of geochronologic results from near the WIPP site.

Potassium–Argon Dates		Reference
K-sulfates		
Langbeinite	245 ± 10 MYBP	Schilling (1973)
Impure langbeinite–sylvite	137–146 MYBP (n=2)	Schilling (1973)
Pure polyhalites	195–216 MYBP (n=10)	Brookins (1981e)
Impure polyhalites (undisturbed evaporites)	154–187 MYBP	Brookins (1981e)
Impure polyhalite (from dike contact zone)	21.4 ± 0.8 MYBP	Brookins (1981e)
Sylvites		
Sylvite	18 MYBP, 74 MYBP	Schilling (1973)
Sylvite	37.9 ± 1.3 MYBP	Brookins (1981b)
Sylvite	180 ± 4 MYBP	Brookins (unpub.)
Lamprophyre dike		
USGS dates	32.2, 34.4 MYBP	Calzia and Hiss (1978)
Kerr–McGee Mine	43.4, 34.7 MYBP	Brookins (1981c)

Rubidium–Strontium Dates		Reference
Evaporite minerals		
Selected mine samples	120–129 MYBP	Tremba (1969)
Mineral composits	230–240 MYBP	Tremba (1969)
41 mineral isochron	214 ± 15 MYBP	Register and Brookins (1980)
Clay minerals in evaporites-rubble chimney	330 ± 77 MYBP	Register and Brookins (1980)
Initial 87/Sr/86-Sr Ratios		
Intercept for 41 data isochron	0.7076 ± 0.0014	Brookins (1981c)
Anhydrite data, polyhalite data	0.7084 ± 0.0014	Register and Brookins (1980).

source of water could be argued, but this is not the case). More important, since no elemental transfer occurred between dike and evaporites, then these igneous intrusions were isolated and are of no consequence in addressing the suitability of the WIPP site evaporites for repository consideration.

Clay seams, actually seams of clay minerals mixed with other silicates and some oxide–hydroxide minerals, are found throughout the evaporites. These seams are of special interest as they will, in effect, be natural barriers to radionuclide loss from a breached radioactive waste canister. It is a common practice to advocate loss of radionuclides due to movement in a brine that may be too corrosive for the backfill clay minerals to maintain their high sorptive capacity. We have argued that the minerals present in the clay seams were deposited under aeolian conditions and acted on by the late-stage brines present during initial compaction and diagenesis. As such, the clay minerals can be rigorously tested for their ability to withstand brine exposure, again using Rb–Sr technique. The clay minerals are detrital and may represent material only slightly older than the Permian Age or possibly ranging in age back to the Precambrian. Regardless of age, if the clay minerals have been thoroughly affected by the brine they will lose much of their ^{*87}Sr and fix Rb, both processes that would help to completely reset the minerals so that they would yield an age of no greater than 225–235 MYBP. Yet we (see Brookins *et al.*, 1980) have shown that the clay minerals yield an approximate age of 390 ± 77 MYBP. The 390 MYBP date does not necessarily hold any real significance for age of provenance or any other presedimentation event, as the possible source areas for the clay minerals are not known, yet it does show that the clay minerals that define it have not been totally reset due to chemical reactions with the brines, thus lending support to the ability of these natural barrier clays, as well as engineered backfill clay minerals, to effectively sorb and retain the alkali and alkaline earth elements. Further, the clay seams contain a normal budget of U, Th, and the lanthanides, thus also showing that the brines did not cause any movement of these elements by chloride or other complexes. These data are also favorable for the stability of the clay minerals at the WIPP site.

Rubble chimneys occur in many places near the WIPP site. These may be the result of local collapse due, possibly, to brine pockets. Alternatively, they may connect at depth to aquifers and signal deep salt dissolution, indicating that large volumes of water pass along the chimney conduits into the evaporites, dissolving them. Although the latter alternative is unlikely based on geologic reasoning, it can nevertheless be tested by radiometric age study. The chimneys contain large blocks of evaporites, which, if the chimneys have served as conduits, should be newly precipitated minerals with radiometric dates near 0 MYBP. Study of the salts in the rubble chimneys show that they are very angular, commonly heteromineralic, of several generations (i.e., if due to precipitation from waters flowing through the chimney the order of crystallization would be very regular, like that of salts from seawater), and probably of xenolithic origin. Further, the beds near the rubble chimney are mildly folded, with possible channels from the chimneys into the contorted areas noted. Many of the evaporite blocks are halite and anhydrite, which cannot be dated by themselves by the Rb–Sr or K–Ar methods, but some chimney exposures contain polyhalite, and in one mine polyhalite is present not only in the chimney but in contorted beds nearby. I have dated polyhalites from these occurrences, noting ages of 205 and 209 MYBP, which attests to the xenolithic nature of the chimney polyhalite and to the fact that the host rock polyhalite was not exposed to any source of water.

These data collectively argue that the rubble chimneys are isolated Pleistocene events that did not affect the surrounding evaporites.

The presence of these rubble chimneys has been linked to possible deep dissolution of very large amounts (i.e., perhaps 10^{12} m^3) of halite (Anderson, 1981), which has taken place in the last million years or so. No data or calculations are offered in support of the statements, however. More recently, Lambert (1983) has discussed the rubble chimneys and the problem of deep dissolution in a rigorous geologic, hydrologic, and geochemical framework. His explanation offers surface waters, transported in part along bedding surfaces, as promoting local collapse and formation of the rubble chimneys. Further, what is interpreted as a removed amount of halite may actually be due to a lack of precipitation of halite originally. The collapse structure in Nash Draw (see Lambert, 1983) has, based on geomorphic evidence, taken some 600,000 years to form—in short, of no consequence to buried waste at 600–800 m.

It is unfortunate that uranium and thorium are not concentrated in evaporites, otherwise U–Pb and Th–Pb dating could be used to further test the evaporites. Yet the Rb–Sr and K–Ar dates show the following:

(1) The primary minerals, sylvite, polyhalite, anhydrite, yield a radiometric age of 214 ± 15 MYBP by the Rb–Sr method, in agreement with the age assigned based on geologic arguments.

(2) Polyhalite, a hydrous mineral, yields K–Ar ages of 210–226 MYBP, equal to the age by the Rb–Sr method and to the age assigned. Polyhalite is thus an early mineral in the evaporite sequence, in agreement with thermodynamic and petrographic arguments.

(3) The Rb–Sr and K–Ar ages show that the minerals mentioned above have retained their isotopic and chemical integrity since the approximate age of sedimentation. No younger events of meteoritic water intrusion into the rocks is evident; thus the arguments of more or less continuous polyhalite formation throughout the Mesozoic and Cenozoic are wrong, and the new sources of water needed for the polyhalite formation are not necessary nor real.

(4) The presedimentation age of 390 MYBP on clay minerals in clay–oxide seams shows that the evaporitic brines did not reset the clays to the age of sedimentation by ion exchange or any other process. This information supports the use of the clays in natural barriers as well as in engineered backfill to sorb and retain radionuclides if necessary.

(5) Polyhalite from a rubble chimney yields an age of 205 MYBP by the K–Ar method, not a Pleistocene date, thus showing that these chimneys have not acted as conduits for large intrusions of meteoric water into the evaporites in the last million years or so. The polyhalite is a xenolith like the other salt minerals found in the chimneys.

Groundwater Hydrologic Studies

Sandia National Laboratories, with assistance of the U.S. Geological Survey, is investigating the groundwater hydrology of the WIPP site rocks. The presence and quality of the groundwater, as well as its hydraulic conductivity and other properties, are being determined. When complete, the data will allow a thorough assessment of the groundwaters of the area and their potential for reaching and affecting the WIPP site rocks.

Regionally, the groundwaters of the Delaware Basin are of very poor quality, and total dissolved solids exceed 3000 ppm. Potable groundwaters are only found in aquifers along the Pecos River, well removed from the WIPP site. Good aquifers, the San Andres Limestone and the Capitan Limestone, with associated reef limestones, are not in communication with the WIPP site rocks.

For convenience, the formations and their hydrologic characteristics are given (cf. SAND, 1978) as related to the Salado Formation, since the Salado Formation will house the repositories.

The thick anhydrite and halite beds of the Castile Formation, which underlies the Salado Formation, are effective in isolating the Salado from the sandstones beneath the Castile Formation that occur at a depth of about 1400 m. The extremely low hydraulic conductivity of the Salado and Castile Formations prevents vertical groundwater flow and thus isolates these rocks from underlying, as well as overlying, aquifers. The more deeply buried aquifers known in the area include rocks ranging from earlier Permian through Devonian. These formations are not important to the WIPP site and will not be discussed.

Hydrologically important units present at great depths below the Salado Formation include the Bone Spring Formation, which has only limited potential as an aquifer, and the Guadalupian Age Rocks, which include shelf aquifers, the San Andres Limestone, and the Artesia Group; the Capitan aquifer, which includes the Capitan Reef and the Goat Seep Limestone; and basin aquifers, which include the sands found in the Delaware Mountain Group. Hiss (1975) has thoroughly discussed these aquifers.

The Salado Formation itself has extremely low porosity, which is explained in part by the plasticity of the salt that seals any openings that might have been formed. Thus the hydraulic conductivity is also extremely low. In addition to the very small amounts of water along grain boundaries, larger pockets of brine are found in many parts of the evaporite sequence. These brines are halite saturated and are so rich in nitrogen gas under pressure that they may blow out when tapped by drilling. Yet these brines are not in contact with each other nor with meteoric sources, and because they are saturated with halite, they cannot dissolve any of the surrounding salt.

Overlying the Salado rocks is the Rustler Formation, which contains two poor-quality aquifers, the Magenta and Culebra dolomites. The Magenta is a finely crystallized, dense dolomite while the Culebra is a vuggy dolomite with some anhydrite. Communication between the proposed repository site in the Salado Formation and these aquifers has not been established and, even if established, would show that too little water is transported to effectively cause dissolution.

In summary, the hydrology of the site is very favorable. The volume of water in the Rustler Formation above the WIPP horizon and in the Delaware Mountain Group below it is too low, and has too low a head, to be a threat to the WIPP site.

Fate of Radionuclides Released from Hypothetical Critical Assemblies of Actinides

The probability of critical mass assembly of actinides is discussed elsewhere in this book (Chapter 3). Although this is very unlikely, I will consider the hypothetical release of radionuclides. A constraint imposed, however, is the extremely low power outage of such an hypothetical assembly based on 1 to 4 canisters that have released all their HLW. A further assumption requires initial segregation of one or more actinides from fissiogenic isotopes and other elements that will act as poisons and thus prevent criticality. A previous inspection of this problem has led the writer (Brookins, 1978e) to argue that only uranium meets the transport criteria to allow its segregation from other actinides, fission products, and canister materials. For sake of argument, it will further be assumed that the uranium released does not interact with the canister and is transported as either uranyl ion, UO_2^{2+}, or uranyl dicarbonite ion, $UO_2(CO_3)_2^{2-}$ (UDC). Point concentration will then be assumed, where U(VI) can either be sorbed in a clay trap or else reduced to U(IV) as the oxide. In comparison of this point source with Oklo (Chapter 11), it must be emphasized that at Oklo several hundred cubic meters of ore sustained some damage of criticality, whereas for the hypothetical WIPP scenario the assumed total hypothetical volume will be ~ 1 m^3.

Radioactive species generated from such a low-level source will be those of concern to the initial waste form, viz. ^{90}Sr, ^{137}Cs, ^{99}Tc, ^{123}Sb, the REE, the actinides, and the actinide daughters. Other isotopes will be commented on as necessary.

Local fixation of ^{90}Sr can be argued based on evidence from Oklo (Frejacques *et al.*, 1976; Brookins, 1981a) either in clay minerals or by exchange with sulfate minerals. The solubility product constant, K_{so}, for $SrSO_4$ is smaller than that for $CaSO_4$, hence small amounts of celestite will form at the expense of anhydrite (or gypsum). At 100°C, the reaction:

$$Sr^{2+} + 2NaCl = SrCl_2 + 2Na^+$$

will favor retaining Sr in the aqueous phase, but other halides and sulfates (sylvite, carnallite, polyhalite) will act as more efficient getters for Sr^{2+}. Since these minerals are common in the halite-rich host rock then Sr^{2+} removal from an aqueous phase is likely.

In the case of Cs^+ even more efficient removal from solution is argued based on, first, the reaction:

$$Cs^+ + NaCl = CsCl + Na^+,$$

which at 100°C favors exchange of Cs^+ (aq.) for Na^+ (halite). When the essentially infinite ratio of $NaCl/Cs^+$ is considered, then this reaction becomes even more efficient.

Rb–Sr studies indicate closed-system conditions for most of the evaporite minerals in/near the WIPP site. Since the retention of Cs is higher than that of Rb, the rocks will enforce closed-system conditions for Cs as well. When clay seams are encountered, Cs^+ will be efficiently scavenged as well, and preferentially with regard to Rb^+ and K^+ (Kharaka and Berry, 1973). In short, there is no readily apparent way for transport of Cs^+ in such a system.

The behavior of Tc is more difficult to assess, but even in chloride-rich brines, TcO_2 should form and thus remove isotopes such as ^{99}Tc from solution. The presence of Fe^{2+} in WIPP and nearby samples indicates sufficiently low Eh for formation of TcO_2.

Antimony will possibly be transported short distances as $HSbO_2^\circ$, or possibly as SbO^+, depending on local pH. Removal as Sb_2O_3 or by sorption on Fe_2O_3 is highly probable. No evidence indicates oxidation potentials sufficient to oxidize Sb(III) to Sb(V). Chloride complexes of Sb(III) should not be of importance at pH > 3.

The lanthanides can be discussed by analogy with ferromanganese nodules in seawater. The lanthanides are extremely insoluble in seawater and coprecipitate with the nodules or with other oxyhydroxide phases (see discussion in Brookins, 1978e). Chloride and phosphate complexes for possible lanthanide transport are unimportant in halide-rich rocks; the chloride complexes are stable only at extremely low pH values and the total dissolved phosphorus content ($\Sigma P < 0.01$ ppm) is sufficiently low that Ln^{3+} ions are stable with respect to $Ln_2P_2O_7^{2+}$ complexes. Thus lanthanides should not actually be transported from their origin, and in turn will act as poisons in causing criticality to cease. If 500,000-yr operational lifetime is assumed for 1000 m^3 of Oklo ore, and criticality stopped by lanthanide accumulation (Naudet, 1974), then for 1 m^3 of hypothetical critical mass in evaporites, the criticality should be stopped in less than 500 years (maximum). Further, early armoring of uranium oxide by $Ln(OH)_3$ and other secondary minerals should impede escape of the more mobile isotopes. While speculative, 1 m^3 cannot be expected to sustain criticality for more than a few dozen year without some mechanism for (1) removal of fissiogenic nuclides and (2) replenishment of uranium. The removal is difficult to envisage for reasons stated above, and the armoring minerals should prevent further point concentration of uranium. The reason for large accumulations of U-rich ore at Oklo was the presence of very large amounts of low-grade uranium ore as, essentially, an infinite source for U supply. No such large supply of U exists in the HLW at the WIPP site.

The fate of the actinides other than uranium is straightforward. None form chloride complexes, and carbonate complexes are present only at extremely high oxidation potentials. In short, any Np, Pu, or Am generated will remain fixed in host MO_2 without loss. In all probability, the neutron flux will be insufficient for more than minute quantities of Np, Pu, and especially Am to form. Any Cm present must be introduced from the original waste form, yet this is difficult to imagine as it should be segregated from U prior to formation of the MO_2 point source.

The fate of Pb and other actinide daughters can be commented on in semiquantitative fashion. Pb may be removed with Cs^+ or possibly trapped in sulfates (i.e., anglesite-equivalent). Only Ra of the other actinide daughters poses a significant threat to the environment, and, as in the case of Sr, the K_{so} for $RaSO_4$ is smaller than that for $BaSO_4$, which will be efficiently removed from solution (see below). Ra^{2+} can also be sorbed on clay minerals, on K-bearing evaporite salts, and on carbonates, in addition to removal by ion exchange with sulfates.

The noble gases Kr, Xe, and Rn cannot, in likelihood, be retained from the MO_2 source; yet the amounts generated will be extremely small. Further, the rock column above the hypothetical MO_2 site will be sufficient to ensure dissemination of these gases and their short half-lives thus suggest zero environmental risk.

Iodine poses special problems, but removal in chlorides seems plausible (see earlier discussion).

Barium will be generated directly (^{138}Ba) and by decay from Cs (^{135}Ba, ^{137}Ba). Removal as the sulfate is predicted based on its low K_{so} or, alternately, scavenging of some by K-bearing salts or clay minerals.

Formation of Secondary Minerals

The formation of secondary minerals in the approximate vicinity of waste plus canister at the WIPP site will be dominated by availability of elements of ions in the encroaching aqueous phase and in the host rocks, with ions/elements from the waste form of secondary importance. The corrosion of the canister proper will, in all likelihood, not release significant amounts of dissolved matter to the aqueous phase with the possible exception of Fe^{2+}. Secondary minerals throughout the evaporite sequence have been listed in SAND, 1978.

Brines A and B (Table 7-2) are typical of solutions from potash and halite-rich zones, respectively, while solution C is typical of groundwater from the Culebra and Magenta aquifers (i.e., of groundwaters in general). Brine B, by definition, is most likely to play a major role in the formation of secondary minerals from interactions with the waste form; but both Brines A and B should be considered for reactions in halide-rich rocks. For reactions away from the repository proper, Solution C may be the most appropriate.

Table 7-2. Representative brines/solutions for WIPP experimentation.

Ion	Brine A (mg/L) ($\pm 3\%$)	Brine B (mg/L) ($\pm 3\%$)	Solution C (mg/L) ($\pm 3\%$)
Na^+	42,000	115,000	100
K^+	30,000	15	5
Mg^{2+}	34,000	10	200
Ca^{2+}	600	900	600
Fe^{3+}	2	2	1
Sr^{2+}	5	15	15
Li^+	20	—	—
Rb^+	20	1	1
Cs^+	1	1	1
Cl^-	190,000	175,000	200
SO_4^{2-}	3,500	3,500	1,750
$B\ (BO_3^{3-})$	1,200	10	—
HCO_3^-	700	10	100
NO_3^-	—	—	20
Br^-	400	400	—
I^-	10	10	—
pH (adjusted)	6.5	6.5	7.5
Specific gravity	1.2	1.2	1.0

Source: Modified from SAND (1978).

Complex oxyborohalides, such as boracite ($Mg_3B_7O_{13}Cl$), ericaite (($Fe,Mn)_3B_7O_{13}Cl$), chambersite ($Mn_3B_7O_{13}Cl$), and parahilgardite ($Ca_2B_5O_8(OH)_2$), have been reported from bedded and dome salts and may well form in the near field. Extensive solid solution of Na^+ for Ca^{2+} is predicted from Brine B. Should these minerals form, the zinc from the waste form will be incorporated into Mg, Mn, and Fe sites. Formation of $ZnSO_4 \cdot nH_2O$ ($n = 0$ to 7) is unlikely due to the high concentrations of Mg in the system.

Potassium-bearing salts, such as langbeinite, kainite, carnallite, glaserite, polyhalite, and sylvite (second generation), will, if formed in the near field or at any other local source (i.e., such as in a zone of re-accumulated actinides), act as efficient getters for Cs, Sr, and other ions with ionic radii greater than about 1.2 Å. Calcium- and/or sodium-bearing phases (e.g., anhydrite, gypsum, boedite, glauberite, polyhalite, thenardite) and Mg- or Mg-K-bearing phases (in addition to those mentioned above) (kainite, kieserite, leonite) will scavenge the actinides and lanthanides (U,Ln for Ca,Na), and zinc (Zn for Mg), while newly formed silicates, such as the Mg-phyllosilicates montmorillonite, illite, and chlorite, will remove Si, Zn, Sb(?), and Ti, as well as some Fe, B, I, Cs, Ba, Sr, and Rb. Unfortunately, there exist insufficient data to make quantitative statements as to which secondary phases will dominate the system and, once these are known, which will selectively remove the greatest amount of radionuclides.

Near Field Effects: Canister Metals plus Uranium

The iron–carbon–titanium metals being considered for canisters have been predicted to be resistant to corrosion from hundreds to thousands of years, even that resulting from exposure to aqueous fluids in excess of those that may logically be expected to come into contact with the metal due to brine migration from pockets, fluid inclusions, or other local sources. Since the temperature in the canister vicinity will be on the order of $60°$ to $100°C$, then, with H_2O in the liquid phase, simple thermodynamic relationships can be considered that will allow semiquantitative comment on the fate of actinides from waste in the near field. For convenience, one-molal concentrations are assumed (unless otherwise stated) for aqueous species and activities are taken as proportional to concentrations.

Uranium is considered to be the actinide most likely to be present after approximately 0.25 million years, and the species of importance in a silica-deficient system such as that at the WIPP site will consist of the soluble U(VI) species: UO_2^{2+}, $UO_2 (CO_3)_2^{2-}$ (uranyl dicarbonate ion, UDC), $UO_2 (CO_3)_3^{4-}$ (uranyl tricarbonate ion, UTC), and $UO_2 (CO_3)_3^{0}$, and solid species: UO_2CO_3 (rutherfordine) and UO_2 (uraninite; here analogous to any uranium (IV) oxide). The relationship between the various soluble and aqueous species is shown in Fig. 11-1. Halide, sulfate, and hydroxyl complexes are not thought to be important in the Eh–pH range indicated in Fig. 11-1, especially since a reasonable pH for near-field reactions is thought to be that indicated by fluid inclusion pH = 5.2. If waters from the Magenta or Culebra Formations are considered as the media for interaction, then the system will be carbonate buffered at pH 7.4–9, and UDC or UTC instead of UO_2^{2+} will be the species of importance.

For the past few years I have been applying thermodynamic calculations to various problems of radioactive waste–container reactions, as well as supporting studies (see Brookins, 1978e, 1980b). Some of my conclusions from these studies are commented on below. Native iron, available from canisters, is a sufficient reductant to effectively remove any U^{6+} species present by reducing it to U^{4+}, whereby it will be precipitated as $U(OH)_4$. This reaction effectively scavenges all U in the pH range 3–14, and forms small amounts of $Fe(OH)_3$ as well (note: it makes little difference if $FeO(OH)$ is produced as both will age to hematite). Further, the fact that pitchblende–hematite is a most common assemblage in uraniferous rocks attests to the probability of the $Fe + U + H_2O$ reaction going to $U(OH)_4 + Fe(OH)_3$.

Metal–Oxyhalide Reactions and Naturally Occurring Halides and Oxyhalides

Free-energy data for the oxyhalides of the REE, several actinides, and a few other elements have been tabulated by Krestov (1972), Schumm *et al.* (1973), and Robie *et al.* (1978). All MCl_3 compounds are readily soluble to

$M^{3+} + 3Cl^-$. The oxyhalides are presumably also highly soluble, but the mechanism of dissolution (i.e., congruent versus incongruent) is unknown. Bismoclite (BiOCl) does occur naturally, as does laurionite (PbCl(OH)), but in very different types of environments than expected at the WIPP. Bismoclite forms as an alteration product from native bismuth or from bismuthinite and laurionite from seawater interaction on native lead slag. Similarly, the other Pb, Fe chlorides, bromides, and iodides form from alteration of sulfide ores or by sublimation in volcanic terrane. Of the minerals listed by Krestov (1972), only halite, sylvite, and chlormagnesite are common to evaporite deposits. The oxyhalides are not important products in natural environments, even when a metal phase (i.e., canister) is present.

Other Reactions

The oxidation of halides to form HaO^- or HaO_3^- can be examined by Eh–pH considerations. Brookins (1980b) has shown that Cl^- and Br^- are stable with respect to ClO^-, BrO^-, ClO_3^-, and BrO_3^- in the stability field of water. Further, even oxidation of I^- to IO_3^- takes places at an extremely high oxidation potential. Finally, none of the species ClO^-, BrO^-, or IO^- are stable with respect to the reduced halide form. It is unlikely that HaO^- or HaO_3^- species will be of importance except as early formed, and rapidly destroyed, metastable species, or that these species plus possible small amounts of Cl_2 (g) [or Br_2 (g), I_2 (g)] will be reduced to the appropriate halide. A further consequence of this is that should these species be generated, they will easily be removed by titanium:

$$Ti^\circ + 2Cl_2 \text{ (g)} = TiCl_4,$$

$$\Delta G_R^\circ = -176.2 \text{ kcal}/M,$$

where, as mentioned earlier, $TiCl_4$ formed in this fashion will be metastable with respect to TiO_2.

Of the halides, that which poses the most potential for migration is fissiogenic $^{129}I^-$. For exchange of one-molar amounts of dissolved halides, $NaCl + I^-$ is stable with respect to $NaI + Cl^-$. By increasing the ratio of $NaCl/I^-$ greater than 10, I^- should be scavenged by NaCl as Na(Cl,I) or as a mechanical mix of $NaCl + NaI$. The point is that in an essentially infinite amount of NaCl that I^- will be scavenged from solution with concomitant release of small and essentially insignificant amounts of Cl^-. Further, the seams of clay minerals will also sorb some I^-, thus further ensuring its retention close to the point of release Brookins (1980b).

Titanium–uranium reactions in the presence of sulfate-bearing solutions have been postulated to be of possible importance. When written as half-cell reactions, not only are very large amounts of SO_4^{2-} required, but the E_0 terms fall well above the sulfate–sulfide boundary. Further, if $TiS_2 + O_2(g)$ are written as products, then the reactions indicate Ti metal to be stable in the presence of SO_4^{2-}. This is borne out by the calculated partial pressures of $O_2 = 10^{-65.3}$ atm (25°C) and $10^{-38.2}$ atm (200°C), which plot well above the stability field for TiS_2.

Further, if TiO_2 is formed from $Ti°$ as documented by experimental work, then in the presence of H_2S, equilibrium $P_{H_2S} = 10^{11.9}$ (25°C) and $10^{14.4}$ (200°C) atm, respectively, which are inconsistent with P_{H_2S} values measured and calculated from fluid inclusions and brines.

Finally, TiO_2 is stable with respect to TiS_2; $TiCl_4$ is stable with respect to TiS_2; and TiO_2 is also stable with respect to $TiCl_4$. Thus the thermodynamics of reactions involving $Ti°$, TiO_2, $TiCl_4$, TiS_2 favor formation of TiO_2 by $Ti°$ interaction with aqueous media and, should any TiS_2 form in some unexpected reaction it will be strongly metastable to $TiCl_4$ and metastable with respect to TiO_2. Reactions involving formation of TiS_2 plus either H_2S or O_2 (g) as reactant or product, respectively, are highly unlikely.

Concluding Statements on WIPP Criticality Studies

Essential to the accumulation of a critical assemblage of actinides away from a breached canister are the following parameters:

(1) Method of transport, presumably due to dissolution following oxidation.

(2) Segregation of actinides from each other and from neutron poisons.

(3) Point, or zone, of accumulation in both evaporite and carbonate rocks as well as oxide–hydroxide–clay mineral seams.

(4) Isolation of newly formed mass from additional, noncritical actinides.

(5) Larger point source actinide concentration in newly formed zone than in original canister.

The previous sections allow each of these parameters, as well as general statements concerning criticality, to be evaluated. Each of the above will be discussed in order. For (1), a scenario allowing MO_2 to oxidize to soluble $M(VI,V)$ is necessary. Yet Wang and Katayama (1981) have shown that armoring will preserve much of the MO_2 due to MO_3-hydrate coatings (due to irradiation effects). Further, the $Fe(II)$ and $Ti(0)$ have been shown to be sufficient to reduce Pu, Np (Bondietti and Francis, 1979), and U (as shown elsewhere in this book). Am and Cm will presumably behave in the same fashion. With regard to (2), in the pH range of 5–6 (assumed for distances greater than 0.5 m from canister wall), carbonate complexes may be important if (1) abundant dissolved CO_2 is available, and (2) a $M(CO_3)_x^{y\pm}$ complex is stable with respect to other species (see Langmuir, 1978). If dissolved CO_2 is less than 10^{-2}, only U is likely to be transported as UDC and then only near the canister as at lower temperatures UO_2^{2+} will dominate, and UO_2^{2+} will not be segregated from other MO_2^+, MO_2^{2+} species. Further, for UO_2^{2+} transport, other fissile ions, such as SbO^+ and TcO_4^-, may be transported with all actinides. In short, segregation of one actinide from the other actinides and other elements that might act either as neutron poisons or diluents is unlikely. Of interest concerning this point is the recent study of the abovementioned lamprophyre dike that intrudes into the evaporites near the WIPP site. Brookins (1981c) has found no evidence for migration of the lanthanides from the dike into the contact zone evaporites, which suggests 100% retention in the dike. Data for U and Th support the lanthanide data. Further, where lanthanides are noted in the evaporites, it is in the vicinity of oxide–clay mineral zones and the REE distribution patterns are normal, indicating no segregation of the light rare earth elements (LREE) from the heavy rare earth elements (HREE) of segregation of Ce(IV) or Eu(II) from the M(III)REE. Brookins (1978c,e) used the argument of lanthanide–actinide similarities to argue in a generic sense against actinide segregation, and the experimental lanthanide data for the WIPP rocks supports his hypothesis.

With regard to (3), a point source for an actinide demands that condition (2) be first met or, alternately, that segregation by selective sorption or precipitation occur. Since the work of Allard and co-workers (Allard, 1979; Allard and Beall, 1978) show similar sorptive properties for the actinides, separation of one actinide from the others seems unlikely. An exception might be mechanically carried MO_2, which settles out away from M(V,VI) soluble species, but only Th meets this criterion and it is not of concern. Selective precipitation of one actinide relative to the others is inconsistent with solubility data, Eh–pH diagrams, and natural analog (Oklo) data.

For (4) to be a reality, not only would conditions (2) and (3) have to be met, but the problem of overnucleation could, as at Oklo, cause dilution and thus prevention of criticality. More important, a point source is likely where there is a structural constriction in the migration path, and such sites will not be single-element specific. Fracture and vein fillings in remobilized evaporites tend to be heteroelemental, even if largely monomineralic, hence it is unreasonable to argue for segregation and preservation of one actinide in a specific site.

Parameter (5) is perhaps the most revealing of all in arguing against a critical reassemblage of an actinide mass. The actinides are more or less randomly distributed throughout the waste forms (whether TRU, HLW, or SURF is irrelevant). For (5) to be met, the critical actinide would first have to be concentrated by some unspecified (i.e., in essence unknown) mechanism and survive reactions involving canister and overpack + backfill materials. Since the few diffusion parameters for the actinides are fairly similar, to argue that D/a^2 [diffusivity/(radius of diffusion)] values will vary widely in a breached canister is unrealistic. Irradiation effects may allow some segregation of ^{238}Pu from other Pu isotopes (Fleischer and Raabe, 1978), but this is not important to the problem of critical assemblage. To then postulate that an actinide will be selectively segregated and further isolated in a monoelemental point source is totally inconsistent with natural observation on U, Th behavior, lanthanide behavior, and, for Oklo, Pu–U–Th–Np (–Am?) systematics (see Frejacques *et al.*, 1976). If a monoelemental mass should form, it will in all probability be a much smaller amount (by any orders of magnitude) than that originally present in the canister.

The Register *et al.* (1980) study of clay material–brine interaction shows that clay minerals, and associated oxide–hydroxide phases, act as systems isolated from the evaporite, with the exception of Mg-addition to the clay mineral assemblage. Their work, plus that of Loehr (1979), for the contact zone of the dike mentioned earlier, argues for heteromineralic–heteroelemental as opposed to monoelemental–monomineralic assemblages. In the case of the 3-m contact zone adjacent to the dike, while the mineral assemblage changed as expected, the bulk and trace element chemistry did not; thus convincingly demonstrating no transport of trace elements from the dike into the evaporite plus clay mineral plus oxide–hydroxide contact zone. Since the temperatures in the contact zone start at about 700°C at the dike wall and reach ambient (60°C) by 2 m, the thermal effects have been insufficient for transport of elements from the dike outward. Further, despite solution effects along the immediate contact zone, no enrichment of the REE, Co, U, or Th is noted in the sulfates or oxide-silicate phases present. Applied to a waste canister breached at much lower temperatures, the dike analog argues against transport of fissile elements away from the canister; this is especially true since the canister and overpack will retain the actinides and other elements while the dike was emplaced without benefit of such barriers.

The Basalt Waste Isolation Plant Site

The Columbia River Basalts were formed some 17–6 MYBP, and consist of the largest volume of basalts in a single series of flows in the United States. These flows underlie the Hanford Operations of the U.S. Department of Energy and Department of Defense (Fig. 7-3). The Columbia River Basalt Group can be subdivided into the oldest Imnaha Basalt overlain by the Grande Ronde Basalt, which is in turn overlain by the Wanampum Basalt and the youngest basalt, the Saddle Mountain (Table 7-3).

The Grande Ronde Basalt consists of some 35 flows, making up about 800 m thickness (Table 7-4). Most of the flows range in thickness from 7 to 70 m with an average flow thickness of 16 m. The Grand Ronde Basalts contain several flows of interest for repositories, especially the Umtanum Basalt, which occurs at about the 800- to 1400-m depth range. The Grande Ronde Basalts range in age from about 15.5 to 14.5 million years. They are aphyric to fine grained, and consist of microphenocrysts of plagioclase (An_{40-60}) and augite in a groundmass of plagioclase, augite, ilmenite, pigeonite, titanomagnetite, olivine, and some orthopyroxene. Individual flow chemistry is

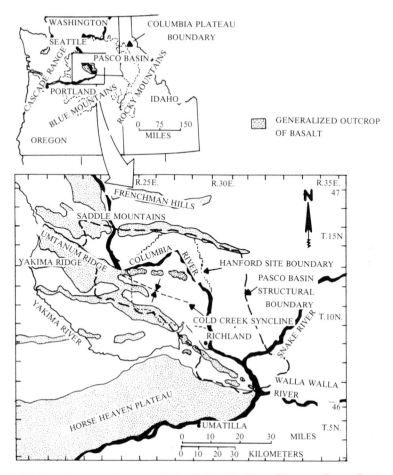

Fig. 7-3.(a) Location map for part of the Columbia River Plateau, Pasco Basin, Hanford site (BWIP). *Source:* Modified from Myers and Price (1981).

often repeated in the Columbia River Basalts, and it is difficult to differentiate between flows based on chemistry alone. This is important, though, for it shows that the rocks are, for modeling and other purposes, fairly homogeneous. Interlayer volcaniclastic sedimentary units occur between some of the basalt flows. These interlayer rocks are more porous and permeable than the flows, yet are apparently not in communication with waters from the basalts.

The basalts are commonly fractured, with a range of about 100 μ to over 1 cm common. Clay minerals and other secondary minerals fill these fractures. The order of formation of the secondary minerals appears to be clay minerals, followed by zeolite and asome chalcedony, and a new generation of clay minerals. The clay minerals are all montmorillonites, although the

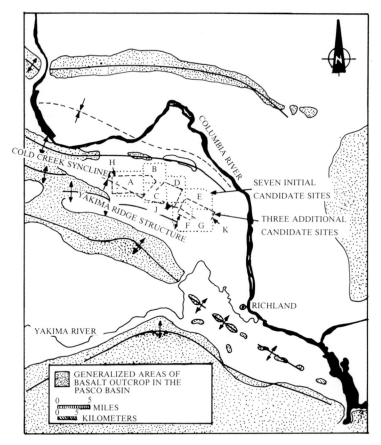

Fig. 7-3.(b) Location of candidate sites in the Cold Creek syncline. *Source:* Modified from Myers and Price (1981).

earlier generation is more iron rich than the younger material. The zeolites are mainly clinoptilolite, $(Na,K,Ca)_{2-3}Al_3(Al,Si)_2Si_{13}O_{36} \cdot 12H_2O$, which also contains some Fe and Mg. Silica is present as opal or quartz, tridymite, or rarely, cristobalite. Pyrite is a common minor phase in the basalts, and K-feldspar, muscovite, and calcite are less-abundant secondary minerals.

The primary basalt phases will, ultimately, control the chemistry of waters in the rocks. *Plagioclase* makes up from 25 to 50% of the basalt, commonly as microphenocrysts. Melt inclusions and inclusions of ilmenite are common in the plagioclase. *Augite* makes up 20 to 45% of the basalt; it ranges in composition from $Wo_{30}-En_{38}-Fs_{32}$ to $Wo_{33}-En_{45}-Fs_{22}$ (Wo = wollastonite, $CaSiO_3$; En = enstatite, $MgSiO_3$; Fs = ferrosilite, $FeSiO_3$), and is commonly intergrown with pigeonite. *Pigeonite* makes up 0 to 10% of the rock, and varies in composition from $Wo_{13}-En_{50}-Fs_{37}$ to $Wo_9-En_{64}-Fs_{29}$. It is intergrown with augite and forms rims on other pyroxenes. *Orthopyroxene* is a minor constituent of the basalt, varying in composition from $Wo_4-En_{61}-$

Fs_{35} to $Wo_6-En_{75}-Fs_{19}$. *Titaniferous magnetite* makes up from trace to 7% of the rock. Its titanium content varies from 28 to 32%. *The mesostasis* makes up from 15 to 70% of the rock, and it varies in SiO_2 content from 60 to 74%. Compositional zoning is common, as are immiscible liquid blebs. *Apatite* varies from trace to 2%, and is found as discrete crystals and as minute crystals in immiscible liquid blebs. *Olivine* (Fo_{67} to Fo_{47}) is not present in most of the Grande Ronde flows. When it is present, it is usually found restricted to the glassy rims of some flows.

The secondary minerals include various smectites. The most common are the dioctahedral varieties: montmorillonite, beidellite, montronite. Saponite is the most common trioctahedral clay mineral. Clinoptilolite is the dominant zeolite, but heulandite, mordenite, phillipsite, harmatome, chabazite, and erionite are all found. Several varieties of silica are found, including quartz, tridymite, cristobalite, and opal. Miscellaneous secondary minerals include gypsum, calcite, and pyrite. Long and Davidson (1981) have described the mineralogy of the Grande Ronde basalts in detail.

The secondary minerals may locally be more important than primary phases in controlling the chemistry of groundwaters in the basalt because they are found primarily along fractures and coating primary minerals in contact with fracture or vein systems. As such, they can, to a large extent, chemically react with any waters in the rock. In vesicles, smectite commonly makes up 32% (by volume), clinoptilolite 43%, and silica 25%. Along fractures the volume percents are: smectite 75%, clinoptilolite 20%, and silica 5%. The smectites are earliest in the order of crystallization, followed by clinoptilolite and silica. Even younger generations of zeolite and clay may form after the silica.

The lateral variation of internal structures within the Umtanum Flow is fairly regular (Fig. 7-4). A flow breccia is found at the top of the Umtanum, a feature which is common for such flows. From the drilling to date in the area, this flow breccia has apparently not served as an access unit for water to penetrate the underlying basaltic rocks. The origin of the flow breccia is problematic (Long and Davidson, 1981), but its regular distribution and more or less uniform mineralogy and chemical composition indicate that it has not directly or indirectly reacted with, or promoted reactions within, the underlying basalts.

Several types of flows are noted in the Grande Ronde basalt, but the significance of the flows is their uniform mineralogy, chemistry, and texture, again indicating more or less closed systems. Cooling by conduction, with minor convective cooling at the top or bottom of the flows, is probable (see Long and Davidson, 1981).

Zones of pillow lava are common in the Grande Ronde, indicating that extensive, shallow lakes were present during the emplacement of the Grande Ronde flows. Pillow zones are rocks with higher permeability than interiors of flows or even flow breccias. It is important to note that (cf. Long and Davidson, 1981) pillow zones are not encountered in drilling in the Grande

Table 7-3. Modified stratigraphic nomenclature, Columbia River Basalt Group, Pasco Basin, Washington.

Period	Group	Formation	K-Ar Age (MYBP)	Member or Sequence	Sediment Stratigraphy Or Flows Or Beds
Quaternary				Surficial Units	Loess Sand Dunes Alluvium Landslides Talus Colluvium
				Gravels	Ringold
Tertiary	Columbia River Basalt Group	Saddle Mountain Basalt	8.5 10.5 12.0	Ice Harbor Member Elephant Mountain Member Pomona Member Esquatzel Member Asotin Member Wilbur Creek Member Umatilla Member	
		Wanapum Basalt	13.6	Priest Rapids Member Roza Member	
			14.5	Frenchman Springs Member Sentinel Bluffs Sequence	Vantage Interbed
		Grande Ronde Basalt	16.5	Schwana Sequence	Intermediate Mg-flow Low-Mg flow above Umtanum Umtanum Flow High Mg-flow below Umtanum Very high Mg flows Many Low-Mg flows

Table 7-4. Groundwater composition in the Pasco Basin.[a]

	(A)	(B)	(C)
pH	9.7–10.1	7.6–8.8	7.3–8.3
SiO$_2$	105–120	54–75	39–50
Cl	100	4–81	3–6
Na	180–240	30–150	5–21
HCO$_3$	2–51	144–277	85–186
CO$_3$	100–127	ND–15	ND
K	3.3–5.9	7.7–17	1.5–5.2
Mg	ND–2	ND–11	7–13
Ca	0.6–1.3	1–29	23–24
SO$_4$	10–96	ND–27	10–18

[a]Data in mg/L (except for pH. ND = not detected.

Note: (A) = Deep groundwaters, from depths greater than 750 m; (B) = Intermediate depth groundwaters, from 300 to 500 m; (C) = Shallow groundwaters, from less than 300 m. Data from Benson *et al.* (1980).

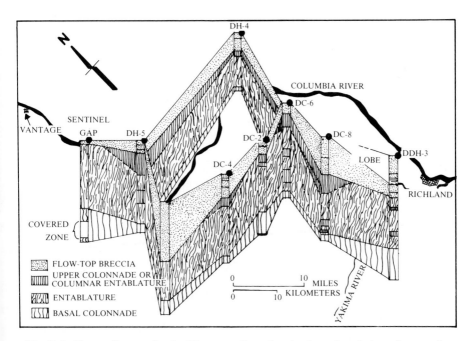

Fig. 7-4. Fence diagram for the Umtanum flow showing lateral variation of some of the flow. *Source:* Modified from Long and Davidson (1981).

Ronde basalt, and they are not expected to be encountered as the main drainage, for the area was to the west and northwest of the present-day Pasco Basin. Consequently, pillow zones are not expected to have any effect on the proposed repository in the basalts. However, communication between the flow breccia and the colonnade may be possible due to fanning joints in the entablature formed during the basalt cooling history. Since outcrops of the Umtanum Flow contain fanning columns, typically spaced at 150-m intervals, any large repository is likely to encounter fanning columns at depth. These columns must be hydrologically tested to ensure that they will not serve as conduits for water encroachment on the repository area. Yet the fracture porosity of the rocks is very small. The total volume of fractures is small, and only 0.1% of this small volume is unfilled fractures. This means that the fractures do not, for the most part, interconnect, and thus are not conduits for groundwaters moving in the rocks.

The entablature part of the basalts contains abundant mesostasis and secondary minerals. Not only will these materials greatly reduce the permeability (and porosity) of the rocks, but these materials are more susceptible to alteration than the colonnade part of the basalt. This means that the entablature will react with, and remove, radionuclides from any waters that have reacted with a waste canister. How these reactions work is discussed below.

The secondary minerals are of extreme importance in addressing the overall safety of the BWIP site rocks, since it is these minerals along fractures that any escaping radionuclides from a breached canister will contact. Vesicle-filling minerals are more complex (Benson and Teague, 1979), but are less important in addressing water–rock interaction, as the vesicles are usually isolated from the flow, which is fracture controlled for the most part. Distribution of the secondary minerals with depth shows (Benson and Teague, 1979) clinoptilolite appearing at a depth of nearly 400 m and some mordenite (after clinoptilolite?) at depths greater than 850 m. The montmorillonites show no general pattern as a function with depth as do other minerals.

Rock–water interaction using basalt–water plus simulated HLW, glass, calcine, and SURF have been studied experimentally by researchers at Pennsylvania State University. These studies (see McCarthy et al., 1978b) are of interest because of the nature of the minerals formed during the experiments and their implication, along with the secondary minerals already present, concerning radionuclides.

In several experiments U^{6+}-bearing minerals, such as weeksite or boltwoodite, formed. This is important, as it indicates that even under chemical environments not reducing enough to remove dissolved uranium as some U^{4+}-bearing phase, U^{6+}-bearing minerals will form. Brookins (1981e) has shown that U^{6+}-bearing minerals in nature are remarkably stable in the natural geochemical environment. In addition, the weeksite or boltwoodite will serve as a nucleation site for many fissiogenic elements, including Rb, Sr, Ba, U, Th, and REE.

Other minerals formed from the simulated HLW–basalt interaction include willemite (Zn_2SiO_4), powellite ($CaMoO_4$), tincalconite ($Na_2B_4O_7 \cdot 5H_2O$), pyroxene (($Na,Ca)(Fe,Zn,Ti)Si_2O_6$), and REE phosphate, plus magnetite plus RuO_2 plus native palladium. These experiments demonstrate that a wide variety of minerals, all with sites favorable for one or more fission products, form readily. The willemite will prevent escape of zinc from the waste package. The borate mineral is a soluble phase, and its presence shows that boron will be fixed near/at where formed. Boron is also a neutron poison, so its presence with U-bearing phases is an additional safeguard against any critical reaction being allowed. Any molybdenum (from CW or fission products) will be incorporated into a powellite-like mineral. This is important, for Mo^{6+} is readily soluble as MoO_4^{2-} or $HMoO_4^-$ in many waters and thus could conceivably be transported away from where Mo was formed were it not for the formation of the powellite. The identification of the RuO_2 shows that fission-produced Ru will remain where released, and that this phase should accomodate technetium as well (see discussion on Oklo, Chapter 11). The native palladium is predicted from theoretical considerations (Brookins, 1978b) and any fission-produced silver and rhodium should be mixed with it.

Experiments in which SURF were exposed to hydrothermal solutions yielded $SrZrO_3$ and pollucite in most runs. The $SrZrO_3$ compound is the host for any fission-produced Sr and Zr that may be present. Ba and Ra should also be effectively scavenged by the $SrZrO_3$. Formation of ZrO_2 or $ZrSiO_4$ in place of the $SrZrO_3$ will take place in the absence of Sr, in agreement with theory (Brookins, 1978c). The pollucite will effectively scavenge fission-produced Cs as well as some K, Ba, Rb, and Sr and Ra.

The naturally occurring minerals in basalts should serve as good barriers to radionuclide escape. The clay minerals and clinoptilolite should serve as effective getters for Rb, Sr, Ba, Cs, Ra, REE, Cd, Mo, Co, Zn, Zr (?), Tc, Ru, and others, while pyrite will help to keep the oxidation potential low so that U^{6+} will be reduced to U^{4+}, thus allowing the formation of some insoluble U-oxide. This phase will also contain Tc and Ru as well as other actinides Th, Pa, Np, Pu, Am, and Cm. In addition, the rock-forming minerals pyroxene and plagioclase may locally interact with the radionuclides but the amount of fission products sorbed or in other ways removed by these minerals will be small compared to the secondary and hydrothermal minerals.

The nature of the alteration products noted in the Columbia River Basalts is of direct importance in addressing both the effect of infiltration of the repository and surrounding rocks by meteoric water and the ability of these minerals to act as barriers to released radionuclides. At the Lawrence Berkeley Laboratories Benson *et al.* (1980) and Apps (1978) have

Table 7–5. BWIP basalt-groundwater reaction (simplified).

Reactants (moles)	Products (moles)
$0.55\ SiO_2 + 0.09\ Al_2O_3 + 0.10\ FeO$ $+ 0.12\ MgO + 0.11\ CaO + 0.025\ Na_2O$ $+ 0.005\ K_2O + 1.26\ H^+ + 0.47\ H_2O$	$0.55\ H_4SiO_4 + 0.18\ Al^{3+}$ $+ 0.10\ Fe^{2+} + 0.12\ Mg^{2+}$ $+ 0.11\ Ca^{2+} + 0.05\ Na^+$ $+ 0.01\ K^+$

Source: Modified from Carnahan *et al.* (1978).

investigated these problems by computer study. In their work, they use several computer codes to predict the order and amounts of mineral precipitates from assumed amounts of hypothetically dissolved basalt (FASTPATH and WOLERY Codes). If basalt of the Pamona flow is allowed to dissolve in water, the equation shown in Table 7-5 applies. In their approach, both CO_2-bearing and CO_2-absent waters were used; both started with pH = 7. This approach is justified, because, for the basalts in question, the deep groundwaters are meteoric waters that have been modified by contact with the rocks and minerals. The regular distribution of the alteration products indicates an approach to, if not actual, chemical equilibrium between the circulating waters and the rock. In the case of CO_2-bearing waters, the ratio of reacted basalt to CO_2 was 9:1 (Carnahan *et al.*, 1978), which assumes much more CO_2 than is actually present. The computer-simulated reaction, which assumes dissolution of the basalt, calculation of the degree of saturation or supersaturation of the waters with respect to dissolved species, and the predicted order of crystallization of new minerals in order of a reaction progress variable, $\bar{E}$, was first used by De Donder and Van Rysselberghe (1936). To each unit of $\bar{E}$, is added one mole of any reactant to a reacting system containing dissolved species and some initially produced mineral. The predicted order of the various minerals for CO_2-bearing and CO_2-absent waters reacted with basalt is given in Carnahan *et al.* (1978) and it is different in each case. For the CO_2-bearing waters the pH is buffered at about 8.6, which might be expected where calcite is present in the basalt fractures. In the other water pH changes from 7 to about 11, which is consistent with pH values measured for deep groundwaters in the basalts. Apps (1978) has reported additional data for simulated reactions at temperatures of $25°C$ and $60–300°C$; the data reported here (Table 7-5) are for $60°C$ (from Carnahan *et al.*, 1978). For the CO_2-bearing system, calcite and siderite are predicted as precipitates, as well as talc, kaolinite, Ca-montmorillonite, and Na-montmorillonite. Phases common to both systems include fayalite, laumontite, adularia, and low albite. Chlorite was noted in both systems but was consumed in the CO_2-bearing system to form talc at higher values of $\bar{E}$. Tremolite and prehnite appear in the CO_2-free system but not in the other. One important point of the computer-simulated reaction is

Table 7–6. Composition of Granitic Rocks, NTS.

Mineral	Composition	Weight Percent
Plagioclase	$(Na_{2.3}Ca_{1.8})(Al_{5.4}Si_{10.5})O_{32}$	46
Quartz	SiO_2	28
Microcline	$(K_{3.9})(Al_{4.1}Si_{11.97})O_{32}$	15
Biotite	$K_{2.2}(Mg_{2.61}Fe_{3.4})(Fe_{0.7}Al_{0.45}Ti_{0.4})(Si_{5.7}Al_{2.3})O_{24}(OH)_2$	10
Pyrite	FeS_2	

that no mechanism for precipitation is indicated; the precipitates and their order are simply a function of the degree of saturation. Further, the minerals in Table 7-6 do not correspond to those actually observed as alteration products, but this is primarily due to a lack of thermodynamic data for many of the zeolites and montmorillonites. Thus laumontite precipitation indicates clinoptilolite or heulandite, while fayalite–talc–chlorite substitute for the Fe- and Mg-bearing montmorillonites. One difficulty is apparent: in the natural-alteration products both K and Na minerals appear throughout, but in the computer-simulated reactions they do not appear until values of $\bar{E} = 0.1$ to 1 are reached. However, the importance of the computer-simulated reactions is that, basically, the minerals observed are those predicted, hence discussion of these minerals as to their ability to retain or retard fission products and actinides can be directly addressed.

The Climax Granite of the Nevada Test Site

The granitic rocks of the NTS are being investigated to assess their potential as a radioactive waste repository candidate. The rocks actually consist of two principal units, granodiorite and quartz monzonite; the Climax Stock is located at the north end of the NTS. The composition of the granitic rocks is given in Table 7-6.

A series of heater experiments and chemical studies are presently being conducted at the Climax Granite site (Fig. 7-5). The heaters have been placed in a new tunnel some 420 m below the surface. Spent-fuel elements have also been placed along with the electrical heaters so that a comparison of heat from HLW and from non-HLW can be made. The goal is to test the integrity of the granite in response to these new thermal loads; both the HLW and the electrical furnaces generate approximately 2 kW of thermal energy, and the HLW will decrease to about 1 kW after 5 years. The electrical heaters will be adjusted so that they match the HLW canisters. This experiment has been designed so that a repository with about 44 W/m^2 thermal loading is simulated. Such a repository will reach its maximum temperature 50 years after emplacement. The experiment calls for maximum

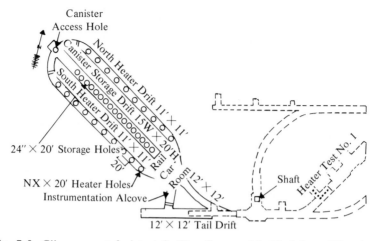

Fig. 7-5. Climax spent fuel test facility. *Source:* Modified from Klingsberg and Duguid (1980).

temperature to be reached in only 2 years so that the data can be directly used in evaluating the granitic rocks as a repository candidate. Near-field effects of HLW in granite can then be addressed.

Preliminary heater experiments have already been conducted in Climax stock. Since the granite is highly fractured, the effect of thermal loading on the thermal conductivity across fractures can be studied. The preliminary experiments (see Klingsberg and Duguid, 1980) reveal that the thermal conductivity is greater across fractures (by about 10%) than values obtained parallel to the major fractures. The thermal conductivities measured in the mined excavation are also about 10–20% greater than values obtained in the laboratory. Of even more interest, the rock permeability as determined by injection of nitrogen tends to decrease with increasing rock temperature, thus indicating that more fracture annealing occurs as temperature increases. This implies that during the high-temperature period of the radioactive waste canister any escaping radionuclide will be impeded by the sealed fractures. However, any fluid escaping from the breached canister may be more reactive than fluids present at lower temperatures.

How fracture-filling material behaves in response to fluids at high temperatures is moderately well known (see McCarthy *et al.*, 1978b); these fluids can be considered as hydrothermal fluids containing the fission products and radioactive elements released from the HLW canister. Under such hydrothermal conditions, fracture filling may occur by cooling of the fluid, redox reactions, precipitation by concentration to saturated conditions, and other factors. The studies of the Fenton Hill, New Mexico hot dry rock project site are of potentially great value in studying the Climax stock. These rocks are also highly fractured and the fractures sealed with calcite, muscovite, feldspar, quartz, and other minerals. The rocks have been subjected to a high heat flow, which can be used to determine how these

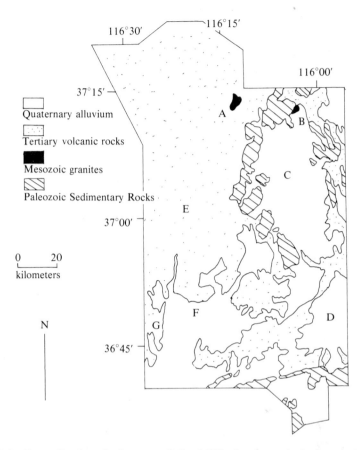

Fig. 7-6. Generalized geologic map of the NTS showing principal rock types. *Source:* Modified from Ramspott and Howard (1975).

rocks will behave in the presence of a new heat source. As mentioned earlier, the fracture fillings at the HDR site have been derived in large part from the host rocks and not from meteoric sources. This means that even if some fluid movement takes place, the fluids do not represent surface sources, nor have they been in communication with fluids far from their sites of formation.

This, in turn, implies that any radionuclide-bearing fluid will likely be trapped in the rocks surrounding the canisters. The actinides would precipitate in vein-type accumulations much like the well-described uranium veins in many granitic rocks (see Rich *et al.*, 1977) and/or thorium veins in various rocks (see Brookins, 1978c). The lanthanides, barium, strontium, and radium would be scavenged into the sulfate/or carbonate minerals likely to be present, and the elements Tc, Ru, Cd, Pd, Ag, In, Sn, Rh, Te, and Sb would be fixed under more reducing conditions, either with pyrite or other sulfides. Similarly, Rb, Cs, and some Sr will be scavenged by the usual clay

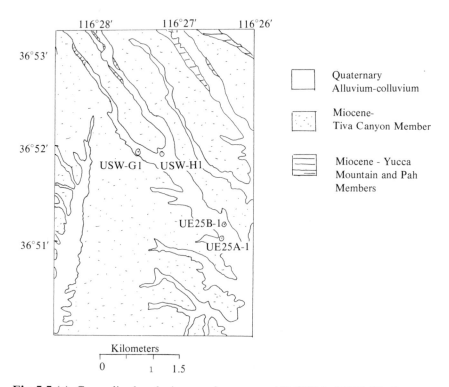

Fig. 7-7.(a) Generalized geologic map of area around Drill Hole USW-G1. *Source*: Modified from Sykes and Smith (1980).

minerals. Iodine also should be sorbed by clay minerals, or possibly by sulfides. Only Kr and Xe will not be readily incorporated into newly formed minerals or sorbed by older ones, but the overall amounts of these gases are so small that they will be so disseminated that their radioactivity is about background long before the biosphere is reached.

Studies of the Climax stock granitic rocks (Beall *et al.*, 1980) have been made to assess their ability to remove actinides by redox reactions involving Fe^{2+}-bearing minerals present in the granite. Bondietti and Francis (1979) earlier showed that minerals, such as magnetite ($FeFe_2O_4$), in various rocks contained sufficient quantities of Fe^{2+} to reduce Np^{5+} and Tc^{7+}, which are highly soluble, to Np^{4+} and Tc^{4+} species, which are highly insoluble. Beall *et al.* (1980) worked with oxidized species of U, Np, Pu, and Am in solutions that were allowed to interact with Climax stock granite, and become fairly uniformly distributed. Any radionuclide-bearing fluid that escaped entrapment in fractures would come into contact with the biotite in the granite, which would result in removal of the actinides by reduction and sorption caused by biotite. Other elements that should also be concentrated in the same chemical environment include the REE, Zr, Nb, Y, Tc, Ru, and Rh. Since pyrite is a common accessory mineral in those rocks, sites for fixation

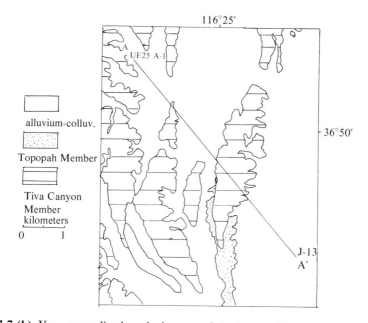

Fig. 7-7.(b) Very generalized geologic map of the Yucca Mountain area, NTS. *Source:* Modified from Sykes and Smith (1980).

of the chalcophile elements include Cd, Mo, Ag, Tc, Pd, In, Sb, Te, and Sn. Behavior of Rb, Sr, Cs, and Ba is also predictable. The alkalis (Rb, Cs) should be sorbed by biotite or perhaps plagiolase, and especially by clay minerals present as deuteric or later alteration. The alkaline earths (Sr, Ba) will be preferentially fixed in carbonates or sulfates if the fluids penetrating the rocks contain dissolved CO_2 and sulfur oxides; if they do not, then sorbtion into the same sites as the alkalis is probable. Iodine behavior is again problematic. Fixation is the hydroxyl site of apatite is possible, but simple sorption may be more likely. Even if not effectively removed from the granite, iodine escaping the repository will quickly be so diluted that it will not pose any threat to the biosphere. The noble gases, Kr and Xe, again, may not be removed by simple precipitation or sorption, but they will be disseminated in and diluted by the granite surrounding the repository.

Tuffaceous Rocks of the Nevada Test Site

The general geology of the NTS, which includes granite, argillite, and tuffaceous rocks, is shown in Fig. 7-6. The focus of NWTS program has shifted somewhat to place a great emphasis on tuffaceous rocks. For much greater detail the reader is referred to Cornwall (1972).

The tuffaceous rocks at NTS are extremely thick and variable in composition, mineralogy, and physical properties. Examination of these

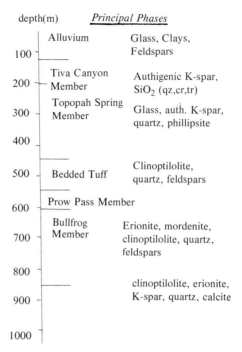

Fig. 7-8. Generalized stratigraphic section of test well J-13 of major authigenic phases. *Source:* Modified from Sykes and Smith (1980).

rocks has focused on Yucca Mountain in the southwestern part of the NTS (see Fig. 7.7a); several drill holes have penetrated these rocks, and their locations are given in Fig. 7-7b.

At Yucca Mountain the tuffaceous rocks have been described by numerous investigators, and summarized by Sykes and Smyth (1980). Although their work summarized results from drill holes J-13 and UE25a-1, additional drilling has been carried out (drill holes UE25b-1 and USW-G1; see Fig. 7-7a). The general lithologic and mineralogic description to follow is modified from the report of Sykes and Smyth (1980), as it is directly applicable to the NTS rocks and for silicic tuffs in general. The stratigraphic section for J-13 is given in Fig. 7-8, and variation of authigenic phases in this section is given in Fig. 7-9. A cross section between holes J-13 and UE25a-1 is shown in Fig. 7-10.

As Sykes and Smyth (1980) point out, there is considerable lithologic variation between these holes, but this is not unexpected considering the dynamic origin of tuffaceous rocks. The oldest unit encountered is a nonwelded ash-flow tuff below the Crater Flat Tuff, of probable thickness in excess of 200 m. Phenocrysts make up 10–14% of the rock and consist of sanidine, plagioclase, resorbed quartz with minor biotite, hornblende, and opaques. Glass is replaced by authigenic K-feldspar and quartz and some

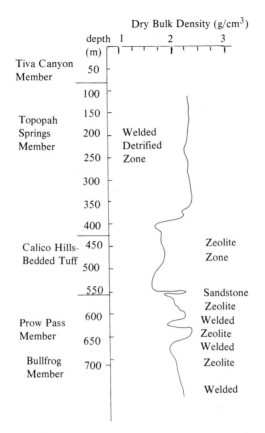

Fig. 7-9. Generalized stratigraphic section (Drill Hole UE25A-1) with major authigenic phases. *Source:* Modified from Sykes and Smith (1980).

clinoptilolite and analcime. It is not known precisely how this tuff unit correlates with other tuffs in the area.

The Crater Flat Tuff contains the Bullfrog Member and the Prow Pass Member. The Bullfrog Member is a welded ash-flow tuff containing abundant biotite. The thickness of the Bullfrog Member is about 140 m (Lappin *et al.*, 1981). Phenocrysts make up some 10–20% of the unit and consist of sanidine, plagioclase, quartz, biotite, and magnetite. Lithic fragments are common at the base of the unit, but are otherwise scarce. Fibrous to spherulitic quartz and K-feldspar replace pyroclasts in the basal units. Erionite and analcime are minor. No pronounced zeolitization is noted (Sykes and Smyth, 1980).

The Prow Pass Member is a vitric crystal tuff in which the lower part consists of nonwelded air-fall and ash-flow units with an upper welded ash-flow unit. It thickness varies from 50 to 150 m. Phenocrysts make up

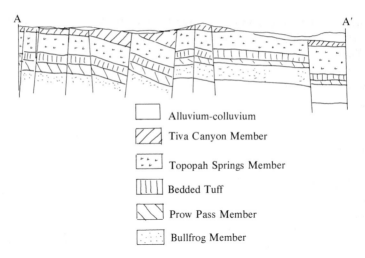

Alluvium-colluvium

Tiva Canyon Member

Topopah Springs Member

Bedded Tuff

Prow Pass Member

Bullfrog Member

Fig. 7-10. Generalized cross section A–A'. *Source:* Modified from Sykes and Smith (1980).

18–20% of the rock and include sanidine, resorbed quartz, plagioclase, and some anorthoclase, biotite, magnetite, and pyroxene. In drill hole UE25a-1 the nonwelded part is zeolitized, with clinoptilolite dominant, whereas in J-13 the nonwelded tuffs have altered to analcime, quartz, and K-feldspar with only minor clinoptilolite (Sykes and Smyth, 1980).

The Bedded Tuff of Calico Hills is a sequence of air-fall, ash-flow tuffaceous rocks and volaniclastic sedimentary rocks whose thickness varies from 95 to 140 m. The volcaniclastic sedimentary unit includes lithic fragments of tuff, perlite, flow rocks, and allogenic phenocrysts. Nonwelded bedded tuffs contain 2–16% phenocrysts of sodic plagioclase, sanidine, quartz and minor biotite. Zeolitization to clinoptilolite (in UE25a-1) is common in the upper part, and silicification accompanies zeolitization with increasing depth.

The Paintbrush Tuff includes the basal Topapah Springs Member (370 m thick), then the Pah Canyon Member (33 m thick), followed by the Yucca Mountain Member (8 m thick) and the uppermost Tiva Canyon Member (12 m thick) (Lappin *et al.*, 1981). Only the Topapah Springs Member has been described in detail (Sykes and Smyth, 1980). It is a zoned ash-flow tuff, with nonwelded basal units grading upward into welded vitrophyre, which is in turn overlain by devitrified welded tuff. Phenocrysts increase to near 17% at the top and consist of sanidine, plagioclase quartz, biotite, and opaques. Lithic fragments are common near the base but less frequent upward in the unit.

Alteration is variable in the unit, and both the basal vitrophyre and the upper welded zone are essentially unaltered. Much of the upper welded zone, however, is detrified to intergrowths of cristobalite and K-feldspar. Lower nonwelded tuffs are highly zeolitized to clinoptilolite, and heulandite occurs as fracture fillings. Minor amounts of erionite and phillipsite also occur.

The youngest tuffaceous unit in the area is the Tiva Canyon Member, which contains a basal bedded air-fall tuff and an upper ash-flow tuff. The lower part of the unit has 11–14% phenocrysts of plagioclase, sanidine, biotite, magnetite, orthopyroxene, while the upper part has only less than 5% sanidine, magnetite, plagioclase, hornblende, and sphene. Authigenic montmorillonite is common in the nonwelded parts of the unit, along with opal/cristobalite, while the welded parts have devitrifeid to intergrowths of cristobalite an alkali feldspar.

The presence of zeolites in the NTS tuffaceous rocks has received a great deal of attention (see discussion in Smyth and Sykes, 1980), as these minerals may control processes such as dehydration and other reactions. The dominant zeolite is clinoptilolite, which is the Na–K analog to Ca-bearing heulandite. Water loss from the zeolites by heating yields 7–10% water loss to 200°C with a volume decrease of 1.3% (Smyth and Sykes, 1980), while at 400°C an additional 2 wt.% water was lost with a volume decrease of an additional 1%.

Numerous sorption–desorption studies have been carried out on NTS tuffs. Some of the data (from Johnstone and Wolfsberg, 1980) are shown in Table 7-7. These data are reported as a sorption factor, R_d, in ml/g. A positive value indicates a high degree of sorption on the solid phase, and a value of ≤0 incorporation into the liquid phase. It is important to note that, with the exception of C, all R_d values are positive and usually large. Further, where data have been taken under both reducing and oxidizing conditions, the larger R_d values are reported for reducing conditions, which again shows the importance of chemically reducing environments to retain or retard possible radionuclide movement.

Bedded Salts of the Paradox Basin Area

The Paradox Basin bedded salt region consists of approximately 15,000 ft (4570 m) of evaporites and clastic sedimentary rocks lying nonconformably on a granite and metamorphic rock basement complex. The basement rocks are Precambrian (Hite et al., 1972) and the overlying sedimentary rocks range in age from Cambrian to Tertiary, with an absence of Silurian and Ordovician rocks. The stratigraphic section (from Ritzma and Doelling, 1969) is like other bedded salts. The evaporites are found in the Pennsylvanian System. The lowermost Pennsylvanian rocks are the Molas Formation, which consists of red, calcareous shale and sandstone with interbeds of limestone. This is overlain by the Pinkerton Trail Formation of limestone and shale (Wengerd and Matheney, 1958). The Paradox Formation overlies the Pinkerton Trail. It consists of a lower member of black shale interbedded with siltstone, gypsum and dolomite, and limestone, with some arkose

Table 7-7. Empirically derived R_d (ml/g) values from batch experiments at 25°C. 0.1 MPa for selected radionuclides in order of ranking.

Element	Salt	Basalt	Tuff	Granite
Tc	2^a	$20,^b$ O	$10,^b$ O	4^a
Pu	$500,^b$ 50	$200,^b$ 100	$500,^b$ 40	$500,^b$ 100
Np	$30,^b$ 7	$50,^b$ 3	$50,^b$ 3	$50,^b$ 1
I	0	0	0	0
U	1^a	6^a	4^a	4^a
Cs	1,800	300	100	300
Ra	5	50	200	50
Sr	5	100	100	12
C	0	0	0	0
Am	300	50	50	200
Sn	$1,^b,$ 50	$10,^b$ 100	$50,^b$ 500	$10,^b$ 500
Ni	6	50	50	10
Se	20; $100,^b$ 20	$20,^b$ 5	2	2
Cm	300	50	50	200
Zr	500	500	500	500
Sm	50	50	50	100
Pd	3	50	50	10
Th	50, 100	500	500	500
Nb	50	100	100	100
Eu	50	50	50	100
Pa	50	100	100	100
Pb	2	25	25	5
Mo	0, $5,^b$ 1	$10,^b$ 4	$10,^b$ 4	$5,^b$ 1

[a]No significant difference beween value measured in oxidizing and reducing Eh's.

[b]Reducing conditions, second-value oxidizing conditions. First value for dome salt, second for bedded salt.

interbedded with the dolomite and gypsum on the periphery of the basin. The upper member is similar to the lower member, and they are separated by a halite-rich evaporitic member. The evaporite sequence of the Paradox Formation is not homogeneous. Several different evaporite cyclothems are noted (Hite, 1960). A typical cyclothem consists of halite, anhydrite, dolomite ± potash zones, and impure shale, all deposited under stagnant conditions (Hite and Lohman, 1973). Each evaporite cyclothem ranges from 8 to 150 m. The Paradox Formation is overlain by Pennsylvanian limestone (Honaker Trail Formation) and limestone interbedded with sandstone (Rico Formation). These, in turn, are overlain by the Permian Cutler Formation and then by Mesozoic and younger rocks.

Several salt anticlines are known in the Paradox Bedded Salt Region. These are on the average 50–115 km in length and 50–60 m in thickness. The core evaporites of these anticlines ranges from 1250 to 4270 m. Halite makes up 70–80% of the anticlinal core rocks with interbedded potash zones and anhydrite marker beds, and some dolomite and black shale. Dissolution of halite at the tops of the anticlinal crests has resulted in gypsiferous cap rocks and collapse structures on the tops of the crests. The salt anticlines originated as shallow burial features (Cater, 1972), unlike the Louisiana–Texas coastal salt domes, which are the result of deep diapiric movement.

These bedded salts meet the criteria for repository considerations on first scrutiny, and are under investigation by the Office of Nuclear Waste Isolation in conjunction with subcontractors (see ONWI, 1980; and ONWI, 1982).

Salt Domes of the Gulf Coastal Region

In the Gulf Coast region, including parts of Texas, Louisiana, and Mississippi, some 263 known or suspected salt domes have been investigated as to their potential for the storage of radioactive wastes. Of these, 36 were selected for further study (see ONWI, 1981) based on their fairly shallow depth (less than 2000 ft) and lack of utilization by industry for oil, gas, salt, brine, or sulfur production. Care was taken to investigate those domes with no petroleum potential as well. The reader is referred to Ledbetter et al. (1975) for a discussion of some of the salient characteristics of dome salts, which are much like bedded salts discussed earlier in this book. They conclude that there is no valid geologic or technical reason for not considering dome salts for radioactive waste storage, especially if the dome is surrounded by shale, is below the lower level of actively circulating groundwater and is in a tectonically stable area.

Late Triassic block faulting led to the formation of the Gulf Basin, accompanied by some igneous activity. During the Late Triassic through the early Jurassic, isolated basins were formed in which salts were deposited by evaporation of sea water. These evaporites were then buried by younger sediments. Some salt movement started in the Late Jurassic, but more pronounced movement followed, culminating in the Mesozoic for the interior salts and in the Late Tertiary for salts on the coastal plain, and movement is still ongoing for the offshore salt domes. Many of the domes, especially in the interior region, are hydrologically stable (ONWI, 1981); little or no dissolution of salt is taking place in the salts. Domes where the dissolution is pronounced are listed as hydrologically unstable and are not under consideration for waste storage. As pointed out in ONWI (1981), the interior salt domes are overlain by aquifers containing fresh water, therefore seeps from the domes result in brackish water. The existence of the fresh water argues against a rapid exchange between the two waters, and the saline water is interpreted as stationary for the most part. It is not known if all these saline waters are saturated or not. If so, then they cannot promote dissolution; if

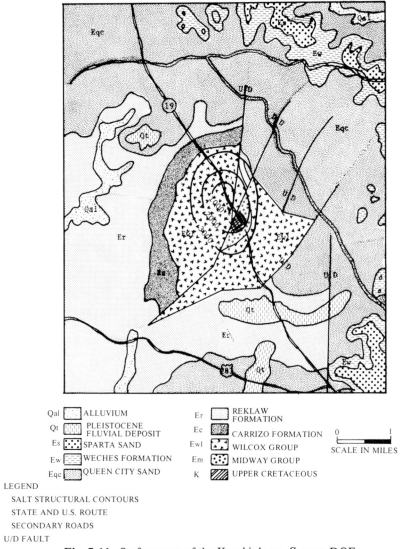

Qal ▨ ALLUVIUM
Qt ▨ PLEISTOCENE FLUVIAL DEPOSIT
Es ▨ SPARTA SAND
Ew ▨ WECHES FORMATION
Eqc ▨ QUEEN CITY SAND

Er ▨ REKLAW FORMATION
Ec ▨ CARRIZO FORMATION
Ewl ▨ WILCOX GROUP
Em ▨ MIDWAY GROUP
K ▨ UPPER CRETACEOUS

0 _____ 1
SCALE IN MILES

LEGEND
  SALT STRUCTURAL CONTOURS
  STATE AND U.S. ROUTE
  SECONDARY ROADS
U/D FAULT

Fig. 7-11. Surface map of the Keechi dome. *Source*: DOE.

not, then they can dissolve salt only to the point of saturation unless new, fresh water is allowed to infiltrate the rocks.

The caprock present in the domes is problematic, and several theories on origin have been suggested (see ONWI, 1981). Early dissolution accompanying diapirism is one such attractive hypothesis. One attempt at calculating a dissolution rate has been made (ONWI, 1981), giving a value of 0.005 mm/yr, i.e., even if dissolution were promoted, it would take one million years for 5 m of salt to be dissolved for the conditions given.

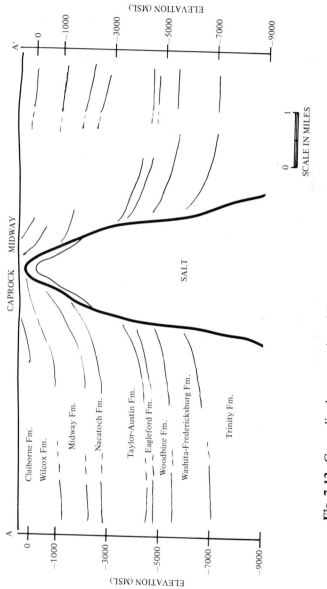

Fig. 7-12. Generalized cross section of Keechi dome. *Source:* Modified from **DOE.**

Several domes are considered better targets than others. The Keechi, Oakwood, and Palestine domes in Texas, the Vacherie and Rayburn's domes in Louisiana, and the Cypress Creek, Lampton, and Richton domes in Mississippi have all been earmarked for more detailed study. In Figs. 7-11 and 7-12 a surface geologic map and cross section of the Keechi dome are shown, which within limits, is typical of most of the domes. The disturbance of the beds in the vicinity of the dome is shown in Fig. 7-12.

The geologic criteria used to evaluate the dome salts are (ONWI, 1981): (1) minimum thickness and lateral extent for containment, (2) adequate depth for containment and for repository construction, (3) in an acceptable area based on low seismicity and tectonic activity, (4) acceptable hydrologic characteristics, and (5) no significant valuable mineral resources present.

Basically, the attractive features of bedded salt deposits (see Chapter 6) are the same as those for the dome salts. In addition, the dome salts, and especially those in the interior regions, are even more H_2O-poor than many bedded evaporites. The in-depth work on dome salts is only now being carried out, and more commentary on the suitability of such a medium for the disposal of radioactive wastes must await the results of these studies.

Some Conclusions

Detailed studies are underway at several sites within the United States to assess their suitability as repositories for radioactive wastes. Of these, the WIPP site in southeastern New Mexico has been designated, pending approval, for disposal of TRU defense wastes. The other sites, including the BWIP site in Washington, the NTS, the Paradox Basin bedded salts in Utah, and salt domes in Louisiana and Texas, are being investigated in terms of HLW disposal. If necessary, the WIPP site may be used for experiments to further investigate its potential use for HLW disposal. The studies at the various sites are in different stages. Very comprehensive work has been carried out at the WIPP site, considerable work at the BWIP site, and an increasingly significant amount of work at the NTS. Investigations in the Paradox Basin and in the salt domes is also progressing nicely.

To date, the collective studies have not revealed any geologic, hydrologic, geochemical, or other data or observations to preclude their possible use for disposal of radioactive wastes. In fact, the data, discussed earlier in this chapter, offer strong support for use of these rocks for radwaste purposes.

CHAPTER 8
Alternate Plans for Waste Disposal

Introduction

Several alternate plans for radwaste disposal have been presented over the last two decades, and the major alternatives are presented herein. Subseabed disposal is again receiving a high degree of interest, and will be examined in more detail than the other alternate methods.

Subseabed Disposal

The idea of disposal of radioactive wastes on the ocean floor, or, more specifically, burial in ocean bottom clays, is an attractive one. Seventy percent of the earth is covered by the oceans, and much of the ocean bottom is not considered for economic purposes. The ocean can be divided into near-shore, continental-shelf, and deep-sea reservoirs.

The near-shore environment includes the intertidal zone of open coast as well as those areas that are partly enclosed by land, such as bays, harbors, and estuaries. Currents and various mixing processes are very intense in this region relative to the other parts of the ocean. Ocean currents, in response to very large-scale oceanic patterns, tidal effects, or wind, generally follow coastlines, and they may reach velocities of several meters per second. Local variations in sea level occur due to tidal and wind action and result in a high degree of horizontal and vertical mixing with associated complicated flow. The effect of these processes in the near-shore environment is the prevention of rapid dilution, and for this, and many other, reasons the near shore environment is not favorably considered for a waste disposal site.

The continental shelf is the submarine rim of the continental land masses. It runs from the shoreline to about a depth of 400 m, although this may vary widely from place to place, and where active subduction is taking place the effective continental shelf is essentially nil or very thin at best. Mixing in the continental-shelf environment is quite variable, with some areas highly mixed and others largely unmixed. Eddy currents are common phenomena that occur over many shelf areas, and, while these currents are good for dispersion of newly introduced elements into them, their overall effect on the water and on the ocean-bottom sediments, especially when turbidity currents are considered, makes the continental-shelf environments unfavorable for repository consideration.

The deep-sea environment makes up about 75% of the total ocean volume. This is a cold-water reservoir, with temperatures commonly ranging from 1 to 4°C. The upper few meters, to a depth of some 200 m, make up a zone of relatively active mixing resulting from seasonal temperature changes, wind action, and possibly, minor tidal effects. There is a marked density change at the thermocline, the barrier between the more shallow, mixed zone, and the cold, unmixed zone. This barrier helps impede elemental transfer between the two zones, and helps explain why the salinity of the deep sea is quite constant at about 34.7 o/oo. The surface layers mix so readily that any material introduced to them will be dispersed extremely rapidly. In one interesting experiment, Folsom and Vine (1957) showed that for 1000 c (curies) introduced as a point source in surface waters, where mixing persists to a depth of 50 m, within 40 days the initial activity was spread over 40,000 km^2 such that the resultant activity was only 1.5×10^{-10} μc/ml (microcuries per milliliter). Movement and mixing in the deep seas below the mixed layer is slight, and horizontal dispersion is more pronounced than vertical movement (see Strommel, 1957). This means that any point source contamination in the very deep ocean will be so dispersed horizontally through a large volume of water that only below-background amounts of the material will reach the surface layers, where even further dispersal will occur.

On the floors of the oceans is commonly found reddish clay, which persists to depths of hundreds of meters. These reddish clays are common on the abyssal plains of the deep ocean bottom, where no human accidental entry is likely and where the resources are, by inspections to date, very few. From experiments, the clays themselves are known to readily sorb most radionuclides and also permit only limited pore water movement. Preliminary study (DOE/ET-0028, 1979) shows that the clays have remained stable for millions of years, which thus would allow sufficient time for radioactive species to safely decay to stable daughter products.

Again geochronologic studies on ocean-floor materials are of importance. Study of aeolian clay minerals on the Bermuda Rise by Hurley *et al.* (1963) shows that the K–Ar systematics of the clays have not been disturbed by their long residence times on the ocean floors. Thus the illitic clays of the Sharan Desert Province that have been transported to the deep sea have retained their original K–Ar systematics, which shows that the seawater has not caused any gain or loss of either K or Ar. Further, basalts from either side of the Mid-Atlantic Ridge can be dated by K–Ar techniques, which also shows that even basalts are not greatly affected by the deep-sea environment.

Yet a description of deep-sea-floor sediment in the abyssal plains as simply "reddish clay" is not sufficient for our purposes. Deep-sea sediment consists of biogenic and nonbiogenic materials. Biogenic materials include $CaCO_3$-secreting foraminifera, coccoliths, and pteropods, with the foraminifera being dominant. SiO_2-secreting life includes diatoms, radiolaria, sponges, and silicoflagellates, with the diatoms and radiolaria being the most important. Silica is fairly soluble in seawater, hence the distribution of the forams and radiolaria is variable, with zones of concentration occurring at high latitudes and in the equatorial zones.

The nonbiogenic clay minerals include material transported from non-ocean sources, bottom-transported material, and authigenic clay minerals. Continental sources are more important than ocean sources. Also, because of weathering effects in source areas, the abundance and distribution of clay minerals are variable as well. Kaolinite is common in zones of tropical weathering, and the equatorial zones of the deep-sea sediments are kaolinite rich, whereas a higher ratio of chlorite to kaolinite is found in higher latitudes. Illite is virtually all detrital, yet it does not react with sea water (as mentioned above). Montmorillonite is mostly authigenic and is derived largely from alteration of glassy parts of ocean-floor basalts along with the zeolite, phillipsite. Quartz is largely land derived as are the small amounts of feldspar, hornblende, and other rock-forming minerals noted in deep-sea sediments. Gibbsite occurs with kaolinite in the equatorial zones and is scarce elsewhere. As expected, there are numerous volcanic materials found in the deep-sea sediments, including basalts, volcanic glasses, and zeolites.

Of special interest are the manganese-rich nodules that cover extensive parts of the ocean floors. These nodules contain an average of 16–24 wt% manganese, high iron and silicon, and high concentrations of many metals, including Cu, Co, Ni, Cr, Pt, and the REE. The nodules attest to the fact that these elements are removed effectively from the oceans in which they are found.

The fairly complex mineralogy of the ocean-bottom sediments is useful in considering the selection of such sites for a possible repository. The clay minerals found are similar to those being considered for backfill and thus will sorb and remove most alkali and alkaline earth elements as well as much of the transition metals and, presumably, the actinides. Most of the transition metals and the actinides will be controlled by inorganic precipitation akin to nodule formation). For radioactive HLW, Co, the REE, Pd, Ru, Tc, Ag, Cd, Sb, Te, Th, In, Sn, Mo, Zr, and Y should be scavenged in the oxy-hydroxide (nodule equivalent) sediment and Rb, Sr, Ba, Cs by the clay minerals. Uranium and the transuranics will be sorbed in part on the clays and in part

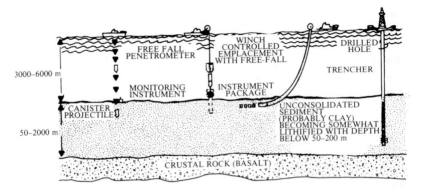

Fig. 8-1. Sediment emplacement concepts for seabed isolation. *Source:* Modified from DOE (1979).

removed in the oxy-hydroxide fraction. Iodine behavior is hard to predict, but even if it is not scavenged by clay, it will be easily dispersed in the seas.

Some of the ways in which radioactive wastes may be entrenched into the ocean bottom are shown in Fig. 8-1. No firm endorsement has yet been made of any of these particular methods, however.

There are several sites being considered for possible subseabed disposal (see Fig. 8-2). The Seabed Working Group (SWG) comprises Canada, the Commission of European Communities, France, Japan, The Netherlands, Switzerland, the United Kingdom, Federal Republic of Germany, and the United States. In addition, Belgium and Italy are involved as observers in the

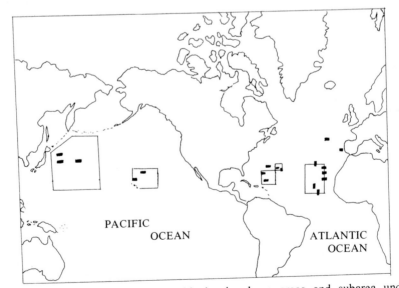

Fig. 8-2. Map of part of the world showing large areas and subarea under investigation for subseabed disposal of radwaste. Modified from DOE (1979).

seabed research. The subseabed research is being actively pursued by the above-named countries, especially since some may not have direct access to a continental repository site suitable for HLW storage. The subseabed option is attractive because of the very stable parts of the ocean floor. The model for the entire subseabed program is shown in Fig. 8-3. It varies from the continental repositories primarily in that the ultimate barrier is the ocean, not the rocks. Yet any stored radwaste that is attacked by bottom ocean water will release an entirely predictable amount of species, and these will be essentially infinitely diluted in the ocean. As shown in Fig. 8-3, and assessment of dose to biota and, ultimately, dose effects on man, will be made as an integral part of the program.

The sites selected for initial scrutiny include clay-rich sediments of the abyssal hills and the edges of abyssal plains. These are under intensive study at the present time.

Other Alternate Waste Disposal Proposals

In addition to the subseabed disposal alternative, several other methods for the disposal of radioactive waste have been proposed. These include the very deep hole waste disposal concept, the rock melt disposal concept, the island-based geologic disposal concept, the ice sheet disposal concept, the well injection disposal concept, the transmutation concept, and the space disposal concept. These are discussed below.

Very Deep Hole Waste Disposal

The very deep hole waste disposal concept would require the placing of nuclear waste at great depths, about 10,000 m, where it is presumed that it would remain isolated from the biosphere. An immediate question is, "How deep?" At a depth of 10,000 m in tectonically stable areas, rocks such as granite and other crystalline rocks and shales are suitable candidate rocks. These rocks should remain sealed and stable, despite the heat and radiation from the radioactive waste. Yet data on deep groundwater and its ability to possibly interact with deep-hole waste are not known. Even natural analogs would be of limited use here as only at- or near-surface rocks have been studied (see Chapter 11). Presumably at a depth of 10,000 m or so any radioactive waste attacked by deep circulating groundwater would only migrate on local levels and not reach the biosphere before being so diluted and radioactivity poor (i.e., the short- to intermediate-level radionuclides would have decayed away) that their hazard would be nil. However, some rocks, like limestones and some other sedimentary rocks, have relatively high porosity and permeability, even at great depths. Crystalline rocks, on the other hand, may possess very low porosity and permeability at depths of only 3000 m or so, hence depths of 10,000 m would not be required. Once a depth has been agreed on, the limiting factors would basically be those of drilling, emplacement, and sealing.

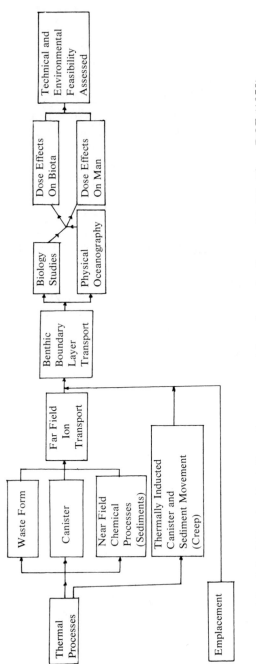

Fig. 8-3. U.S. subseabed disposal program models under study. *Source:* Modified from DOE (1979).

A major disadvantage of the very deep hole concept is that the stored waste would not be retrievable. Environmental factors include the usual drilling, surface facilities, waste emplacement, and assurance of long-term confinement of the wastes.

Transportation would be by rail, and wastes would be disposed of on a regular basis. The U.S. Government (DOE 1980) has estimated that 140 km^2 would be required for a 40-year period of radwaste disposal.

Possible sites have not been designated, due in part to the waiting for decision on mined repositories. The geologic criteria to be used in evaluating potential candidate sites include lithology, tectonic setting, structural setting, hydrology, knowledge of mechanical rock properties at depth, the thermal regime of the rocks, and the rock geochemistry.

The sealing would be done by placing a series of seals above the buried waste canisters. The practice proposed is that used by the oil and gas industries, which have used seals at depth for many decades. Temperature at depths of 10,000 m would be on the order of 300° to 400°C or so, and as such, the heat from the buried canisters would not be pronounced. The actual effect of the additional heat is unknown, but, in the absence of a reaction medium, such as water, thermally driven reactions would be nil. Study of drill core from deep holes drilled as part of the hot dry rock (HDR) research projects (Los Alamos National Laboratory) shows that the rocks at depth are usually fractured but that the fractures are mostly filled with sealing, secondary minerals. Further, the source of much of the fracture-filling minerals is the surrounding rocks, not from more shallow rocks in communication with meteoric waters. It is assumed that at depths of 10,000 m similar filled fractures will occur, but these have not yet been investigated.

Rock Melt Disposal

The rock melt concept would require direct emplacement of HLW and some TRU as reprocessed liquids or slurries into underground cavities. The concept is based on the realization that after water has evaporated from the initial transport phase, the heat from the remaining radioactive waste would cause melting of the surrounding rocks, with later dissolution of the waste. With time, the melts would mix and homogenize at sites well below the surface, and solidifcation would be completed in about 1000 years. Containment of the more long-lived radionuclides is expected from this method. One drawback is that SURF, CW, and some TRU wastes cannot be handled by this method without laborious treatment and preparation of slurries. This would increase the potential handling toxicity of the wastes, and therefore is not desirable.

The mix of rock melt with dissolved waste will presumably be leach resistant, so that its chemical integrity will be preserved for a lengthy period of time. Several analogs to assess and evaluate the rock melt method are known, including emplacement of igneous rocks, with partial melting of the intruded rocks and some high-temperature hydrothermal mineral deposits, but these have not been studied for their potential application to the rock melt concept.

The waste isolated by the rock melt concept will be irretrievable, although, in theory, the rock melt–dissolved waste mixture could be recovered by deep mining. In practice, however, this type of recovery would not be economical nor practical.

The actual placement of the waste would require excavation of a shaft about 2000 m deep and a cavern to handle the waste about 6000 m^3 in size. The waste would be emplaced over a period of about 25 years, after which the water and steam present would be pumped out, the excavations sealed, and temperatures would cause melting. After about 60–70 years solidification would start.

The rocks selected must be those whose melting characteristics are well known, like granite. The size of the rock melt cavity will depend on the size of the mined cavity, the thermal conductivity and diffusivity of the rock, heat of fusion, amount and chemistry of rock and contained water, and mineralogy and chemistry of the rock. Once these and other relevant parameters are known, predictions on rock melting can be made from knowledge of potential radioactive heat from the buried waste.

It should be clear that the rock melt concept demands knowledge of the rocks involved in a specific task, and while no prototype experiment has been carried out, the extensive work on experimental petrology, mineral deposits, and hydrothermal alteration, HDR research, and so on, suggests this concept can work. At the same time, it is equally apparent that there are major disadvantages to this concept, including irretrievability of the wastes, necessity of finding multidisposal sites for the non-HLW, handling times and predisposal chemical treatment, and handling of water and steam pumped from the mined cavity. Further, data on waste–rock interactions are meager, and there no previous technical or engineering design of these types of facilities has been attempted. For these reasons, it is the author's opinion that this concept is less practical than other alternatives.

Island Disposal

The island disposal concept would involve placing the radioactive wastes in deep, stable geologic formations on islands. The repository requirements are like those for mined repositories on continental land masses (Chapters 6 and 7), only the wastes must be transported to the islands. The natural barriers at such a site would include the repository rock plus the oceans. The oceans surrounding the island would ensure rapid dilution of any escaped waste (see Subseabed disposal, above). The status of island disposal is not clear. Investigation of the island disposal concept is not being actively pursued by the U.S. Department of Energy, although the United Kingdom (Royal Commission on Environmental Pollution) has pointed out the advantages of having an island disposal site both in view of favorable geology and for possible use as an international repository. Many islands contain very little groundwater, and exchange with the oceans is slight. Thus little water would be available for possible attack on stored waste, and if such unlikely reactions were to take place, the surrounding ocean would so dilute the products of the reactions that they would be rendered harmless. Basically two different below-ground sites are available for inspection. The near-surface waters are fresh waters, which do not readily exchange with the deeper, saline waters, and whose shape is that of a lens sitting in heavier, saline water.

Three types of islands are available for repository considerations: continental islands, oceanic islands, and island arcs. The continental islands are found for the most part on the continental shelves and are made up of igneous, metamorphic, and sedimentary rocks. The oceanic islands are located in oceanic basins and are made up of volcanic rocks, usually basalt. The island arcs are those isalnds located at plate boundaries near continental masses. They have originated by tectonic processes and frequently are the sites for andesitic volcanism.

There are serious disadvantages to this method. First, it is logistically impractical. In addition to the secure ocean vessels, there would have to be seaport and handling facilities both on the mainland and on the islands, and land transportation both on the mainland and the islands. This and the fact that handling time and transportation times are much greater than those for land repositories make this method economically less attractive than other alternatives. The U.S. Department of Energy has estimated (DOE, 1980) that an island repository would cost at least twice as much as a land repository. A possible attractive feature would be use of an island site for the storage of international waste. This may not seem necessary now, but, in view of the fact that 35 nations now rely to some degree on the nuclear option, such an international site may be necessary in the not-too-distant future.

Another serious disadvantage for many of the islands of the earth is possible volcanism. Volcanism is a risk on too many islands and ocean islandic provinces to unequivocally say that it will not occur on a particular island. When only continental islands are considered, the volcanism risk is reduced, but these islands are perhaps less well situated geographically than the oceanic islands in order to be considered for the disposal of inter-nationally generated wastes.

Ice Sheet Disposal

The ice sheet disposal concept would involve placing radioactive wastes into continental ice masses in isolated areas, mainly in the Antarctic. Vast amounts of internationally held land in cold, remote areas of the Antarctic are, at first glance, attractive for the disposal of radioactive waste, as the waste should be isolated for at least several thousand years in order for the short-lived radionuclides to decay. In theory, the waste would be placed in the ice at shallow depths where it would melt its way down into the ice sheet. But there are problems with ice sheet disposal. First, the transportation of wastes will be very expensive, as it involves the usual (or more complex) land transportation, two sets of seaports and handling facilities, and a land–ice transportation system. Second, evolutionary processes in ice sheets are not well-understood. It is assumed that the wastes that melt their way into the ice will remain safely isolated from the surface, yet different rates of flow in the ice do exist, and the distribution and numbers of these areas of high flow are not well understood. Further, in the Antarctic, especially, the amount and distribution of volcanic centers are not well known. Still further, the ecology of Antarctica has not been extensively studied, and therefore the impact of radioactive wastes in ice on the environment needs more detailed assessment.

Three types of emplacement in ice have been proposed. One method calls for simple melting of the radioactive wastes into the ice without any chance for retrievability. Melting would occur most efficiently if special canisters were used for the wastes; if carried out, the rate of descent of the waste would be about 1 to 1.5 m per day, or, in other words, it would take about 5 to 10 years for the wastes to reach the bedrock (for a 3000-m-thick ice sheet). A second method calls for anchored emplacement, in which the canisters would be attached by cable to the surface so that retrievability could be carried out, if desired. In a third method, a surface storage facility would be constructed in which a specified amount of waste would be stored, after which the entire facility would be allowed to melt into the ice. This third method also rules out retrievability.

The main advantage of the ice sheet concept is that, in the case of the Antarctic, remote, international land is available, temperatures are low, and the presence of the ice assures isolation. Yet the disadvantages, which include high transportation and operational costs, the uncertainty of ice dynamics, and the possibility of severe changes in global climate, outweigh the advantages. An additional disadvantage is that, somewhat like the rock melt concept, temperatures high enough to melt the ice are required. This means that most TRU will not, by itself, possess enough heat for this to be possible; hence the TRU will have to be mixed with some HLW. The disposal of SURF and CW in ice sheets does not appear to be possible, although, in theory, SURF would melt ice if not stored on site for a few years. Yet this argues against the conventional policy of allowing most wastes to cool above ground for 5 to 10 years, by which time the SURF and CW would be too cool to effectively melt ice. TRU wastes, because of cost and volume factors, cannot be considered realistic for ice sheet disposal, nor can SURF and CW, because of cost and melting parameter factors. Even if HLW could be safely disposed of in ice sheets, the transportation factor would be further hindered by the fact that the sea lanes are open for only 1 to 3 months per year.

The author again concludes that there are too many uncertainties of a geologic nature and undesirable economic factors that make the ice sheet disposal concept less desirable than other alternatives.

Well Injection Disposal

The well injection concept is based on the experience of the oil and gas industry, whose practice has been to get rid of oil and gas field brines by pumping them back into the oil or gas reservoirs. Their success has lead to the use of well injection for both natural and industrial wastes, although the methods used in the oil and gas industries are for liquid waste only (because solids can cause blocking of natural passage ways). Two different injection techniques have been developed: the deep well injection method and the shale grout injection method.

In the deep well method, acidic liquid wastes would be pumped into rocks isolated from the biosphere to depths of 1000 to 5000 m. These rocks would presumably be somewhat porous and permeable, but for candidate sites would be overlain by impervious rocks, such as shales, thus isolating the injected wastes. Questions (DOE, 1980) have been posed as to the possibility of differential radionuclide migration such that, possibly, plutonium or some other actinide may be isolated from other elements and concentrated to the point where criticality could occur. The writer has assessed this problem (see discussion in Chapter 4) and finds it to not be significant for two major reasons. First, an actinide-selective process to allow this type of segregation and reconcentration is unknown. Second, a local critical reaction at a depth of over a 1000 m would not affect the rocks beyond a few meters. The studies of the Oklo natural reactor convincingly show that even where criticality continued for some 500,000 years, the fission products largely remained in place.

In order for deep well injection to be carried out, the various wastes would first have to be shipped to a processing plant, presumably at the well site, where the wastes would be dissolved in an acidic solution separate from the hulls. The resultant acidic solutions would include both the actinides and the fission products. Candidate rocks best suited for the well injection method are sandstones, especially those in synclines, where the injected rocks are overlain by an impervious rock, such as shale. The chemistry and lithology of the injected rocks must be known so that accurate predictions about radionuclide behavior can be made. By definition, groundwater flow in the injected formation would have to be slight and with essentially no vertical parameter. This concept has the advantage that, due primarily to the extensive experience and investigation of the oil and gas industries, the rock characteristics and hydrology of many such rocks are known.

The shale grout method would require fracturing of the candidate shale by high-pressure water injection, after which a mixture of wastes and clay and cement would be injected into favorable shale at depths of 300 to 500 m. The injected slurries would then be allowed to solidify *in situ*. Since shale is not highly porous and possesses low permeability, migration of the injected material is not likely. Further, the high sorption potential of the shale would further hinder any possible radionuclide movement. It is important to this method that fractures created will be parallel to bedding planes so as to minimize vertical transport while ensuring that any possible radionuclide flow would be restricted to the shale horizon in which it was emplaced. To date, the Oak Ridge National Laboratory has demonstrated the success of the shale grouting method for the disposal of TRU wastes. For the shale grout method, neutralized liquid waste or a slurry of irradiated fuel would be mixed with cement and clay, and then injected into hydrofractured shale at depths of 500 to 1000 m. For shale in which the horizontal stresses are greater than the vertical stresses, the fractures will be parallel to the horizontal, thus not allowing vertical fractures and possible movement of radionuclides, to occur.

A distinct advantage of this method is the abundance of shale in the United States (see Fig. 6-6). Widespread areas, many tectonically suitable, are underlain by shales, including shales that are flat lying, of low porosity and permeability, and those in which the horizontal stresses are less than the vertical stresses. These parameters, coupled with the ability of the shale to easily adsorb and fix actinide and fission product elements of concern, makes this method attractive. In addition, there has been considerable work done on the disposal of nonradioactive toxic and nontoxic wastes, and disposal of radioactive wastes has been investigated by the Oak Ridge National Laboratory. In all cases, successful waste isolation has resulted.[a] Wastes that have been removed by these methods include oil field brines, uranium mill tailings, steel industry waste, refinery and chemical wastes, and, by the Oak Ridge National Laboratory, nearly 2 million gallons of ^{137}Cs-bearing (524,000 Ci) and ^{90}Sr-bearing (36,800 Ci) wastes over a period of 10 years. Some useful references are given in Pinder and Gray (1977), Warner and Orcutt (1973), ERDA (1976), Pickett (1968), Burkholder and Koester (1975), Maini (1972), Spitsyn *et al.* (1973). The studies at Oak Ridge National Laboratory (Moore *et al.*, 1975) show that use of ash as a substitute for some of the cement yields a more efficient sorbing agent for Sr, and is cheaper.

Some advantages of the method are that the methods do not rely on multibarriers for isolation, although the shale for the shale grout method and shale for the rock overlying sandstone for the deep well injection method both provide an infinite number of barriers *if* flow is considered vertical to the source of accumulation. HLW studies for behavior in both methods are lacking, and, while attractive, must await much more research.

An advantage of these methods is that the shale grout method has already

---

[a] It is a recognized that the longest duration of isolation is for only several decades, but no loss from the storage areas has been noted.

been demonstrated for TRU wastes, hence a method is at our disposal, with necessary emplacement technology, to make this method practical. For HLW, however, the picture is less clear.

Economically, the methods, especially the shale grout method, are cost competitive for handling of TRU wastes. In fact, the shale grout method may be cheaper than mined repositories. Storage and handling facilities are expensive, but so are facilities for HLW from production.

Natural events that might affect the well injection methods are those for virtually all repositories, including meteorite impact, volcanism, seismic events, and gross climatic changes. These are not significant for well injected waste unless the hydrology of the site is changed so that greater releases of stored radionuclides is possible. Only infrequent earthquakes are thought to be of any realistic consideration, and, even here, areas underlain by shales where numerous earthquakes have occurred could be studied to determine just how many elements have been mobilized due to the tectonic activity. Drilling of the site for non-waste-storage purposes would have to be assessed. If a high number of drill holes have been made into the candidate rock, then these, as well as natural penetrations, would have to be studied to see if they present possible migration paths higher for water into the waste-injected rocks or avenues for radionuclide escape. Further, the rocks will have to be studied to see if the combination of hydrofracturing and waste-injection could promote local earthquake activity (e.g., storage of liquids at the Rocky Mountain Arsenal injection facility was related to an increased amount of seismic activity resulting in earthquakes of small scale).

Flow by groundwater through the shale or sandstone remains the most probable route by which radionuclide migration might occur. While there is admitedly a paucity of relevant studies in the area of radionuclide behavior and flow in waters in nature, the studies at Oak Ridge National Laboratory have shown that Sr and Cs would be fixed by shale surrounding the grouted area, that Pu and U would be sorbed essentially where injected, and that most, if not all, radionuclides would have decayed away before reaching the biosphere and those that did would be so diluted that they would pose no threat to the environment.

Transmutation

The transmutation concept involves elimination of long-lived elements by transmutation to short-lived isotopes. Thus the residence time for waste before decay of radioactive species would be shortened appreciably, and the potential danger from long-lived radionuclides would be greatly diminished. In order for the transmutation concept to work, it is assumed that Pu and U would be separated from spent fuel and remaining HLW separated into a fission product component and an actinide component. The fission products would then be treated separately by freezing them into a glass or ceramic waste form, and the waste actinides mixed with the Pu and U recovered (above) from the spent fuel plus the actinides that are recovered from TRU wastes. A new fuel rod would then be fabricated from the total actinides and irradiated in a reactor. If this is carried successively during a year, almost all the actinides would be converted to short-lived radionuclides that could be separated out during the fission product: actinide separation. Only about 5 to 7% of the actinides are transmuted in any irradiation, hence numerous irradiations would be necessary to completely remove the actinides.

This method serves one purpose: the removal of long-lived actinides, although additional fission products are produced. In addition, problems are presented by the necessity of additional facilities and handling, which increase the risk of possible releases to the environment, occupational exposure, possible accidents and, finally, greater cost. Final disposal of the fission product is identical to that for HLW generated during routine reactor operation.

Transmutation technology is straightforward but cumbersome. For this method to work efficiently, valuable reactor time must be spent on transmuting irradiations, and the pretransmutation processing and other chemical steps are time consuming. Existing systems would have to be modified to handle activation of fissile isotopes, and additional containment systems would have to be built to prevent loss of radioactivity from the core. Further, additional shielding would have to be built for the processing, irradiations, and newly formed HLW fission products. Finally, decommissioning would be more time consuming and unwieldy due to the new shielding, multichemical treatments, and waste separations.

It must also be emphasized that this concept involves only a partial solution to the waste disposal problem. The HLW, original and newly formed, would still need to be disposed of separately from the actinides. Further, the separation of actinides from HLW prior to transmutation would cause a net increase in the total volume of wastes generated. It should also be pointed out again that nature has successfully been storing actinides in rocks for over 3.8 billion years, and that actinides mixed with HLW can be technologically handled in a mined (or alternate) disposal site. Finally, a very large work force, and a new and very large (twice normal size) physical plant facilities, would be required, thus greatly increasing the cost for waste treatment.

Space Disposal

The space disposal concept was designed to permanently remove radioactive wastes from earth. It is intended that encapsulated HLW would be processed into a cermet matrix, packaged, and placed in a rocket, and sent into solar orbit for one million years. In theory, the waste package would be carried by space shuttle to an orbit near earth from which it would be sent into the deep solar orbit by a transfer rocket with a supplementary solar orbit

Table 8–1. Radioactive species in the oceans.

Nuclide	Concentration (g/ml)	Specific Activity (d/sec/ml)	Total Amount in Ocean (millions of tons)	Total Activity in Ocean (millions of curies)
^{40}K	4.5×10^{-8}	1.2×10^{-2}	63,000	460,000
^{87}Rb	8.4×10^{-8}	2.2×10^{-4}	118,000	8400
^{238}U	2.0×10^{-9}	1×10^{-4a}	2800	3800
^{235}U	1.5×10^{-11}	3×10^{-6a}	21	110
^{232}Th	1×10^{-11}	2×10^{-7a}	14	8
^{226}Ra	3×10^{-16}	3×10^{-5a}	4.2×10^{-4}	1100
^{14}C	4×10^{-17}	7×10^{-6}	5.6×10^{-5}	270
^{3}H[b]	8×10^{-20}	2.5×10^{-5}	1.5×10^{-9}	12

Source: Modified from DOE (1979).

[a] Activity is for nuclide plus daughter products.

[b] Top 50 m of ocean only.

rocket. The transfer rocket would return to the shuttle craft, while the supplementary rocket would carry its payload into solar orbit.

This concept was designed to handle some HLW and long-lived actinides to permanently remove them from the biosphere. Much TRU, much HLW, and all of SURF and CW would not be treated in this fashion, but would be stored in mined geologic repositories.

The risks, which principally involve launch pad operations and near-earth space path, must be compared to the possibilities of breaching a waste canister in a mined repository. Special handling facilities separate from normal HLW–TRU facilities would have to be constructed, and wastes would have to be encapsulated on a regular basis. The economics boil down to the simple question of to whether or not removal of some wastes by rocketry would be worth the expense. A mined repository has a very large initial expense, but once completed, has relatively low operating costs. Space disposal has a lesser initial cost, but very high costs afterward. Noneconomic factors include any of a number of accidents associated with treatment of wastes, waste form fabrication, ground transportation, handling, launch preparations, shallow orbit, and deep orbit. Accidents that might occur on the launch pad include intense fires and explosions, while accidents from orbit(s) could involve high-velocity impact and spreading of payload over part of the biosphere. Further, international aspects involve advising other nations of the mission.

Conclusions

Subseabed disposal of radioactive wastes is certainly a viable alternate to land disposal. As more and more data are gathered, subseabed disposal looks very favorable indeed. It is realistic to predict that even more favorable response to subseabed disposal will be forthcoming. One major problem is simple economics, however, in that transportation costs and some handling costs may be extremely high. Further, retrievability of radwaste is not feasible. Yet despite these factors, some radwaste may be disposed of in subseabed environment as safely as in land mined repositories.

The other alternate plans offer a combination of some merits and some uncertain factors, and extremely high economic costs in some cases. At present, the plans are to more fully investigate the land repository sites and to further explore the subseabed options.

CHAPTER 9

Low-Level Radioactive Wastes and Their Siting

Introduction

Low-level radioactive wastes are, by volume, the largest of the radioactive wastes generated. They are also benign relative to HLW, SURF, and TRU. Siting of the low-level wastes is a job requiring cooperation between Federal and State Agencies, in accord with 10CFR61 (1981), and different types of sites are currently under investigation in the United States.

Low-Level Radioactive Wastes

Low-level radioactive waste is difficult to categorize. It consists of all waste material having some radioactivity that is not classified as high-level waste, spent fuel, transuranic waste, or uranium mill tailings. In short, everything else generated by, or involved in some aspect of, use of radioactive materials is low-level waste. Thus medical materials, animal carcasses containing some radioactive materials, contaminated gloves and packaging materials, and non-HLW, TRU, SURF, and CW wastes from the nuclear power industry and from the military and other government operations, all are lumped together as low-level radioactive waste. The radioactivity of many of these wastes is so slight that, in theory, they may be of no hazard to the environment or may possibly be at risk from contained nonradioactive elements relative to radioactive species. Still, this vast volume of waste is slightly radioactive and must be treated as such as directed by law. Low-level radioactive waste generation is grouped into the following broad sectors: institutional wastes, industrial wastes, commercial power reactor wastes, and government and military wastes. Of these, the low-level wastes from commercial nuclear power plants provide the largest volume and activity. Some data are summarized in Table 9.1. A discussion of how low-level

Table 9–1. Volume and total activity of low-level radioactive wastes in the United States for 1978.

Sector	Volume (m³)	% of total	Activity (Ci)	% of total
Commercial nuclear power plants	35,600	42.4	473,600	53.4
Educational/hospital	21,300	25.4	2500	0.3
Industry (other than nuclear power)	20,400	24.3	404,500	45.8
Government and military	6600	7.9	5400	0.6
Total	83,800		886,000	

Source: Modified from DOE (1980).

radioactive wastes are described, treated, packaged, transported and disposed of is presented below.

Only three of the original six commercial sites available for low-level waste disposal are now open: Barnwell, South Carolina; Beatty, Nevada; and Hanford, Washington. The closed sites are located at West Valley, New York; Sheffield, Illinois; and Maxey Flats, Kentucky. Some relevant information concerning these sites is given in Table 9-2.

Based on current data, commercial nuclear power plants wastes include spent resins, filter sludges, evaporator bottoms, dry compressible wastes, contaminated equipment, and irradiated components. Typical radionuclides include ^{51}Cr, ^{54}Mn, ^{59}Fe, ^{60}Co, ^{58}Co, ^{65}Zn, ^{134}Cs, ^{136}Cs, ^{137}Cs, ^{140}Ba, ^{141}Ce. Sixty-six power reactors generated this waste in 1978.

Waste from medical facilities and universities include biological wastes, scintillation vials, solidified and absorbed liquids, and dry trash for the 2390 reporting units. Dry trash and scintillation vials make up about 90% of the total volume, and 92% of the total activity. Typical radionuclides include ^{3}H, ^{14}C, ^{32}P, ^{35}S, ^{51}Cr, ^{67}Ga, ^{99}Tc, ^{125}I, and ^{131}I. By volume, the dry trash and contaminated equipment make up about 48% of the total and the resins, sludges, and evaporator bottoms about 47%. The irradiated components make up about 90% of the total activity, although they represent the smallest volume, (5%).

Waste from the military and U.S. Government includes biological wastes, dry solids (almost 100% of the total), solid sludges, and other minor wastes. The individual activities for these are not available. Neither the types of wastes nor their activities are available for industrial wastes.

The handling of low-level radioactive wastes by states is in accord with the Atomic Energy Act of 1954 as amended by the Energy Reorganization Act of 1974, and subsequent laws and regulations. States that have entered into agreement with the U.S. Nuclear Regulatory Commission (or the Atomic Energy Commission before 1975) are known as Agreement States. These are Alabama, Arizona, Arkansas, California, Colorado, Florida, Georgia, Idaho, Kansas, Kentucky, Louisiana, Maryland, Mississippi, Nebraska, Nevada, New Hampshire, New Mexico, North Carolina, North Dakota, Oregon, Rhode Island, South Carolina, Tennessee, Texas, and Washington. The NRC will work directly with states and provide them with assistance in locating and licensing sites for low-level radioactive waste disposal, and for plans for operating, maintaining, and closing sites.

The low-level radioactive waste sites at West Valley, New York and Maxey Flats, Kentucky opened in 1963. Operations were stopped at West Valley in 1975 and at Maxey Flats in 1977. The Sheffield, Illinois site opened in 1967 and was filled to capacity by 1978.

In the cases of both West Valley and Maxey Flats, problems with water control forced the closing. The caps used to seal trenches leaked. At West Valley, normal precipitation infiltrated the soil used as a cap over the trenches causing the water table in the waste-containing trenches to rise and subsequently spill downslope. Although no significant increase in radioactivity was found in the area affected by the spill, the site was closed due to the unknown possible risks to the public, compounded by the confusion between site and operator as to liability for the spill. The on-site water has been pumped and treated since the summer of 1980, but no present plans call for reopening the site.

The West Valley site was selected because of its proximity to nuclear power plants in the Northeast, its then-assumed favorable geologic conditions, and the low population density in the area. The trenches are located in glacial sediments about 120 ft thick, of which the upper 40 to 60 ft consist of low-permeability sediment underlain by 32 to 50 ft of fine sand that is in turn underlain by impervious clays. Annual precipitation in the area of about 35 in is greater than the rate of evaporation of about 15 in per year.

Table 9-2. Storage of low-level radioactive wastes at active and closed sites.

Site	Storage Area (in acres)	Storage Volume	Amount Stored to 1980 (m^3)
Active Sites			
Beatty, Nevada	46	318,250 m^3	87,000 m^3
Hanford, Washington	100	891,200 m^3	61,000 m^3
Barnwell, South Carolina	256	1,600,000 m^3	324,000 m^3
Closed Sites			
West Valley, New York	25	67,000 m^3	67,000 m^3 (to 1975)
Sheffield, Illinois	20	90,000 m^3	90,000 m^3 (to 1978)
Maxey Flats, Kentucky	300	135,000 m^3	135,000 m^3 (to 1975)

Source: Modified from DOE (1980).

The problems at West Valley have been addressed by new covers and water management, thus stabilizing the area.

Water problems also plagued the operation at Maxey Flats. Water from waste-filled trenches has infiltrated a permeable sandstone some 100 ft from the trenches, although still on site. The radioactivity measured from the sandstone affected by the trench waters was slightly elevated, although considered to be of no threat to public health, and the contaminated water had to be processed prior to discharge. The site probably could have remained open with a new and better water control plan, but the State of Kentucky imposed an excise tax of 10 cents per pound of waste, which made operation of the site economically unattractive, and the site was closed.

U.S. Ecology, Inc. (formerly Nuclear Engineering Company, Inc.) operated the Maxey Flats site. The site is underlain by shale, siltstone, and sandstones, and the surface soils possess low permeability and low capacity for water storage. The site is in a humid region, receiving about 35 in of rain each year. The site was selected because the shales underlying the site were considered impermeable, which would make seepage to even deeper aquifers unlikely. Materials used to cover the trenches at Maxey Flats were more permeable than the surrounding soils and the underlying shales. The trench capping material was further weakened by subsidence, which was in turn caused by compression of the deeper waste due to corrosion of metallic containers and scrap and decomposition of organic wastes. Water collected in the trenches due to the surface subsidence coupled with high rates of precipitation, and the waste was exposed to leaching of radionuclides. To overcome this, the trench standing water was pumped to storage tanks where the liquid was evaporated and the residue buried. Some lateral, on-site migration of radionuclides and some vertical flow has occurred. The closed site offers no radiological hazard, but water management will continue to be of concern.

The site at Sheffield, Illinois was closed after capacity had been reached and because of delays and changes in federal regulatory policy. Coupled with problems of zoning for site expansion and other State of Illinois concerns, the operator withdrew the license renewal application. Expansion of the site was thus not undertaken. Subsequently, the U.S. Geological Survey, as part of its investigation of the site (see Robertson, 1980), revealed a layer of permeable sand below the site that would be a potential aquifer for radionuclide transport, although the rate of water accumulation at Sheffield comparable to that at West Valley and Maxey Flats has not risen. Water has infiltrated the burial soil in the trench caps, however, and small amounts of tritium have been measured off the site (Robertson, 1980).

There has been no threat to public health from the three closed sites, yet with proper advance planning, the conditions that led to the closings of Maxey Flats and West Valley could probably have been avoided. Better site selection and water management should have been used; hopefully they will serve as a lesson to planners of sites selected elsewhere. Further, the state regulations were inadequate, leading to confusion and delays on the operation of the sites, and the public, with extensive media help, professed great concern over the health aspects of the sites.

Water problems at West Valley and at Maxey Flats have continued since their closing. Tritium has been measured off the site in small amounts, although none has been measured in groundwater. Other radionuclides have migrated onsite at Maxey Flats.

The Barnwell, South Carolina site is operated by Chem-Nuclear Systems, Inc. on land leased from the state. The facility, which opened in 1971, consists of 230 acres of a total 300 acres marked for waste disposal. The site is bounded by the Barnwell Fuel Facility and is close to the Savannah River Plant. Factors that led to the selection of the Barnwell site included its proximity to nuclear power plants, good weather, good highway system, suitable hydrology, and good water supply. The population density of areas surrounding the site is low. Further, it had been demonstrated that waste could be stored in the general area by the Savannah River Plant.

The geology of the site consists of flat-lying, largely unconsolidated clays and sands. The nearest stream is about one mile away and empties into the Savannah river. The stream water is unused and the river is not used for about 80 miles downstream. The water table is 35 ft below the surface.

The Barnwell site can handle a large amount of waste (see Table 9-2) yet, as of October 1, 1981, only 1,200,000 ft^3 annually will be accepted. This is about one-half that previously accepted. Solid low-level radioactive wastes and solidified or adsorbed liquid radioactive wastes only are accepted, although incinerated organic-bearing residue is accepted.

The trenches at Barnwell are 8 ft deep and 100 ft wide by 100 ft long. The trenches are filled with waste (usually in drums or metal containers) and covered with 6 to 10 ft of soil, which is then compacted. Monitoring of the site is done by a well-trained technical staff. The plans for site treatment and closing coupled with a plan for extended care and reclamation have been made, and are funded by a small amount of the disposal fees. The State of South Carolina maintains the funds.

The low-level waste disposal site in Washington is located near Richland on the Hanford Reservation of the U.S. Department of Energy. It is operated by U.S. Ecology, Inc. The facility opened in 1968, and accepts solid and solidified low-level radioactive wastes, plus contaminated tools, clothing, etc. As of December 1982, organic liquids are no longer allowed at the facility. About 100 acres are designated for low-level radioactive waste storage, although a larger area, in theory, could be made available.

The site is underlain by dense basalt, although an aquifer exists at a depth of 300 ft. The surface materals consist of layers of sand and silt with local zones of sand, gravel, cobbles, and clay-like material. Surface waters do not penetrate the rocks below due to their extremely low permeability. The climate of the area is very dry, which helps to prevent water seepage to the aquifer.

Trenches at the Richland site are typically 30 ft deep by 70 ft wide by 400 ft long. The trenches are filled with 8 ft of soil over the containers of waste, and a gravel topping is used to prevent wind erosion. The site is monitored by constant analyses of air, soil, water, and vegetation.

The site at Beatty, Nevada is also operated by U.S. Ecology, Inc. Opening in 1962, it was the first site licensed by the U.S. Atomic Energy Commission for commercially generated wastes. At present, it is licensed by the State of Nevada as an NRC Agreement State. The site is on land leased from the Nevada Department of Human Resources. The site can be expanded, if that is the decision reached in the near future, as abundant land of essentially no other surface use bounds the 100 acres of the site.

Solidified and solid low-level radioactive wastes are accepted at the Beatty site, as well as organic liquids if they are still contained in scintillation vials. The shipments to the site are certified in compliance with acceptance criteria in use at the site.

The climate in the area is extremely arid. The rocks in the area consist of about 115 ft of impermeable clays through which no water passes. The surface is covered by an indurated, cemented sand and gravel. The rate of evaporation in the area is about 100 in per year, far in excess of the 4 in or so total precipitation per year.

Trenches at the Beatty site are typically 24 to 40 ft deep by 30 to 90 ft in width and 300 to 800 ft in length. Twenty of the trenches have been filled, capped, and marked with a permenent monument that describes the contents of the trenches. Air, soil, water, and vegetation at thirty locations at the site are monitored.

LLW Handling and Treatment

Since the range materials described as low-level radioactive wastes is extremely broad, there exists a variety of technologies to handle and treat the wastes as well. Low-level radioactive wastes from the different sources, institutional, commercial power plants, industrial, and military/government, all have different chemical, physical, and radiological characteristics (see Table 9-1). In determining the treatment for a particular waste, its total hazard, not just from its radioactive component, must be considered in order to determine the most efficient method for treatment and disposal. Other factors that will influence the treatment to be used include government regulations, environmental considerations, economics, and political factors.

Universities and hospitals produce waste that contains very little radio-activity, consisting essentially of paper, rags, combustible wastes, glass and ceramic ware, and other noncombustibles, scintillation vials, and others. Institutional practices for handling wastes include limited storage, shipment, and contained waste burial in shallow land sites. Some is diluted and released to common refuse systems, and some wastes are incinerated. The wastes are not just generated for medical research, but include radiopharmaceutical wastes, tracers for analytical soil and water testing, and neutron activation analysis.

Commercial nuclear power reactors generate a large volume of radioactive wastes, including combustible paper and rags, filter sludges, resins, and concentrated solutions. The combustible reactor waste is handled differently from institutional combustible waste as it is more radioactive. Hence it is only compacted prior to shallow land burial. Future processes will include improved solidification of wetted resins and other liquid-containing waste followed by incineration. The higher-level wastes will be buried at intermediate depths.

Low-level radioactive wastes are so varied that it is necessary to comment on the different factors that dictate just how hazardous various types are. For example, some low-level radioactive wastes may be enriched in one or more short-lived radioactive species, while others may contain virtually no radioactive species but potentially hazardous amounts of nonradioactive species. Were low-level radioactive wastes to be classified, presumably prior to treatment for disposal in shallow burial sites, the following factors should be considered: half-lives of radioactive species, toxicity, chemical and physical form, radiogenic and other new isotopes or compounds formed during storage, volume, concentration, possibility of release, and mobility.

The Conservation Foundation (1981) has proposed a new classification scheme (Table 9-3) for low-level radioactive wastes. In it, the attempt is made to separate those wastes with very little, short-half-life waste from those containing more, long-half-life waste. The geotoxicity of the various wastes must also be considered. In brief, the proposed plan calls for temporary storage of the small quantity, short-half-life material until radioactive decay has progressed to the point where the waste is essentially nonradioactive, at which point it can be disposed of as common, nonradioactive refuse.

The responsibility for any new classification scheme, however, rests with the NRC.

The technologies for the handling and treatment of LLW are varied. The Conservation Foundation (1981) has concluded that "... the technical capability currently exists to select sites properly and to package, handle, transport and isolate wastes safely." Their report stresses the need for not only more public participation, but for an educated public as well, so that the usual fears associated with any form of radioactive wastes can be put in proper prespective.

Levin (1981) has pointed out that "the most cost-effective way to reduce the volume of low-level waste, however, is not to generate it." Very simply, means to greatly reduce the volumes of LLW on-site are now feasible. Incineration to remove the bulky combustible parts of the LLW may reduce the volume by 40% alone, and smelting to separate different materials may also be used to reduce the volume. Further, for LLW with short-half-lived

Table 9-3. New classification scheme.

	Radioactive Waste Activity Level		
	High[a]	Greater than De Minimis, Less than High Level	De Minimis[b]
High-level site			Nonradiologically controlled disposal
Low-Level[c] Site	Long-Term (greater than 2 years)	Greater than Short-Term, Less than Long-Term	Short-term (e.g., less than 14 days)
Half-life		20-year moderate term storage facility	Holding facility

[a] "High" simply refers to an activity level above a newly established definition for low level. In all probability, it will be different than the current understanding of high-level radioactive waste. TRU wastes in designated amounts would be considered as high level.

[b] Refers to other screens (e.g., toxicity, concentration). If the other screens detect a nonradiological hazard for a *de minimis* waste, the disposal method would be determined by the nonradiological hazard presented. For example, if the waste was radiologically *de minimis* yet toxic, it would be disposed of according to the applicable regulations for hazardous wastes. Similarly, for wastes that have a half-life greater than two years, other characteristics may demand that the wastes receive more restrictive disposal than shallow-land burial.

[c] Site criteria and operating requirements must be established for the low-level waste sites. While not intended to be all inclusive, examples of some of the characteristics discussed by the group include all waste packaged in a prescribed manner, specified uniform depth for burial, requirements for proper handling, effective monitoring and documentation of disposal activities, designated hydrological features to ensure the ability of the site to contain wastes and prevent off-site migrations, and soil and geologic stability.

radioactive components, storage for a time roughly equal to 10 half-lives would reduce the overall radiotoxicity so that the LLW could be disposed of as non-LLW. All these possibilities are being looked into at the present time. LLW is the least hazardous type of radwaste to deal with, but it is also voluminous, and ways to dispose of it in accordance with 10CFR61 (1981) are imperative.

Use of Poisoned Land/Inland Seas for Low-Level Radioactive Waste Disposal

Brookins *et al.* (1983) have discussed some of the many aspects of poisoned land/inland seas for consideration for LLW studies. The following section is modified from their work.

It is widely known that certain areas of the earth are characterized by such high concentrations of toxic elements that their use by plants or animals is precluded. A waterhole whose water contains potentially toxic quantities of As is a common example of such a naturally poisoned area. The dry salt basins of the Basin and Range Province in Nevada and western Utah are largely relict inland salt seas and, because many of the salt basins have total interior drainage of both surface and ground water, these basins represent unique geochemical environments for the entrapment and concentration of mobile elements. Through leaching, transport, and ultimate deposition from throughout the watershed, the closed salt basins are hydrologic sinks in which toxic materials such as As, Sb, Hg, and other heavy metals may be concentrated in a relatively confined area.

Successful low-level radioactive waste retention in a burial medium requires isolation of the waste for a very long time, perhaps tens of thousands of years. Toxic material must be confined in a relatively small area, and migration out of that region must not be greater than allowable limits. The total interior drainage of closed hydrologic basins and the natural toxicity of many dry salt basins make them possible candidates for the siting of low-level radioactive waste storage facilities. Further, such sites will also be suitable for nonradioactive waste storage as well. This section is directed at identification and description of such naturally occurring poisoned areas, at compilation and analysis of factors that influence natural toxic material availability, and at compilation of key factors to be analyzed in assessment of the potential of a poisoned area for use as a low-level waste site. Specific effects of toxic-element concentrations on life are an important aspect to be sutided in future work.

Toxic metals and nonmetals regarded as active down to very low doses include Be, Cr, Ni, As, and Cd (Westermark, 1980); many biologists believe that the toxic effects of these elements are due to an interference with enzyme repair of DNA (Westermark, 1980). The chemical state of the element affects its genotoxic properties. Pb, Hg, and Cd are elements giving rise to concern of neurotoxic actions or other effects at very low level (Westermark, 1980).

The arid to semiarid Basin and Range Province of the western United States is characterized by numerous basins separated by interbasin mountain ranges. The principal reason for the large number of individual basins in this area, which covers Nevada, western Utah, and parts of California, Idaho, and Oregon, is Cenozoic block faulting. This faulting created many hydrologically closed basins, many of which contained pluvial lakes from Miocene through Pleistocene time, with a climate more humid than at present (Noble, 1972). Hydrologically closed basins typically contain playa "lakes" in the lowest part of the basins. In the Basin and Range Province, the principal reason for the abundance of playas is that all streams drain to permanent lakes or playas and none drains to the ocean. Playa is a general term for a variety of desiccated former lakes and topographic depressions found in arid and semiarid regions. Playas are flat and barren, and are frequently flooded during heavy rainfall, creating playa lakes. The playas in hydrologically closed basins are commonly large, with the water table at or near the surface: part of the playa surface is generally soft or damp from the high water table. Pluvial or rain water lakes on the other hand are permanent lakes. Playa sizes range from a few hundred square meters to hundreds of square kilometers in areal extent. Nevada has over 90 playas, most less than 100 km^2 in area.

The typical arid land hydrologic system includes a recharge area in mountains and a discharge area in the basin. In the Basin and Range Province the precipitation over the mountain ranges is greater than over the intermontane basins. Mountain streams commonly become ephemeral at the mountain fronts, and a considerable part of the flow is lost to the ground water reservoir in the upper reaches of the alluvial apron. During infrequent heavy rainfall, surface runoff may reach the playa, producing a temporary playa lake. Otherwise, ground water recharge in mountain aquifers, alluvial fans, and sediment surfaces will move toward the lowest part of the basin. In hydrologically closed basins, water discharge in the playa area occurs as springs (generally near the playa edge), through evaporation via capillary action from beneath the playa surface, and through plant evapotranspiration. The playa discharge area in a closed basin is probably the only place within the hydrologic flow system where permanent waste disposal cannot result in damage to the whole ground water reservoir, because in the discharge zone of a closed basin, no large-scale water movement occurs except discharge by evaporation.

The concentration of certain elements in the playa environment is well documented. The earliest and most common use of playas has been as a source of salt and other evaporite minerals. Borax, soda ash, and sodium sulfate have been produced from playa lake deposits for many years, and recently ancient lake deposits have become a uranium exploration target. Playa sediments and interstitial waters have also been sources of Li and zeolites. Trace elements that may be present in Nevada playa brines include Br, F, I, P, K, Rb, and Sr. Due to the resource value of some playa sediments, there could be possible conflict with siting waste storage facilities in the playa environment.

Factors that influence natural toxic material concentration may be divided into three groups: basin characteristics, water characteristics, and trace element/heavy metal characteristics. Basin characteristics include drainage, basin size, background levels of trace elements in drainage source area, stability of drainage basin over time, and playa sediment lithology, which is a function of basin rock lithology. Water characteristics include total volume of water available in the drainage basin (a function of precipitation versus discharge rates), evaporation rates in the playa vicinity, and water chemistry (also a function of basin rock lithology). Parameters reflecting the mobility of trace element/heavy metal ions are also largely parameters that affect water chemistry, and include activities and solubilities of components, adsorptive capacities of solid phases, buffer capacities of system, and complexing ligands present.

Basin characteristics largely affect the ultimate concentration of toxic elements in playas. The effect of overall basin size on concentration of toxic material is unclear. More clearly, high background levels of leachable trace elements in the watershed will enhance potential toxicity of the playa sediments. In sediment samples from Nevada playas in Smoke Creek Desert, Winnemucca Dry Lake, Roach Lake, and Cave Valley, the Roach Lake playa showed the highest Zn concentrations, probably related to the Zn deposits within the basin drainage (Leach *et al.*, 1980). The highest W concentrations were noted in samples from the eastern edge of Winnemucca playa, adjacent to drainages containing known W deposits.

In addition to basin size and background levels of trace elements, two other basin characteristics are important in influencing the eventual concentration of trace elements in playas: the stability of a basin drainage over time is a factor dependent on tectonism; the other factor is the lithology of the playa sediment, determined in part by rock lithologies within the basin drainage. Clay minerals present within playa sediment are partially determined by rock chemistry and mineralogy before weathering and transport have taken place. The presence of precursor carbonate-rich saline lakes would affect playa sediment composition; relative amounts of clays, zeolites, and organic matter are all important variables of playa sediment composition.

Water characteristics have a pronounced effect on the ultimate concentration of toxic trace elements in playas. The total amount of water available within a drainage basin is controlled by the precipitation rates in mountain areas and evaporation/discharge rates in the vicinity of the playas. Potential evaporation for Nevada playas, as an example, varies from approximately 100–200 cm annually; annual precipitation ranges from as little as 7.5 cm in some of the southern valley areas to more than 75 cm in the highest mountain ranges. To maintain conditions favorable for playa existence requires a very high ratio of annual evaporation to precipitation, perhaps on the order of 10:1.

Evaporation rate is an important water characteristic to consider in the concentration of trace elements.

For comparison of element concentrations between playa and nonplaya wet and dry sediment samples, approximate enrichment factors have been calculated. The basins included are: Winnemucca Dry Lake Basin, Roach Lake Basin, Cave Valley Basin, and Smoke Creek Desert Basin. The enrichment factors indicate that, for these four basins, U, As, and Sb are enriched in playa sediment. The median value for As in playa sediments from the four areas averages 20 ppm, several times the 1.8 ppm average crustal abundance value for As (Krauskopf, 1979). The average value for Sb in the playa sediments is 2.2 ppm, with 0.2 ppm the average crustal abundance. Thus, with a favorable source area, these elements are enriched by evaporation in the playa sediments.

Water chemistry is a characteristic which has profound effects on trace element transport and concentration. Important variables include ground and stream water pH, Eh, buffer capacities, and activities of components.

Leach et al. (1980) cite the importance of water chemistry in explaining the contrast in U concentration in Walker Lake, Nevada, and Great Salt Lake, Utah. Walker Lake, a closed-basin lake remnant of Pleistocene Lake Lahontan, has a high concentration of U (130 ppb) compared to other large water bodies. Great Salt Lake, 24 times as saline as Walker Lake, contains only 5 ppb U, (Brookins et al., 1983). Leach et al. (1980) note that Walker Lake has a higher amount of CO_3 (2700 ppm) than does Great Salt Lake (500 ppm), and that under air-saturated conditions (high Eh), it is the carbonate content of the lake water and not alinity that largely determines the partitioning of U between the sediments and the lake water.

The third major set of factors affecting trace element concentration in closed basins are parameters that reflect mobility of the trace metal ions in question. These parameters are summarized in a report by Markos (1979) on geochemical mobility and transfer of contaminants in uranium mill tailings, and include activities of the components, adsorptive capacities of different solid materials, solubilities of species, complexing ligands, and buffer capacities. These parameters are also important for their effect on water chemistry, but their evaluation is beyond the scope of this report.

Future study might best be oriented toward a detailed geochemical and geological study of one or more of the playas and closed basins mentioned earlier, due to their proximity to large metallic ore deposits. The scarcity of playa geochemical studies necessitates a better understanding of both playa water and sediment chemistries. The Buena Vista and Edwards Creek playas, due to their drainage of large Hg, Sb and other metal ore deposits, are among the most favorable targets for such future work. The Roach Lake playa is another possible target, although it is possibly not a true closed basin. Examination of available studies in the literature does indicate that As and other trace metals are present in much of the natural waters of western Utah and Nevada, in amounts exceeding the U.S. Public Health Service standards for "rejection of public water supply."

Characteristics of playas and closed hydrologic basins that make them relevant to low-level waste disposal are listed in Table 9-4. It has been demonstrated that the accumulation of geotoxic elements is enhanced in closed hydrologic systems, and that elements are concentrated and not disseminated in such an environment. Materials comprising sediments of the playas include smectite, illite, calcium carbonate, silica, feldspars, and iron oxides and hydroxides, similar to those proposed as constituents of engineered barriers in waste repositories. Ion exchange, complexation, and precipitation are among the chemical processes active in the playa sediments and water, and are responsible for partitioning of radionuclides between aqueous and solid phases. Water table fluctuations pose no problem in a closed hydrologic system: under high water table conditions, some elemental redistribution will take place, but not removal from the closed system; under lower-water-table conditions, while more local fixation will occur, there will again be no loss of material. Local torrential rains occasionally flood the closed hydrologic basins in arid lands, but the waters do not escape the closed system. By definition, rise in the water table and subsequent dilution

Table 9–4. Characteristics of playas and closed hydrologic basins: relevance to low-level waste disposal.

1. Accumulation of geotoxic elements is enhanced in closed hydrologic systems, in some cases making them unfit for use by animals or plants.
2. In such systems, elements are concentrated and not disseminated.
3. Materials in playa sediment are similar to those being considered for use as radioactive waste engineered barriers.
4. Ion exchange, complexation, and precipitation, processes that partition radionuclides between aqueous and solid phases, are active in playa sediments.
5. Fluctuations in the water table pose no problem, because although some elemental redistribution in playas will take place under conditions of a higher water table, the closed hydrologic system ensures entrapment of low-level waste; under lower-water table conditions, more local fixation will take place, but again, no loss of material.
6. Although local torrential downpours occasionally flood playas, the waters do not escape the overall closed hydrologic system; soluble elements actually become more diluted under such conditions, and thus point concentrations of elements are decreased.
7. The hydrologic flow system of a closed basin causes the playa area, in the discharge zone of the basin, to be an area from which no water movement occurs except discharge by evaporation; in such an area, damage to an entire groundwater reservoir due to waste disposal may be avoided.

of soluble elements result as point concentrations of elements are decreased. As water loss through evaporation then occurs, the elements affected by the flooding will be more or less uniformly distributed within the limits of the basin surface to the extent of the flooding.

Storage of low-level waste in naturally poisoned closed hydrologic basin sites would result in the following: to a degree, overall geotoxicity background would be slightly decreased due to dilution of very geotoxic elements with less-toxic low-level waste, but this is not an important factor. Of greater importance, the closed hydrologic system ensures entrapment of the low-level waste as opposed to loss by hydrologic flow in an open system. Although wind has removed playa sediment in the past, wind erosion would only affect waste buried at very shallow depth, and would probably not transport any playa material over agreat distance. A higher water table would reduce the likelihood of any wind erosion.

Further study of any closed basin should allow evaluation of the basin with respect to numerous factors. A high abundance of geotoxic elements in playa sediments should be established, as should high total alkalinity (or total salt content). These are the characteristics of playa areas that would make them essentially 100% uninhabitable by plant and animal life. Establishment of closed hydrologic system conditions versus open-system conditions could be made through observations of accumulations of varved playa sediments, other layered strata, or possibly by radiometric means (^{14}C and ^{230}Th methods). Isolation of the poisoned area from metropolitan or smaller urban centers is necessary; many poisoned lands in the western United States meet this criteria. Frequency of flooding should be assessed, because this affects the way in which surface elements are mobilized and redistributed. For example, bank deposits may have concentrations different from samples from the last spot of evaporation. A correlation of playa chemistry with mineralogy should be made. Finally, a high background for potentially geotoxic elements in the basin drainage should be established.

Transportation

Transport of all hazardous wastes, including radioactive wastes, is under the regulation of the Department of Transportation. Their regulations are applicable to all aspects of waste handling to final disposal, including initial packaging, loading, transport, unloading, content inventory, vehicle main-tenance, and safety. The Nuclear Regulatory Commission handles the regulation of highly radioactive materials during their transport and for the development of performance standards for shipments involving large quantities of materials. Safety of the waste contained and various safeguards are also the responsibility of the NRC. Economic aspects of waste transport are the responsibility of the Interstate Commerce Commission and some state agencies. Transport of waste over water is the responsibility of the U.S. Coast Guard. Noncivilian vehicular transport is controlled by the U.S. Departments of Defense and Energy.

Low-level wastes from commercial operations are handled as follows: Common carriers are used for the shipment of some wastes, while commercial waste handling firms are used for others. The wastes are packaged according to the U.S. Department of Transportation (DOT) regulations; 55-gallon drums are conventionally used for low-specific-activity wastes. Waste handled by commercial waste handling firms is usually stored at surface facilities for some time prior to ultimate transport to a shallow land burial site. The wastes, due to their low activity, are unshielded.

Decommissioning

Even when a site for lower level radioactive waste disposal has been selected with the utmost care, there remain the possibilities of accidents, new information on waste behavior, and the like. The "best" site must be arranged so that funds have been placed aside for the closing of the site, for monitoring of it for a substantial period of time, and for compensation to third parties who might be inadvertently injured by an unexpected and unspecified event in the future. The LLW disposal facility actually consists of four different phases. Site selection and active operation are the first two, decommissioning or the closing of the site is the third phase, and the fourth phase is the 100- to 200-year period during which the site must be monitored. The third phase, decommissioning, will be discussed here.

Every LLW disposal site will ultimately be closed. After closing, no more wastes will be accepted for burial. No uneventful closings of LLW sites have occurred in the United States, but there have been three closings due to various difficulties (see discussion earlier in this chapter). Regardless of whether or not a site closes on schedule, the site must be prepared for the very long time that it takes for radioactive elements to decay to an environmentally safe level.

Decommissioning is hopefully a simple matter for a site that has experienced no major difficulty during its operative lifetime. It may be necessary, however, to dismantle and bury buildings at the site if they have become slightly contaminated due to handling and disposal of the wastes. The smoothed over, filled trenches will have to be capped with materials (planted soils and riprap) to prevent erosion and accumulation of water. Warning signs must also be placed, and long-term monitoring stations must be built to check for radionuclide migration after site closure. If a site is run properly, the decommissioning should be a simple, straightforward, and relatively cheap process. Funds will have been collected to handle not only a simple closing but for more cumbersome cases as well. Costs for simple decommissioning may run from $500,000 to $3,500,000 (in 1978 dollars) (see Murphy and Holter, 1980), but these costs are only estimates as no U.S. site has ever been decommissioned. For example, in the case of a poorly run operation, it may be necessary to stabilize the site by a series of excavations around the perimeter of the site, plus add cappings over trenches and directional surface modification to control the flow of water. While expensive, the costs associated with such measures are not prohibitively expensive. A crude analog for costs can be taken from the decommissioning of a uranium milling operation. At the Sherwood Uranium Mill near Spokane, Washington, the cost for decommissioning was $1,300,000 and, since this is a normal type of uranium mill, the decommissioning cost may well be reasonable also for a shallow land burial site.

Interestingly, there are no comprehensive sets of federal regulations that control decommissioning. The host state assumes responsibility once it has been established that the operator has fulfilled his responsibilities under the conditions of his lease. The funds gathered for decommissioning during the operations' lifetimes are intended to be adequate for both decommissioning and for long-term care, but it has already been demonstrated that the funds gathered for the three closed sites (West Valley, Maxey Flats, Sheffield) are insufficient for these purposes. The lesson learned from the three closed sites is that a surplus of funds may have to be gathered. Although expensive, this creates a feeling of security for the public in knowing that funds for contingency are indeed available.

Interim Storage of Commercial Low-Level Radioactive Waste

Commercial LLW from all sources reached 92,000 m^3 in 1980 and may reach 129,300 m^3 by 1985 and 184,000 m^3 by 1990. Hopefully space at shallow land burial facilities will be adequate to handle these wastes; if not, then interim storage may become common practice. The U.S. Government has very little experience in the temporary storage of LLW since most is buried directly from the various sources, yet considerable experience has been gained by studies related to the burial of TRU wastes, which are directly applicable to the burial of LLW.

The DOT approves containers for TRU waste packaging of DOE generated wastes. These are stored on asphalt or concrete pads at DOE storage sites. Free liquids are not allowed, nor are combustible materials, explosives, or other materials that could pose a storage hazard. Twenty years retrievability is intended for all materials so packaged, during which time there will be no surface contamination by radioactivity. The containers are stored so that no accidental contact with water is probably prior to their shipment to burial sites. At the Idaho National Engineering Laboratory, some 4000 m^3 of TRU wastes have been stored aboveground on asphalt slabs. A movable building protects the waste from precipitation or wind. Soil is used as a break between the stored waste, primarily for firewall purposes, every 24 m. When stacks are completed, they are covered with reinforced vinyl, plywood, and one meter of soil that is planted with grass (to prevent water and wind erosion when ultimately exposed to the elements).

Table 9–5. Disposal of low-level radioactive wastes in the United States: 1979 data.

State	1979 Volume (%)	1979 Volume (m³)	State	1979 Volume (%)	1979 Volume (m³)
Alabama	5	3672	Mississippi	<1	68
Alaska	<1	<1	Missouri	<1	329
Arizona	<1	54	Montana	<1	3
Arkansas	<1	265	Nebraska	1	801
California	5	4342	Nevada	<1	4
Colorado	<1	225	New Hampshire	<1	77
Connecticut	5	3970	New Jersey	4	3008
Delaware	<1	120	New Mexico	<1	80
District of Columbia	<1	33	New York	12	9572
Florida	3	2592	North Carolina	7	5304
Georgia	2	1261	North Dakota	<1	2
Hawaii	<1	83	Ohio	2	1905
Idaho	<1	7	Oklahoma	<1	21
Illinois	8	6758	Oregon	2	1219
Indiana	<1	27	Pennsylvania	9	6825
Iowa	1	961	Rhode Island	<1	463
Kansas	3	10	South Carolina	10	8089
Kentucky	<1	194	South Dakota	<1	1
Louisiana	<1	19	Tennessee	1	1131
Maine	<1	416	Texas	<1	543
Maryland	1	978	Utah	<1	106
Massachusetts	6	4860	Vermont	<1	370
Michigan	3	2150	Virginia	5	4230
Minnesota	2	1461	Washington	1	779
			West Virginia	<1	40
			Wisconsin	<1	487
			Wyoming	<1	<1

Source: From DOE.

The 1979 figures for disposal of LLW in the United States are given in Table 9-5.

Conclusions

Incineration or other means of volume reduction are being implemented for LLW prior to land disposal. Sites for disposal are varied, but basically require isolation from the biosphere. In addition to sites already in existence, geologic criteria suggest that naturally poisoned lands in the arid western and southwestern United States may be good candidates not only for LLW, but for hazardous chemical wastes as well.

CHAPTER 10

Uranium Milling
and Mill Tailings

Introduction

Uranium milling and mill tailings present problems not inherent to other types of waste disposal. First, the milling process separates uranium from other elements, but elements such as Ra, Se, and Mo may be more or less concentrated in the tailings. Second, while modern milling is much improved over older techniques, and thus less of these elements wind up on the tailings, abandoned mill tailings from some 26 sites pose special problems.

Uranium Milling

Uranium mill tailings represent, by volume, the largest amount of radioactive wastes generated in the United States. Before discussing the mill tailings, however, it is first necessary to discuss the various milling processes in use. Many of the mills have been shut down due to the uranium glut at the present time, although it is reasonable to assume that several if not many of these will re-open in 5 to 6 years. Basically, the main objective of milling is to separate the uranium ore from the gangue (host material). The two most widely used milling processes are the acid and alkaline leach methods, which will be discussed below.

To achieve a finished product of uranium oxide, or yellowcake, a straightforward series of procedures is followed, some of which are unique to the uranium industry. The yellowcake contains about 80 to 90% U_3O_8, which attests to the success of the milling processes, as most starting uranium ores contain only 0.02 to 1.0% U_3O_8. Before the leaching process begins, however, the ore must be sorted according to grade and rock type and crushed to an appropriate size.

The mechanical separation of ore material from gangue can be achieved by one of the following four ways.

(1) Separation by radioactivity—when ore radioactivity reaches a certain level, the counting equipment is linked to a mechanical kicker mechanism that moves the ore to a separate bin. This method is used for spotty, but locally high-grade, ores.

(2) Gravity concentration—due to their high specific gravities, heavy minerals can be separated by spirals, hydrocyclones, and gravity tables. This method is used for copper–uranium ores.

(3) Attrition grinding—ore is ground in horizontal cylinders half-filled with steel or cast-iron balls. As the cylinder is rotated, and the ore fed in one end and discharged from the other, the mixture is suitably ground by the steel balls. The gangue minerals are resistant to grinding and can thus be separated from the uranium ore by screening after the grinding operations.

(4) Flotation—the surfaces of some minerals can be modified by use of chemicals such as xanthate and aerofroth 70. Some particles remain attached to water while others cling to air being bubbled through the pulp. The loaded bubbles rise to the surface, and in the resulting overflow, the concentrated uranium minerals are carried off.

In both acid and alkaline leaching methods, the tetravalent uranium in the ore is oxidized to hexavalent uranium, which is readily soluble. Each leaching method has its own advantages and disadvantages. Acid leaching plants are far more common that alkaline leach plants. Flow diagrams for both are shown in Figs. 10-1 and 10-2.

Acid Leaching

In the acid leaching method U^{4+} is first oxidized to U^{6+} by an oxidant such as MnO_2 or $NaClO_3$ in conjunction with reactions involving iron as shown below. In practice, sulfuric acid and elemental iron, the oxidant, are agitated with a slurry of uranium-rich solids. Some relevant reactions are:

$$U^{4+} + MnO_2 = UO_2^{2+} + Mn^{2+}$$

or

$$3H_2O + 3U^{4+} + ClO_3^- = Cl^- + 3UO_2^{2+} + 6H^+$$

$$Fe + H_2SO_4 = FeSO_4 + H_2$$

and

$$FeSO_4 = Fe^{2+} + SO_4^{2-}.$$

The Fe^{2+} reacts with MnO_2:

$$2Fe^{2+} + MnO_2 = 4H^+ = 2Fe^{3+} + Mn^{2+} + 2H_2O$$

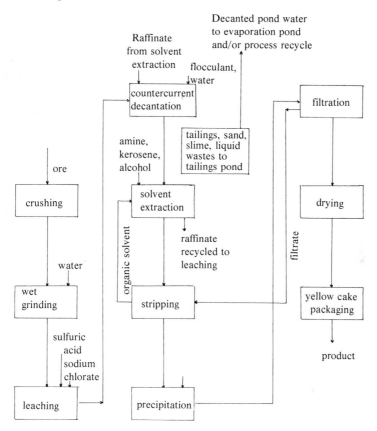

Fig. 10-1. Flow diagram for acid leach process. *Source:* Modified from NUREG (1980).

or with $NaClO_3$:

$$6Fe^{2+} + NaClO_3 + 6H^+ = 6Fe^{3+} + NaCl + 3H_2O,$$

and the following reactions occur simultaneously:

$$UO_2^{2+} + 2SO_4^{2-} = UO_2(SO_4)_2^{2-},$$

$$UO_2(SO_4)_2^{2-} + 2SO_4^{2-} = UO_2(SO_4)_3^{4-}.$$

The uranium in solution thus consists of a mix of UO_2^{2+}, $UO_2(SO_4)_2^{2-}$, and $UO_2(SO_4)_3^{4-}$ after the acid leach. Usually, excess sulfate ion is added to convert all UO_2^{2+} to one of the uranyl sulfate oxyanions. The acid leach method is most commonly used for uranium ores with a low carbonate content, such as many of the ores from the Grants, New Mexico Mineral Belt. Some of the advantages of the acid leach method are: (1) uranium is readily extracted; (2) much of the sulfuric acid can be re-used; and (3) some oxidant (e.g., MnO_2) is already present in the ore, thus facilitating oxidation. Some of the disadvantages are: (1) the sulfuric acid can damage milling

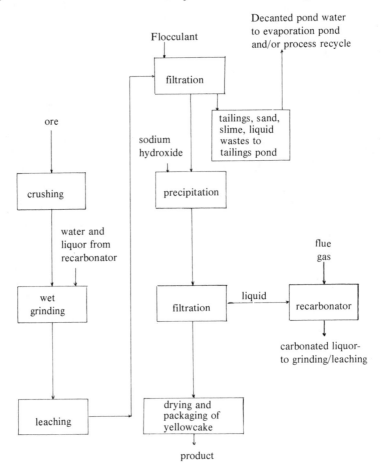

Fig. 10-2. Flow diagram for alkaline leach process. *Source:* Modified from NUREG (1980).

equipment without proper precautions; (2) high-calcite-content ore can not readily be processed by acid leach; and (3) the very low pH of the sulfuric acid leach solution allows radium (as Ra^{2+}) to be put into solution, which is of extreme environmental concern.

Alkaline Leaching

In alkaline leaching, after the initial crushing of the ore the oxidation of U^{4+} to U^{6+} is carried out in mild steel vessels using oxygen from the air as an oxidant in the presence of sodium carbonate:

$$2UO_2 + O_2 + 6Na_2CO_3 + 2H_2O = 2Na_4Uo_2(CO_3)_3 + 4NaOH,$$

but this reaction is not satisfactory as the NaOH will attack the uranyl carbonate complex, therefore $NaHCO_3$ is added so that:

$$2UO_2 + O_2 + 2Na_2CO_3 + 4NaHCO_3 = 2Na_4UO_2(CO_3)_3 + 2H_2O.$$

The leachate is filtered to remove solids and other foreign matter, and the uranium-rich solution (the *pregnant liquor*) and sent to a product-recovery vessel. The following reaction then results in the formation of a first-step yellowcake:

$$2Na_4UO_2(CO_3)_3 + NaOH = Na_2U_2O_7 + 6Na_2CO_3 + 3H_2O.$$

The $Na_2U_2O_7$ (yellowcake) at this step contains about 75% U_3O_8 and 5 to 6% V_2O_5 and 2 to 3% carbonate. The impure yellowcake is roasted at this point to produce a calcined yellowcake, which, when rinsed with water, will be purged of its contaminant vanadium and carbonate. The vanadium is stored and sold as a by-product. The sodium content of the yellowcake is lowered by washing with sulfuric acid-ammonium sulfate, which yields a purified yellowcake with only 0.5% sodium. The resultant yellowcake contains about 87% U_3O_8. This is washed, dried, and packaged.

The advantages of the alkaline leaching method are: (1) oxidation costs are lowered because O_2 is the oxidant; (2) the high calcite content of the ore facilitates dissolution; (3) radium loss is low because it is not soluble in the alkaline leach; and (4) resin is not involved. The disadvantages are: (1) the crushing costs are greater because the ore must be very finely ground; and (2) for low-calcite ores, the alkaline leach does not remove uranium as thoroughly as acid leach.

Agitation leaching processes are carried out in cylindrical vessels with air or mechanical agitation. A pulp is made from the wetted, finely ground ore. This method is used for continuous processing with a high yield in a relatively short retention time, lower chemical consumptions (relative to percolation leaching), little uranium loss from fine material. This method has higher costs, higher water consumption, low-grade pregnant solutions, and a liquid–solid separation process has to be used.

In percolation leaching, a coarser feed is used, which yields a high-grade pregnant solution. Vats with false bottoms are used, in which the pregnant solution is filtered through various materials. This method has lower capital costs, lower amounts of solutions, low water consumption, higher-grade pregnant solutions, and delivery of filtered pregnant solutions. Percolation leaching is not suited for fine materials, and it has a longer retention time, and a higher reagent consumption.

Complex ores are composites of ores of different types, which necessitates use of both acid and alkaline leaching techniques. While the different ores can be separated from each other, this is time consuming and expensive. The alternative taken is to mix high-calcite ore with a larger volume of low-calcite ore followed by acid leaching.

For solution mining of complex ores, an alkaline leach is first used followed by an acid leach for a shorter time. The alkaline leach is used first because it removes fewer metals from the ore and gangue than acid leach, thus ensuring a longer operation. When the ore has an overall low calcite content, the alkaline leach will be followed by a short period of acid leaching to get the alkaline-resistant uranium. Pretesting of the ore is done to determine the amount of acid leach to use so as not to release too much radium to the host aquifer.

Recovery Processes

Uranium is stripped from the pregnant liquor by any of several exchange processes. The most common methods are the resin-in-pulp (RIP) method, the solvent extraction method, and the general ion-exchange process. In each method, advantage is taken of the fact that UO_2^{2+} or other U-oxyanions can be exchanged between sites. If R is an organic fraction, and X the mobile ion, then a general reaction for exchange is:

$$4RX + (UO_2(SO_4)_3)^{4-} = R_4UO_2(SO_4)_3 + 4X^-.$$

Loading refers to the reaction going to the right; *stripping* refers to the reaction going to the left. The pregnant solutions are continuously loaded and stripped in recovery tanks until the uranium is concentrated sufficiently for final recovery.

In the ion-exchange processes, the uranium sulfate complexes have a high affinity for certain anion exchange resins. If R is the resin, the following reaction is typical:

$$4RCl + UO_2(SO_4)_3^{4-} = R_4UO_2(SO_4)_3 + 4Cl^-.$$

The resins are commonly made of ammonium anion salts attached to a cross-lined styrene–divinylbenzene copolymers backbone (see discussion in Brookins *et al.*, 1981b). While effective, this resin releases nitrogen to the mill tailings, which, because of the simplicity of its oxidation, produces nitrates, which move readily in the surface environment.

In the acid leach, the pH is kept below 1.9 so that bisulfate ion (HSO_4^-) is dominant over sulfate ion $(SO_4)^{2-}$. This is done because HSO_4^- competes more strongly for uranium complexes than will SO_4^{2-}. When uranium is saturated on the resin, it can be removed by acid nitrate or chloride ions. Uranium is then precipitated as an uranate with ammonia, sodium hydroxide, or magnesia.

Sands and slimes must be separated from the pregnant leach solution during standard extraction. The RIP method was developed by the U.S. Atomic Energy Commission in the 1950s to speed up operations. Sands are easily separated from uranium-rich solutions although slimes are not. The RIP circuit contains parallel banks of vats connected in series. Each vat is filled with baskets of a strong anion exchange resin. The baskets rise and fall in the vats and the pregnant liquor penetrates the resin, resulting in uranium entrapment. Slime coating, which would prevent uranium entrapment, is prevented by constant agitation of the vats. Both pH and Eh are controlled. Melanterite ($FeSO_4 \cdot 7H_2O$) is used for Eh control to prevent removal of vanadium and other cations, except uranium. This allows uranium to be separated as an oxyion ($UO_2(SO_4)_3^{4-}$). The pregnant liquor cascades from one vat to the next by gravity feed. In a normal 14-vat circuit, 10 tanks are used for loading (by adsorption) and 4 for stripping. When one tank is loaded, it is then stripped and an eluted tank is used for loading. This continues until, from the last vat, a solution rich in uranyl sulfates, sulfate, and ferric cations emerge.

Precipitation is carried out by two steps to purify and clarify the pregnant eluate. First, ferric ions are removed by:

$$Fe_2(SO_4)_3 + H_2SO_4 + 4Ca(OH)_2 + 6H_2O = 2Fe(OH)_3$$
$$+ 4CaSO_4 \cdot 2H_2O,$$

and the gypsum produced is sold. The solution left is filtered, and magnesia is added as a source of OH^- ions to neutralize any acid left and to precipitate uranium as yellowcake.

In solvent extraction organic extractants are used to recover uranium from acid leach liquors. A phosphate compound such as alkyl orthophosphoric acid or alkyl pyrophosphoric acid can remove uranium by cation exchange by (where R is the alkyl group):

$$UO_2^{2+} + 2R_2HPO_4 = UO_2(R_2PO_4)_2 + 2H^+.$$

To reverse this reaction, and to strip the uranium from the organic phase, hydrochloric acid is commonly used, after which the solution is evaporated to recover the HCl. The concentrated uranyl chloride is then diluted and precipitated as ammonium diuranate, which is then converted to U_3O_8. Sodium carbonates can also be used to strip the uranium from the organic phase by:

$$UO_2(R_2PO_4)_2 + 3Na_2CO_3 = 2NaR_2PO_4 + Na_4UO_2(CO_3)_3.$$

After first destroying the sodium uranyl carbonate by sulfuric acid, the uranium can be precipitated by addition of ammonia.

Other Stripping Methods

Uranium extraction from acid leach liquors can also be achieved by addition of organonitrogen compounds:

$$2(R_3HN)_2(SO_4) + UO_2(SO_4)_3^{4-} = (R_3HN)_4UO_2(SO_4)_2 + 2SO_4^{2-}.$$

Uranium extraction from the organic phase can be achieved by (1) addition of acidic nitrate or chloride solutions, (2) sodium carbonate solution stripping, or (3) magnesium oxide stripping.

Recovery of Uranium from Phosphates

All but one of the uranium mills in the United States are located in the western United States. The remaining mill, where uranium is removed from marine evaporites as a by-product of phosphate milling, is located in Florida. Uranium as U^{4+} can substitute easily into the Ca^{2+} site of apatite, $Ca_5(PO_4)_5(F,OH)$, and must be removed prior to use as a phosphate fertilizer and from waste solutions released into the biosphere.

When U-bearing apatite is treated with sulfuric acid for phosphate recovery, the following reaction takes place:

$$Ca_{3-x}(UO_2)_x(PO_4)_2 + 3H_2SO_4 = xUO_2SO_4 + 2H_3PO_4$$
$$+ (3-x)CaSO_4.$$

The uranyl sulfate remains in solution with the phosphoric acid, from which it can be readily removed by standard hydrometallurgical methods. The uranium recovered is not of interest economically, yet thorough uranium recovery must be carried out for environmental protection.

Uranium Mill Tailings

Several radioactive and nonradioactive wastes are generated by uranium ore processing. The bulk of both wastes is the tailings, which account for most of the ore solids, process additives, and water. The U.S. Nuclear Regulatory Commission (NUREG, 1980) has considered the hypothetically generated wastes from a generic model mill in its assessment of uranium milling operations and problems associated with mill tailings. While different from actual mill operations, the generic model mill is very useful in dealing with "average" mill wastes. It is assumed that the model mill will generate about 2000 tons of dry tailings, which are slurried in water to about 50% solids by weight (density about 1.6 g/cm^3) and sent to a tailings retention system. About 30%, as an average, of the tailings liquid is recycled for use in uranium milling. The net consumption of water is about 1400 tons per day.

The components of tailings include sand, slimes, and liquid. Sands consist of solids greater than 75 μ; in many cases these consist of quartz, feldspars, rock fragments, minor amounts of calcite, and other minerals. Some of these gains are coated with clay minerals as well. Slimes consist of solids smaller than 75 μ diameter, and include most ore minerals, organic material, clay minerals calcite, and finely ground rock forming minerals. The liquids contain ore-generated chemicals and process reagents. Most of the radioactivity (about 85%) is contained in the slimes, which make up about 35%, by weight, of the tailings. In Fig. 10-1 is shown a typical flow chart for the model mill. Additives for a hypothetical acid leach process include sulfuric acid (45 kg/MT—kilograms/metric ton), sodium chlorate (1.4 kg/MT), alcohol (0.04 kg/MT), kerosene (0.45 kg/MT), and iron grinding rods (0.25 kg/MT).

At the time of discharge from a mill, typical slurried tailings are pumped through either steel or plastic pipes to a tailings pond, which is typically of a square basin bounded by earthen embankments. The tailings slurry is discharged into the impoundment.

The chemical composition of the sands is largely SiO_2 with minor amounts of complex silicates of Al, Fe, Mg, Ca, Na, K, Mn, Ni, Mo, Zn, U, and V. Minor amounts of phosphate, carbonate, and sulfate are also present. The sands are very complex and vary in composition as a function of types of processes used during the milling. The radiological content is about 26 to 100 pCi/g from radium and from uranium. Small amounts of thorium (about 70 μCi/g) may be present. Slimes contain abundant silicate complexes of Al, Fe, Mg, Ca, Na, K, Mn, and others. They are enriched in the radioactive species, with uranium content about double that of the sands, and radium about double as well. The liquids portion of the tailings produced is about equal in weight to the sands and slimes and is used as the medium to transport the solids to the final place of discharge. The water used can, in many cases, be recycled, with typically 0 to 70% of the amounts recycled. The solutions contain dissolved sulfate, sodium chloride, ammonium, phosphate, minor amounts of flocculants, kerosene, amines, and other organics. Ions in solution include Cu, Mo, U, V, Zn, Mg, Ca, Be, Al, Ni, Sb, and Fe. Concentrations of elements such as Pb, Cd, Hg, Cr, Mo, Ba, and As are generally less than 0.01 ppm. The pH varies usually from 1 to 2, and values for radioactivity are: 20 to 7500 pCi/l of ^{226}Ra, 2000 to 22,000 pCi/l for ^{230}Th, and uranium concentrations vary from 0.001 to 0.01%.

For alkaline leach process wastes, the amounts of sands and slimes are roughly equal to those produced during acid leach processes. A major difference is that almost 100% of the liquids used are recycled, with only minor losses from evaporation and seepage during the transport and short tailings pond residence time. The sands are almost entirely made up of SiO_2 (99%) with small amounts of U (0.01%), Mo (0.001%), and Se (0.002%). About 600 pCi/g for both ^{226}Ra and ^{230}Th are present as well.

Slimes consist primarily of silica with lesser amounts of silicates of Al, Na, Ca, Mg, and Fe, and minor amounts of sulfates, chlorides, and carbonates. The radioactivity of alkaline leach slimes is about the same as for acid leach slimes. The liquids contain, in grams per liter, the following components: CO_3^{2-} (7), HCO_3^- (100), Se (0.04), SO_4^{2-} (7.6), Cl$^-$ (0.7), Mo (0.08), V_2O_5 (0.04), and U_3O_8 (0.04). The pH is usually 10 to 10.5. The radiological content of the liquids is about 200 pCi/l of ^{226}Ra. Thorium is insoluble in the leach process so its radioactivity in the liquids is below detection.

Other Emissions

Materials are emitted from sources other than the mill tailings, including (for an acid leach process) ore dust, uranium oxide from the drying and packaging area, organic solvents, and fumes of sulfur dioxide and sulfuric acid, NO_2. The amounts of these are summarized in Table 10-1. Nonradioactive materials are principally added from the ore storage areas and the tailings disposal area plus from roads and other areas where equipment is used. Radioactive materials consist of particulates and radon gas. Particulates are

Table 10–1. Emissions from mill tailings.

Emission	Source	Daily Rate[a]
Ore dust	Ore storage pads and crushing and grinding	6.3 kg
Tailings dust	Tailings Pile	1080 kg
U_3O_8	Drying and packaging	1.4 kg
Organics (92% kerosene)	Solvent extraction system	70 kg
SO_2, H_2SO_4 fumes	Acid leach tank venting	1 kg
SO_2	Fuel oil	22 kg
NO_2	Fuel oil	5 kg
Sewage	Washrooms, showers, etc.	30,000 L

[a]For a model mill of 1800 MT/day, see (NUREG, 1980).

generated during the ore storage and initial phases of milling, during yellowcake drying and packaging, and from mill tailings. Radon gas is emitted from the stored ore and from the tailings, but not in any significant quantities from the yellowcake operations. The amounts generated from the NRC's model mill are listed in Table 10-2.

Controlling Impacts of Milling Operations

The NRC has decided on three alternatives to mitigate uranium milling impacts, including (1) control of emissions during milling operations, (2) tailings disposal programs, and (3) decommissioning of the milling facilities (exclusive of the tailings disposal area) (see NUREG-0706, 1980).

The control systems now available to help control emissions include the following.

Table 10–2. Radioactive emissions from mill tailings.

Emission Source	Particulates ^{238}U, ^{226}Ra, ^{234}U, ^{210}Pb, ^{230}Th, ^{210}Po (mCi/yr)			^{222}Rn (Ci/yr)
Ore hauling and storage	0.67	0.67	0.67	
Crushing and grinding	0.90	0.90	0.90	68
Yellowcake drying and packaging	150	0.73	0.15	
Tailings pile	8.7	120	120	4400
Dispersed ore and tailings	—	—	—	48

Note: These values are the "worst-case" values calculated by the NRC, and can be significantly lowered by proper technology and management.

(1) Windbreaks around ore unloading areas.

(2) General mill drainage systems to catch the leachate and surface runoff from ore storage areas.

(3) Hooded conveyor belts.

(4) Sprinkling or wetting of ore stockpile.

(5) Sprinkling or wetting of roads.

(6) Storage of ore in warehouse instead of out of doors.

(7) Wet grinding a semiautogenous system would reduce dust.

(8) Wet scrubbers to reduce dust.

(9) Filters around ore areas.

(10) Dust-control spray systems installed at points of dust emission.

(11) High-efficiency particulate air filters to reduce airborne particulates.

(12) Charcoal adsorber delay trap to remove radon and other gaseous species.

(13) Chemical stabilization of dried tailings areas by use of asphalt, plastics, clays, etc., to prevent active wind and water attack.

(14) Wetting of dried tailings surfaces.

(15) Progressive reclamation of tailings areas.

(16) Wet shipment of yellowcake to prevent loss by dust.

The effect of these alternatives on control of milling operations is summarized in Table 10-3.

Uranium mill tailings disposal programs are intended to address the following objectives (see NUREG, 1980): (1) reduce or eliminate airborne radioactive emissions (mainly radon); (2) reduce or eliminate impacts on groundwater; and (3) ensure long-term stability and isolation of the tailings without the need for continued active maintenance.

To achieve these objectives, four general procedures are defined (NUREG, 1980); (1) preparation of tailings for disposal; (2) location of the tailings disposal area: (3) preparation of the tailings disposal area; and (4) stabilization and covering of the tailings.

For tailings preparation, several alternatives are possible, including nitric acid leaching of ore, which will better control the radioactive elements, segregation of sands from slimes for separate treatment, neutralization of the tailings liquids, barium chloride treatment for radium control, removal of toxicants by ion exchange, removal of water by solar *in situ* evaporation, dewatering, and fixation of tailings in asphalt or cement.

Surface emplacement of tailings is the common way of disposal. In addition, disposal could take place in unused open pit mines or in other excavations. Disposal of tailings in mined-out underground mines not only eliminates the radon emission problem, but also provides structural support for the abandoned mine.

Table 10–3. Summary of emission sources from mill tailings.

Milling Activity or Area	Possible Emissions	Potential Controls
Ore stockpile	Particulates, leachate, runoff	1,2,4,6
Ore crushing and grinding	Particulates, radon	3,7,8,9,10,11,12
Yellowcake processing	Product particles, NH_3 gas	8,11,16
Tailings disposal area	Particulates, radon	13,14,15
Roads	Particulates	5

Note: Numbers in third column refer to topics listed in the section entitled "Controlling Impacts of Milling Operation." The reader is referred to NUREG (1980) for discussion of decommissioning alternative.

The tailings area can be left unprepared if the geomorphologic conditions are suitable (i.e., good drainage, solid adequate for chemical stability, and mechanically strong). In other areas, the site can be improved by soil compaction to reduce chances of tailings–soil interaction, use of clay liners (bentonitic clays are especially good for lining purposes), and use of synthetic liners. Each of these has merit in specific cases (see NUREG, 1980).

Tailing stabilization is desirable, both on a temporary basis during mill operations and permanently after mill shutdown. With no treatment, the tailings will dry out and will be attacked by wind and water erosion. Soil covers may be used to prevent this, and vegetative cover, where possible, would further reduce attack by wind and water. In some areas it may be necessary to use gravel, crushed rock, caliche, or riprap cover; this would prevent wind erosion and allow water to infiltrate. Artifical covers can also be used.

When all combinations of all variables for uranium mill tailings disposal are considered, there are literally thousands of alternatives (NUREG, 1980). The NRC has, however, focused on the nine most feasible of these (NUREG, 1980) and the reader is referred to their document for in-depth discussion.

Uranium Mill Tailings Disposal Programs

Any program to deal with uranium mill tailings must address the following: reduction of airborne radioactive species (mainly radon), reduction or elimination of impacts on groundwater, and the long-term stability of isolated mill tailings such that continued active maintenance will not be necessary. The tailings must be prepared for disposal, isolated sites must be selected, the tailings disposal area must also be prepared, and the tailings must be stabilized and covered (see NUREG, 1980).

Of plans to better the environmental impact of tailings processes, several steps could be taken (see NUREG, 1980). Use of nitric acid instead of sulfuric acid or alkaline leach would more effectively remove uranium, radium and thorium, but the method would be very costly as the reagents alone would cost twice as much as sulfuric acid leach and major equipment changes would have to be made as well. Most of the radium and much of the other radioactive species that go to the tailings are found in the slimes. Separation of the slimes from the sands would allow them to be treated separately. Neutralization of an acidic slurry with calcium hydroxide is also attractive, as not only would pH be raised to about 7, but elements such as Ra, Th, U, Fe, Co, V, and As, would be precipitated as a group. This concentrate may actually contain economic concentrations of some of these.

It has been shown (NUREG, 1980) that simple addition of $BaCl_2$ to slurries to the tailings disposal area is not adequate to efficiently remove Ra. Yet if $BaCl_2$ is added to clear Ra-bearing waters the Ra removal is 90 to 99% efficient. Thus some combination of filtering followed by Ra-removal would be an effective way of preventing Ra from reaching the tailings. Ion exchange can also be used for the removal of Ra, but it is more cumbersome and expensive as different processes have to be used for liquids and for particulates.

Regulations proposed and in effect by the NRC address the different problems presented by uranium milling operations and in mill tailings. These are discussed as follows (modified from NUREG, 1980). Airborne radioactive emissions during operations shall be as low as possible; wetting of dry surfaces will be carried out to prevent dusting and to minimize blowing. Dust generation from ore pads is more difficult to control, but the pads can be located to minimize the problem and, as with other sources, constant monitoring will allow proper control.

The site for the uranium mill tailings must be removed from populated areas. Further, the hydrologic and geologic characteristics of the sites must be known, and there must be a minimum potential for loss by erosion, for disruption, and for dispersion by natural forces. The tailings must be isolated for thousands of years so that the effect of ^{222}Rn from ^{226}Ra is negligible (half-life of ^{226}Ra is 1620 years).

Tailings may be placed in specially excavated pits, in mines, or in other suitable sites. Backfilling in mines is one reasonable disposal method, where the sand–slime mixture is used to fill worked-out mines. This method has the advantage that the aboveground siting problems are avoided. It is imperative that the effect of backfilling on mine stability, potential release to aquifers, and variations of these and other parameters with time be known. For example, if an aquifer is in communication with parts of a mine, then backfilling may create more of a problem than storage of the tailings aboveground. Each mine where backfilling is practiced or planned must be assessed on a case-by-case basis. For aboveground sites the following criteria must be adhered to:

(1) The upstream area must not be favorable for rainwater catchment and flood potential.

(2) The topography should be such that wind protection is assured.

(3) Cover slopes and embankments must be as flat as possible after stabilization to minimize erosional potential.

(4) A vegetative cover, or rock cover, must be established to minimize wind and water erosion. In arid areas it may be necessary to use rock covers instead of vegetative covers. If this is the case, then only dense, sound, and abrasion-resistant rocks will be used. Rocks such as shale or other weak or friable rocks will not be used.

(5) The impoundment surfaces must be contoured to avoid channeling or ponding during periods of high precipitation, and surfaces where above-average runoff may occur must be covered with riprap.

(6) The geomorphology of the surrounding area must be investigated to determine the potential for large-scale erosion, such as gullying, which might affect the tailings pile in the future.

(7) The impoundment cannot be located near a fault that is capable of causing structural damage to the impoundment during an earthquake. The earthquake potential of the area and any faults in the impoundment area must be evaluated.

(8) The impoundment should be designed such that deposition is promoted (i.e., a design that allows for deposition from runoff is desirable).

(9) The site should be so selected and engineered that ongoing maintenance is not necessary in order to preserve its isolation.

Radon emanation from mill tailings can be minimized by suitable cover. Use of soil covers of at least 3 m thickness results in a maximum radon emission of less than 2 $pCi/m^2/sec$. Direct gamma radiation must be at background as well. If rock or man-made materials are used as covers, it must be demonstrated that these will not alter or crack with time such that greater radon emission will result. No material, including mine waste or rock, that contains elevated radioactivity, and radium in particular, can be used. This ensures that the radium level in the tailings is minimized.

Toxic materials are present in the tailings piles. It is essential to prevent their release into the surrounding areas, and especially into groundwater. In order to prevent any toxic material from seeping into groundwater supplies, the following steps must be taken: (1) Impermeable bottom liners must be installed. These may consist of clay or artifical materials: both must be evaluated to ensure that they are, and will remain, impervious to impounded liquids. (2) The milling operations must be designed so that maximum recycle of liquids results to reduce the net input of fluid to the impoundment. (3) The tailings must be dewatered. A drainage system can be used for new tailings piles or by *in situ* bottom drainage on a predesigned sloped surface. (4) The solutions must be neutralized to promote immobilization of toxic materials.

At existing sites, if groundwater is being affected by seepage, then remedial actions must be taken to alleviate the original conditions in order to restore the groundwater to precontaminated state.

Further, the applicant or operator of a uranium milling operation must provide information on the chemical and radioactive characteristics of the solutions and on the underlying soil and geologic formations. The potential of such soils and formations for flow after contamination by seepage must be evaluated. Thus the thickness, shape, uniformity, etc., of the underlying strata must be known, and the hydrologic parameters assessed. Further, the present and future requirements of the groundwater in the area must be known. In addition to the tailings area, the same information must be known at the ore storage area so that seepage of uranium and other elements from the ore does not present a health hazard to water supplies.

Other regulatory aspects of uranium milling operations call for plans for decommissioning of the structures and site, a plan for decommissioning and disposal of tailings, environmental review, public participation, financial surety, monitoring both during the mill operation and before and after operation, and long-term control of the disposal sites.

At existing sites, the problems are different because of the large tonnages of materials already there, and the fact that the sites were selected to be convenient to mining operations without a careful evaluation of the sites (i.e., relative to the criteria now used). Further, disposal costs were not built into the original product price plan as tailings were generated. At the active sites future operations will have to handle existing as well as future tailings.

Radon Risk

Although many toxic materials may be locally concentrated in tailings piles, only radon gas is free to emanate easily from them. It is interesting to compare radon emissions from tailings with other sources of radon emission (Table 10-4; see NUREG, 1980, for summary). The assumption is made that mill tailings must be covered, and, if this is done, then the radon emission is assumed equal or less than 2 $pCi/m^2/sec$. At this concentration, about 40 cancer deaths for the period 2000–3000 CE would be expected from the tailings (note: if the tailings are left uncovered, the potential risk is 6000 cancer deaths for a 1000-year period). The amount of radon from tailings is compared with other sources: about 0.002% of radon from soils, 0.02% of radon from evapotranspiration, and roughly 10% of radon from building interiors.

Other Toxic and Hazardous Species

The radioactive and nonradioactive species that may pose hazards to the environment are given in Table 10-5 and the EPA limits for levels of many of

Table 10–4. Comparison of continuously released radon from uranium mill tailings with other sources.

Source	Estimated Annual Release (Ci/yr)	Estimated Annual Population Dose (organ-rem to bronchial epithelium)	Potential Annual Premature Cancer Deaths
Natural soils	1.2×10^8	1.6×10^7	1152
Building interiors	2.8×10^4	2.2×10^7	1594
Evaportranspiration	8.8×10^6	1.2×10^6	86
Soil tillage	3.1×10^6	4.2×10^5	30
Fertilizer used (1900–1977)	4.8×10^4	6.9×10^3	0.50
Reclaimed land from phosphate mining	3.6×10^4	4.9×10^3	0.35
Postoperational releases from tailings generated to the year 2000	3.9×10^3	2.8×10^2	2.9^a–0.02

See NUREG (1980) for details on Table 10–4.

^{a}2.9 for no cover on tailings piles; 0.02 with such cover.

these species in drinking water are given in Table 10-6. In Table 10-6, it is assumed that the NRC Model Mill (see NUREG, 1980) is typical of most mills, although the reader is reminded that any particular mill may have an inventory quite different from that presented here.

In the case of radioactive species (Table 10-5), ^{226}Ra, ^{230}Th, and ^{210}Pb are present in the tailings solutions in amounts greater than those permitted by the EPA. Of these, ^{230}Th is of concern because it is nearly two orders of magnitude above the EPA level. Yet this radionuclide will precipitate out on the bottom or in the tailings pile as pH increases as $Th(OH)_4$ is extremely insoluble from mildly acidic (i.e., pH 4–7) to basic conditions. Hence neutralization of the tailings feed, or, later, milling with surface waters, will stabilize this species. Yet it must be carefully monitored so that if any point concentrations of ^{230}Th occur, they can be identified. The ^{226}Ra, under current plans, will largely be removed prior to transport to the tailings. Where ^{226}Ra is present, however, it may be partially removed by $BaCl_2$ treatment to precipitate it as $(Ba, Ra)SO_4$. The fate of ^{210}Pb is not known with any certainty, yet coprecipitation with Ba–Ra sulfate is likely based on theoretical grounds.

Basically, clay liners or naturally occurring clay or shale should be effective in removing radioactive species from the tailings slurries, leaving only the species present as anions, S, Se, As species, as well as Na^+, Fe-complexes, and nitrates as species, which may move out of the impoundment by aqueous exchange. For a good impoundment structure not in communication with highy surface-drained areas or underground aquifers, these elements should not escape the tailings area.

Risks to Uranium Mill Workers

It is possible to comment on the mortality of uranium mill workers, especially on the risk of occupational radiation exposure, relative to other groups (Table 10-7). The data for Table 10-7, which are for 1975, are perhaps too high because of the decrease in number of mills working and, more important, better working conditions. It is interesting to note that the risk for uranium mill workers of 42 premature deaths/10^5 person-years is higher than for the private sector, but low compared to the other categories shown.

The reader is cautioned, however, to note that the figures in Table 10-7 do not give the actual number of fatalities, but rather a normalized mortality

Table 10-5. Typical concentrations of radionuclides and chemicals in tailings solution.

Radionuclide	Concentration, μc/ml	Maximum Permissible Concentration in Unrestricted Areas[a] (μCi/ml)
U-238	1.7×10^{-6}	4×10^{-5}
U-234	1.7×10^{-6}	3×10^{-5}
Th-230	9.0×10^{-5}	2×10^{-6}
Ra-226	2.5×10^{-7}	3×10^{-8}
Pb-210	2.5×10^{-7}	1×10^{-7}
Po-210	2.5×10^{-7}	7×10^{-7}
Bi-210	2.5×10^{-7}	4×10^{-5}

Chemical	Concentration mg/L	NAS Water Quality Standards for Livestock[b] (mg/L)
As	0.2	0.2
Na	200	—
Fe	1000	—
Al	2000	—
F	5	—
V	0.1	—
Ca	500	—
$SO_4^=$	30,000	250
Cl^-	300	3000
NH_3	500	—

[a]From Rules and Regulations, Title 10, Chapter I, Code of Federal Regulations, Part 20, Standards for Protection Against Radiation, U.S. Nuclear Regulatory Commission.
[b]From "Water Quality Criteria 1972," A report of the Committee on Water Quality Criteria, National Academy of Sciences, National Academy of Engineering, prepared for the U.S. Environmental Protection Agency, 1972.

Table 10–6. U.S. Environmental Protection Agency draft standards for uranium mill tailings disposal (element concentrations in underground drinking water).

Element	Maximum Permissible Concentration in Groundwater
Arsenic	0.05 mg/L
Barium	1.0 mg/L
Cadmium	0.01 mg/L
Chromium	0.05 mg/L
Lead	0.05 mg/L
Mercury	0.002 mg/L
Molybdenum	0.05 mg/L
Nitrate	10.0 mg/L
Selenium	0.01 mg/L
Combined ^{226}Ra and ^{228}Ra	5.0 pCi/L
Gross alpha activity including ^{226}Ra but not uranium or thorium	15.0 pCi/L
Uranium	10.0 pCi/1

Radon Flux Limit from Disposal Site	
Maximum permissible radon flux emitted from residual radioactive materials at the disposal site.	2 pCi/m^2/sec (annual average)

Source: From EPA (1976).

Table 10–7. Incidence of nonviolent job related fatalities.[a]

Occupational Group	Fatality Incidence Rate (premature deaths/10^5 person/year)
Underground metal miners	1244
Asbestos insulation workers	365
Uranium miners	232
Smelter workers	193
Mining	61
Uranium mill workers	42
Transportation and public utilities	23
Services	3
Public sector	10

[a]Data for 1975; sources of data in NUREG (1980).

index (i.e., actual deaths for uranium miners, for example, was 8). In terms of risk, however, the point is made that uranium mill workers are not in the highest risk categories.

Further, it is important to point out that the spontaneous cancer death rate per year in the United States is a staggering 470,000, and the fractional increase in the death rate due to milling is only 1.2×10^{-5} (see NUREG, 1980 for details).

Inactive Mill Sites

The inactive uranium mill sites are listed in Table 10-8. Concern for these sites dates to 1974, when the Vitro tailings site in Salt Lake City, Utah was discussed for remedial action to be administered jointly by the U.S. Atomic Energy Commission and the State of Utah. Since it was pointed out that conditions at the Vitro mill tailings were identical to those from other sites, the AEC and Environmental Protection Agency (EPA) decided that a generic approach to the issue was in order, and, subsequently, 22 inactive sites were identified. These sites were identified and problems for each were published in a series of reports (ORNL, 1980). The U.S. Department of Energy proposed a plan to Congress in 1978 to stabilize and control the mill tailings in a safe and environmentally sound manner. Public Law 95-604 was enacted in November 1978 to authorize the DOE, along with affected states, Indian tribes, and private owners, to establish assessment and remedial action programs at inactive uranium mill sites, UMTRAP (uranium mill tailings remedial action program). The act further states that the DOE will meet the radiation standards of the EPA, and that the DOE will finance 90% of the remedial action costs and the states the remaining 10% (from nonfederal funds).

Major program requirements for UMTRAP are: (1) designation of processing sites; (2) establishment of site priorities for remedial action; (3) establishment of cooperative agreements with affected states and applicable Indian Tribes; (4) acquisition and disposition of lands and materials; (5) reprocessing of residual radioactive materials; (6) compliance with the National Environmental Policy Act (NEPA); (7) remedial action; (8) public participation; and (9) annual status reports to Congress. Three more sites were later identified, bringing to 25 the total number of inactive sites needing remedial action. These are given in Table 10-8. Of the sites listed in Table 10-8, the Baggs, Wyoming and the two North Dakota sites were added to the 22 sites covered by U.S. PL-95-604. The priorities for remedial action are based on possible health effects due to radon emissions as compiled by the DOE. The amounts of tailings (Table 10-8) are in large part due to uranium processing associated with Government needs, roughly in the period 1940–1965. At the Grand Junction, Colorado site the tailings are 80% Government and 20% commercial; at New Rifle, Colorado 99% Government, 1%

Table 10–8. Inactive uranium mill tailings sites.

Location	Remedial Priority	Amount of Tailings (10^6 MT)	Last Year of Operation
Arizona			
Monument City	Low	1.0	1968
Tuba City	Medium	0.7	1966
Colorado			
Durango	High	1.8	1963
Grand Junction	High	2.5	1970
Gunnison	High	0.6	1962
Maybell	Low	2.4	1964
Naturita	Medium	0.6	1963
Rifle (New Rifle)	High	2.4	1972
Rifle (Old Rifle)	High	0.3	1958
Slick Rock (NC)	Low	0.02	1957
Slick Rock (UC)	Low		1961
Idaho			
Lowman	Low	0.08	1960
New Mexico			
Ambrosia Lake	Medium	2.3	1963
Shiprock	High	1.9	1968
North Dakota			
Belfield	Low	0.6	1968
Bowman	Low	0.5	1967
Oregon			
Lakeview	Medium	0.1	1960
Pennsylvania			
Cannonsburg	High	0.2	1966
Texas			
Falls City	Medium	2.2	1973
Utah			
Green River	Low	0.1	1961
Mexican Hat	Medium	2.0	1965
Salt Lake City	High	2.0	1968
Wyoming			
Baggs	Low	0.01	—
Converse County	Low	0.17	1965
Riverton	High	0.8	1963

Source: Modified from NUREG (1980).

commercial; at Falls City, Texas, 34% Government, 66% commercial. Only Government tailings are present at the other sites.

At the present time remedial action plans are being prepared by DOE, with concurrence by the NRC required, for each site. The responsibility for licensing for maintenance and monitoring of the processing sites after the remedial action has been taken lies also with the NRC. Regulations in effect for active mills are complied with at the inactive sites with the DOE and the NRC as overseers.

The amount of mill tailings is enormous. The total for the 25 sites listed in Table 10-8 is 25.28 million MT of tailings. Some of the tailings have been used for various purposes, including some 270,000 MT from the Grand Junction, Colorado site used for fill material for various construction projects. The construction sites have been identified and are being jointly investigated by the DOE and the State of Colorado.

When the DOE visited the inactive sites, it was determined that the stabilization procedures that had been used were adequate only for a relatively short-range holding pattern. Signs of wind and water deterioration, slope instability, and failing vegetative cover were all noted. More thorough, long-range action must be taken. This research is under investigation at the present time.

Of the 25 inactive sites, only the Cannonsburg, Pennsylvania site is located in an area of moderate rainfall. All the other sites are located in more or less arid states in the West, Midwest, and Southwest. The Cannonsburg site thus will require more pains in ensuring site stabilization.

Conclusions

Attempts to improve U milling have been made. For example, Ra removal by sulphate precipitation is efficient. Modern tailings are well designed and monitored. Abandoned mill tailings need to be assessed in terms of potential hazard to the environment. The Canonsburg, Pennsylvania site is the only one in a fairly humid, high rainfall area, and special steps to permanently seal the site have been made. Methods of preventing loss of surface materials at the other sites are under study, although even with extensive wind damage the overall effect on many mill sites may be minimal. Backfilling of abandoned uranium mines with tailings offers a convenient way to remove some of the waste material, although care must be taken to assure no chemical transport between tailings and any aquifers present in the workings.

CHAPTER 11
Natural Analogs

Introduction

Natural analogs provide an excellent means by which to study the ability of natural media to retain radwaste. The Oklo natural reactor provides a remarkable case study for radioactive species in rocks, and use of various contact zone studies for elemental distribution between intrusive and intruded rocks also provides opportunity for such studies. In most of these analog studies, the duration of the naturally conducted experiment is on the order of hundreds to thousands to hundreds of thousands of years. Nature has conducted the experiment and it is up to man to synthesize the data and to make interpretations.

A great deal of research has been conducted on laboratory experiments dealing with radioactive waste forms, waste form–rock interaction, and so on; yet in many cases, even with the most detailed experimental work, it is difficult to make long-range predictions on the order of tens of thousands to millions of years. Yet storage of actinide-bearing radioactive waste must be planned for about 240,000 years—about ten times the half-life of ^{239}Pu. In order to attempt to answer questions concerning rock and mineral stability, and stability of encapsulated radioactive waste in rock–mineral surroundings, natural analogs have been studied on an increasing level for the last few years. The importance of natural analogs is based on the facts that nature has conducted the experiments, and man can make proper measurements, synthesize the data, and attempt an intepretation. Ideally one would pick an analog that would be a very close counterpart to stored radioactive waste in rock; in practice the analog is not as close to the radioactive waste–rock package as desirable. For example, a hypothetical uraninite vein cutting an evaporite sequence would be desirable for a WIPP site (Chapter 7) analog.

But evaporites are impoverished in uranium, let alone uraninite veins, so other analogs must be considered.

By far and away the most important of the analogs is the Oklo "fossil" reactor in the Republic of Gabon, since at this site a fission reaction was sustained for 500,000 years at about two billion years before the present, and many of the fission products were retained in the minerals in which they were formed (see Brookins, 1978a,b,f; 1980c). Other analogs include studies of various rocks intrusive into different rock types representative of some of the radioactive waste candidate rocks. In these analogs elemental behavior in both the intrusive and intruded rocks is studied to see if there has been chemical migration between the two. The assumption is commonly made that a large, high-temperature mass of molten rock of different chemistry from the intruded rocks will act as a heat engine, which, in turn, may promote some elemental migration. If extensive migration of various elements is noted, then the data must be scaled to properly address the problem of a smaller, and less heat-containing radioactive waste canister. If no migration is noted, then the data may, with caution, be interpreted to argue that for the much smaller radioactive waste canister, it is even less likely that elemental migration from waste form into the surrounding rocks, will occur assuming identical thermal field behavior.

Other analogs for some aspects of stored radioactive waste include pitchblende veins in various rocks, sandstone uranium deposits, and volcanic rock–sedimentary rock contacts.

An important facet of analog studies is the proper interpretation of mineral assemblages in the different rocks, and types of equilibrium diagrams that can be applied to such assemblages in order to comment on stable versus metastable assemblages, transportation and precipitation means, and other parameters. In this regard the author has previously argued for the use of Eh–pH diagrams in order to address some of these problems (see Brookins, 1978a, b, f; 1977). For convenience, reference to the Oklo natural reactor setting will be used in discussing the application of Eh–pH diagrams.

Eh–pH Diagrams

The use of Eh–pH diagrams for problems of radionuclide migration and accumulation of actinides has been discussed by Allard et al. (1978), Apps et al. (1977), Ames and Rai (1978), and Brookins (1978a,b,d,f). The Oklo natural reactor has been subjected to numerous investigations (see articles in IAEA, 1976; 1978) and is mentioned here to address some general aspects of radionuclide migration.

In order to use Eh–pH diagrams, some fundamental assumptions concerning the chemistry, temperature, and pressure of aqueous phases is necessary. Parameters such as precise species identification, ionic strength, temperature, pressure, early- versus late-stage epigenetic minerals, and flow parameters are all unknown at Oklo; yet a comparison of predictions of retention versus migration (Brookins, 1979a) based on Eh–pH diagrams compared with measured data is good.

The use of Eh–pH diagrams has proven to be most useful in depicting stability fields for various solids and aqueous species in which the element(s) under consideration exist in more than one valence, species change as a function of either Eh or pH alone, and solid aqueous species boundaries will vary as a function of amount of total dissolved element(s) in solution. The convention here is that used by Garrels (1959) and Garrels and Christ (1965). For oxidation of a metal ion, M^{2+} to M^{4+}, for example, the reaction may be written:

$$M^{2+} = M^{4+} + 2e^-.$$

The free energy for the reaction is:

$$\Delta G_f^\circ(M^{4+}) - \Delta G_f^\circ(M^{2+}) = \Delta G_R^\circ,$$

taking the free energy of formation for the electron (e^-) as zero. The relationship between ΔG_f° and the standard potential, E_o, is:

$$\Delta G_R^\circ = nFE_o,$$

where $n = $ number of electrons and $F = $ the Faraday Constant (23.06 kcal/v/g equivalent).

The expression for Eh, the oxidation potential under natural conditions, is given by the Nernst expression:

$$\text{Eh} = E_o + \frac{2.303\ RT}{nF} \log K,$$

where, if it is assumed that activities $=$ concentrations, for the $M^{2+} = M^{4+} + 2e^-$ reaction written above:

$$\text{Eh} = E_o + \frac{0.059}{2} \log \frac{(M^{4+})}{(M^{2+})} = E_o \quad \text{since } \log \frac{(M^{4+})}{(M^{2+})} = 0.$$

For a more complex reaction such as:

$$M^{2+} + 2H_2O = MO_2 + 4H^+ + 2e^-$$

the corresponding Eh–pH reaction results from the Nernst expression, viz.,

$$\text{Eh} = E_o + \frac{0.059}{2} \log (1/M^{2+}) + \frac{0.059}{2} \log (H^+)^4,$$

$$\text{Eh} = E_o + \frac{0.059}{2} \log (1/M^{2+}) - 0.118\ \text{pH},$$

and the E_o term will be modified by the $(+\ 0.059/2 \log (1/M^{2+}))$ term.

For a more complex reaction, such as one involving oxidation of more than one species, consider the reaction of a sulfide, MS, to a higher-valence oxide, M_2O_3, plus sulfate:

$$2MS + 11H_2O = M_2O_3 + 2SO_4^{2-} + 22H^+ + 18e^-,$$

and the Eh–pH expression is:

$$Eh = E_o + \frac{0.059}{18} \, 10g\,(SO_4^{2-})^2 - \frac{(22)}{(18)} \, pH.$$

At lower pH values it is likely that the Eh-independent dissolution will occur, viz.,

$$MS + 2H^+ = H_2S(aq) + M^{2+}$$

and the equilibrium constant for this reaction is calculated from the expression:

$$\Delta G_R^\circ = -RT \ln K = -1.364 \log K,$$

where:

$$K \text{ for this reaction} = \frac{a_{(M^{2+})}a_{(H_2S)}}{a_{(H^+)^2}},$$

which, with knowledge of, or assumed values for, M^{2+} and (H_2S) allows one to calculate the equilibrium pH. For the oxidation of M^{2+} to M_2O_3 as H_2S oxidizes to SO_4^{2-} we write:

$$2M^{2+} + H_2S + 7H_2O = M_2O_3 + SO_4^{2-} + 16H^+ + 10e^-,$$

and the Eh–pH expression is:

$$Eh = E_o + \frac{0.059}{10} \, \log \frac{(SO_4^{2-})}{(M^{2+})^2} - \frac{(16)}{(10)} \, 0.059 \; pH.$$

For the usual case, the activities of solids, water, and e^- are assumed equal to unity. Further, activity (= concentration) for total dissolved S = 10^{-3} molal. For most dissolved metal ions (simple or complex), an activity of either 10^{-6} or 10^{-8} is commonly assumed. Total dissolved carbonate is taken here as 10^{-2} molal. A further assumption is that halide, sulfate, etc., complexes are important only at low pH values and thus will not be included in the Eh–pH diagrams. The range of natural Eh values for over 10,000 waters has been given by Baas Becking *et al.* (1960) (shown here in modified form as Fig. 11-1) and it is clear that the extreme pH values (i.e., less than 1; greater than 12) can be ignored. At molalities of 10^{-6} to 10^{-8}, conditions approaching infinite dilution are assumed, hence using concentration = activity for those dissolved metal species is justified. The higher values for total dissolved S (10^{-3}m) and C (10^{-2}m) require that concentrations be modified by the appropriate activity coefficients to arrive at proper activities. However, since ideality is assumed and the ionic strength is not known for most natural

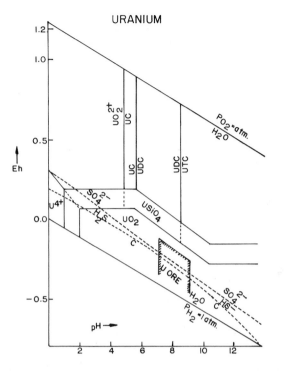

Fig. 11-1.(a) Eh–pH diagram for part of the system U–Si–S–C–H–O.

systems, concentration = activity will be assumed for these species as well.

Stability Limits of Water: 25°C to 200°C

The basic assumption for use of Eh–pH diagrams in depicting mineral equilibria or stability is that the reactions of interest take place in the presence of water. It is assumed that total pressure is 1 bar and that the effects of increased pressure will have only a small (note: but not necessarily insignificant) and presumably systematic effect on all solid and aqueous phases. In this view, it is appropriate to mention that Eh–pH diagrams for many species at temperatures near 100–150°C and pressures to 80 bars are different from the 25°C, 1-bar diagrams only by the slopes of the various curves while the areas for various species are about the same.

For water stability, the conventional approach is to assume the upper and lower limits in the cases where $P_{O_2} = 1$ bar and $P_{H_2} = 1$ bar, respectively. For 25°C and 1 bar total pressure the bounding conditions are thus:

$$\text{Upper Limit: } Eh = 1.23 - 0.059 \text{ pH},$$
$$\text{Lower Limit: } Eh = 0.00 - 0.059 \text{ pH}.$$

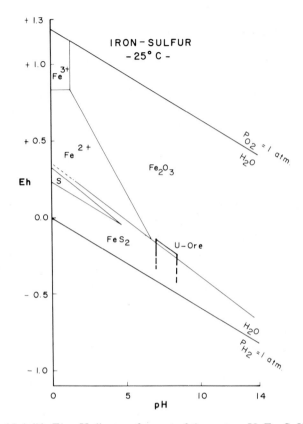

Fig. 11-1.(b) Eh–pH diagram for part of the system U–Fe–S–H–O.

For temperatures in excess of 25°C only the Eh intercept for the upper stability limit and the slope will change due to the temperature effect on the fundamental relationship:

$$\Delta G_R = -RT \ln K.$$

The upper and lower limits for water are:

100°C

Upper Limit: Eh = 1.17 − 0.074 pH,

Lower Limit: Eh = 0.00 − 0.074 pH.

200°C

Upper Limit: Eh = 1.09 − 0.094 pH,

Lower Limit: Eh = 0.00 − 0.094 pH,

where data for H_2O(liq.) are from Barner and Scheuerman (1978) and $\Delta G_f = 0$ for O_2(g) and H_2(g) at both temperatures.

Reference Diagrams for Uranium

The Eh–pH diagram for part of the system U–Si–O–H (most commonly in the absence of Si) has been calculated many times, starting with the work of Hostetler and Garrels (1962). New and revised data have led to the diagrams published by Brookins (1976) and Langmuir (1978). Langmuir's (1978) contains the most comprehensive compilation of uranium data. All these diagrams are, however, for 25°C and 1 bar total pressure. Data for uranium species above 25°C are available for some species of interest (Barner and Scheuerman, 1977; Sergeyeva et al., 1972), but no attempt has previously been made to construct Eh–pH diagrams for temperatures above 25°C.

Species considered to be of importance for Oklo include the hexavalent species UO_2^{2+}, $UO_2CO_3^\circ$ [note: rutherfordine, $UO_2CO_3(c)$, is stable at pH values below 5 and is not treated here; the reader is referred to Sergeyeva et al., (1972) and Langmuir (1978), for discussion of its stability field], $UO_2(CO_3)_2^{2-}$, and $UO_2(CO_3)_3^{4-}$. Tetravalent species of importance include UO_2 and $USiO_4$. At Oklo, pitchblende is common while coffinite is rare (Geffroy, 1975). For ease of depicting Eh–pH relations, it will be assumed that UO_2 and $USiO_4$ very nearly occupy the same Eh–pH space (see Hostetler and Garrels, 1962; Brookins, 1976, for further discussion).

The 25°C data are readily available and the diagram (Fig. 11-1) has been constructed using the data tabulated in Table 11-1. The fields of rutherfordine, schoepite, and other carbonate and oxide species have been omitted,

Table 11–1. Thermodynamic data for uranium species used in this chapter.[a]

Species	$\Delta G_{f(T)}$ (kcal/M)		
	25°C	100°C	200°C
Solids			
UO_2	−246.6	−243.5	−239.3
$USiO_4$	−456.6	—	—
UO_2CO_3	−375.4	−367.8	−357.7
Aqueous species (U)			
UO_2^{2+}	−236.4	−232.8	−227.8
$UO_2CO_3^\circ$	−369.4	−360.4	−348.4
$UO_2(CO_3)_2^{2-}$	−504.9	−495.4	(−469.6)
$UO_2(CO_3)_3^{4-}$	−637.5	−615.5	(−568.0)
Aqueous species (CO_2)			
H_2CO_3	−148.9	−144.5	−139.5
HCO_3^-	−140.3	−133.8	−124.5
CO_3^{2-}	−126.2	−116.3	−100.7

[a] 25°C data from Langmuir (1978) except for $USiO_4$ (from Brookins, 1976). 100°C and 200°C data for UO_2^{2+}, UO_2, and carbonate species from Barner and Scheuerman (1978). 100°C and 200°C data for $UO_2(CO_3)^c$ and $UO_2(CO_3)^\circ$ from Sergeyeva et al. (1972); 100°C and 200°C data for $UO_2(CO_3)_2^{2-}$ and $UO_2(CO_3)_3^{4-}$ calculated (see text).

as they occur in Eh–pH space removed from the partially hatched area designated U-ore. Also shown in Fig. 11-1 are boundaries for carbon (i.e., $C°$): carbonic acid, bicarbonate ion, carbonate ion, and boundaries between reduced (H_2S, HS^-) and oxidized sulfur (SO_4^{2-}) species. For total dissolved $S = 10^{-3}$ mol, a small field of native sulfur occurs between H_2S and SO_4^{2-} from pH = 0 to 5. The pH limits for the zone marked U-ore and the upper Eh limits are derived by considering the works of Brookins (1976), Lisitsin (1969), Harshman (1972), and Lee (1976).

Eh–pH diagrams for temperatures above 25°C have not previously been attempted. This is in part due to lack of data, and in part due to the assumption that the uranyl carbonate complexes are unstable above 125°C at 1 bar pressure (Rafalsky, 1958; see also discussion in Hostetler and Garrels, 1962). More recent work, summarized in Sergeyeva et al. (1972) indicates that the uranyl carbonates are stable at 1 bar pressure in the temperature range 25°C to 150–175°C with the following equilibrium expressions available:

$$UO_2(CO_3)_2^{2-} = UO_2^{2+} + 2CO_3^{2-} \qquad (\log K = -14.14 - 0.0096T),$$

$$UO_2(CO_3)_3^{4-} = UO_2^{2+} + 3CO_3^{2-} \qquad (\log K = -27.84 - 0.0216T).$$

Since data for UO_2^{2+} and CO_3^{2-} are known at 100°C and 200°C, the free energies of formation for the uranyl carbonates have been calculated for 100°C and 200°C (note: the 200°C values are approximate values only). The 100°C diagram is considered more reliable than the 200°C diagram not only because of the uncertainty of ΔG_f^o values, but, as reported by Sergeyeva et al., (1972), for the total dissolved uranium and carbonate concentrations at 200°C, $UO_2(OH)_2^o$ is the observed soluble species rather than either uranyl carbonate. In Fig. 11-2 the range of Eh and pH in which primary uranium mineralization occurs is much less well known that for the 25°C diagram, but the same general pH range is advocated based on mineral assemblages.

Again, the reader is cautioned that the assumption that pressures greater than 1 bar total pressure do not have a pronounced effect on condensed phases may not be valid. This approach is used because where Eh–pH diagrams have been constructed for P > 1 bar, the diagrams are very close to the 25°C, 1-bar diagrams. Since pressure estimates at Oklo vary widely (Brookins, 1980c), data in Figs. 11-2 and 11-3 are more properly used as introductory, working guides only.

Americium

Data for americium have been tabulated by Apps et al. (1977) and Rai and Serne (1978). These sources contain most relevant thermodynamic data for constructing Eh–pH diagrams (Apps et al., 1977) and activity–pH diagrams (Ames and Rai, 1978; Apps et al., 1977). The thermodynamic data in Table

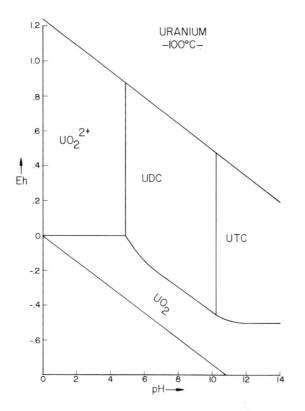

Fig. 11-2. Eh–pH diagram for part of the system U–C–O–H at 100°C, 1 Bar pressure.

11-2 are commonly agreed upon by most investigators. Apps *et al.* (1977) have presented an Eh–pH diagram showing that Am^{3+} and $AmOH^{2+}$ and "hydroxyl" species occupy the entire stability field of water with metastable fields for AmO_2^+ and AmO_2^{2+} above the upper limit for water at 25°C, 1 bar pressure conditions. Further, Brookins (1978f) has presented an Eh–pH diagram in which some older data (i.e., Latimer, 1961) were used to allow inclusion of the fields for $Am(OH)_3$ and AmO_2 as well as Am^{3+}. In view of the more recent compilations cited above, it is necessary to consider $AmOH^{2+}$ as well as to attempt to use better data for the solid species.

In order to obtain ΔG_f° for $AmOH^{2+}$ data (shown in Table 11-3) are available. Because of the much lower ionic strength for the first two reactions, the ΔG_f° calculated at approximately −194.6 kcal/M is preferred to the −191.8 kcal/M value calculated by the last two reactions.

Similarly, for solid species, solubility data exist (Table 11-4) for the reactions (from data in Apps *et al.*, 1977; Rai and Serne, 1978). From these data plus the ΔG_f° data for Am^{3+} and Am^{4+} from Fuger and Oetting (1976),

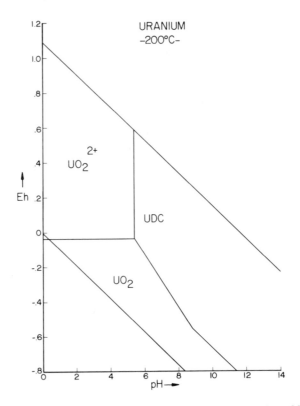

Fig. 11-3. Eh–pH diagram for part of the system U–C–O–H at 200°C, 1 Bar pressure.

the following free energies of formation are calculated: $Am(OH)_3$: -292.7 kcal/M; $Am(OH)_4$: -310.7 kcal/M; AmO_2: -211.2 kcal/M. For the reaction: $AmO_2 + 2H_2O = Am(OH)_4$; $\Delta G_R = +13.9$, thus indicating that the dioxide is stable relative to the tetrahydroxide. Unfortunately, no good data exists for Am_2O_3, hence the Eh–pH diagram must be calculated using $Am(OH)_3$ as a stable solid. The diagram is shown in Fig. 11.4.

Table 11–2. Free energy data for aqueous Am species.

Species	ΔG_f° (kcal/M)	Reference
Am^{3+}	-143.2	Fuger and Oetting (1976)
Am^{4+}	-89.2	Fuger and Oetting (1976)
AmO_2^+	-177.1	Fuger and Oetting (1976)
AmO_2^{2+}	-140.4	Fuger and Oetting (1976)

Table 11-3. Data for $AmOH^{2+}$ reactions.

Reaction	Ionic strength	$T(°K)$	$\log K$	ΔG_R	Reference
$Am^{3+} + OH^- = AmOH^{2+}$	0.005	298	+10.7	−14.6	Shultz (1976); Shalinets and Stepancy (1972)
	0.005	298	+11.3	−14.9[a]	Shultz (1970); Marin and Kikindi (1969)
$Am^{3+} + H_2O = AmOH^{2+} + H^+$	0.1	296	−5.9	+8.0	Shultz (1976); Desiré et al. (1969)
	0.1	296	−5.92	+8.1	Keller (1971)

[a]Based on $\log K$; $\Delta G_R = 15.4$ by calculation.

Table 11–4. Data for $Am(OH)_3$ and AmO_2 reactions.

Reaction	$\log K$	ΔG_R
$Am(OH)_3 = Am^{3+} + 3OH^-$	-19.6	$+26.7$
$Am(OH)_4 = Am^{4+} + 4OH^-$	-52.1	$+71.1$
$AmO_2 + 4H^+ = Am^{4+} + 2H_2O$	-6.23	$+8.5$

Use of the above data for the solid species yields Eh–pH curves that do not intersect as they should. Use, however, of the following data: AmO_2: -231.0 kcal/M; $Am(OH)_3$: -300 kcal/M (both values from Latimer, 1952) with the newly calculated value of -194.6 kcal/M for $AmOH^{2+}$ and the -142.3 kcal/M value for Am^{3+} (Fuger and Oetting, 1976) do not yield curves that very nearly intersect (Fig. 11-4). It is recommended that Fig. 11-4 be used merely to indicate where the stability fields for combined AmO_2–$Am(OH)_3$ occur at the specified activity of total dissolved americium $= 10^{-8}$ molal.

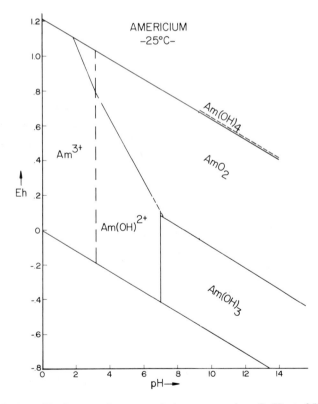

Fig. 11-4. Eh–pH diagram for part of the system Am–O–H at 25°C, 1 Bar pressure.

Also missing from Fig. 11-4 is a field for americium carbonate. Based on carbonate stability for various lanthanides (i.e., $M_2(CO_3)_3$, where $M = La$, Nd, Y), the soluble americium, presumably as $AmOH^{2+}$, should be camouflaged in the lanthanide carbonates at pH's below the stability limits of the oxides–hydroxides.

Curium

Data for curium species are scarce. Fuger and Oetting (1976) report the ΔG_f° for Cm^{3+} to be -124 kcal/M. This is the only reliable value for curium species at this time. For the reaction:

$$Cm(OH)_3 = Cm^{3+} + 3OH^-,$$

Card and Jansen (1975) report solubility data that allow ΔG_R for the above reaction to be estimated at $+4.8$ kcal/mole; i.e., $\log K = -3.49$. From this the ΔG_f° for $Cm(OH)_3$ is calculated at -260.0 kcal/M. In addition, data for the reactions shown in Table 11-5 are known. These data allow the following ΔG_f° data to be calculated:

$$CmOH^{2+} = -194.5 \text{ kcal/M}; \quad Cm(OH)_2^+ = -268.2 \text{ kcal/M}.$$

Unfortunately, the data for $Cm(OH)_3$ are suspect. According to the compilation of Rai and Serne (1978), the species $Cm(OH)_2^+$ should be stable relative to $CmOH^{2+}$ in the vicinity of solid $Cm(OH)_3$; yet the "equilibrium" pH value falls outside the limits of the stability field of water (*Note*: Use of the data from Désiré et al. (1969) yields $\Delta G_f^\circ = -207.1$ kcal/M for $CmOH^{2+}$.)

At this time it is suggested that curium behavior will be similar to americium behavior, but no attempt at constructing even a working Eh–pH diagram can be justified due to the very large uncertainties in the data.

Neptunium

Neptunium species (Fig. 11-5), like the other actinides, are not well known. Apps et al. (1977) and Allard et al. (1977) have presented Eh–pH diagrams for aqueous species, while Bondietti and Francis (1979) and Brookins (1978b) have presented Eh–pH diagrams showing solid species as well.

Table 11–5. Data for Cm reactions.

Reaction	$\log K$	ΔG_R	Reference
$Cm^{3+} + H_2O = CmOH^{2+} + H^+$	-3.4	4.6	Shalinets and Stepanov (1972)
$Cm^{3+} + 2H_2O = Cm(OH)_2^+ + 2H^+$	-9.1	12.4	Shalinets and Stepanov (1972)

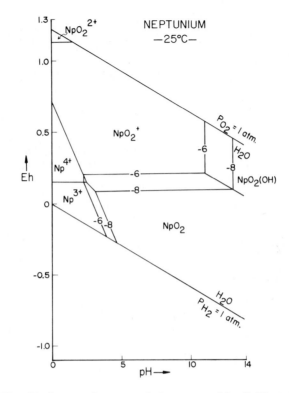

Fig. 11-5. Eh–pH diagram for part of the system Np–O–H at 25°C, 1 Bar pressure.

Frejacques *et al.* (1976) have shown that in the interior of very high grade reactor ore at Oklo the following reactions may have occurred:

$$^{236}U(n,\gamma)^{237}U \qquad \text{and} \qquad ^{238}U(n,2n)^{237}U.$$

both of which would result in formation of ^{237}Np. The amount of calculated ^{237}Np is consistent with the amount of ^{239}Pu predicted to form (Frejacques *et al.*, 1976). Further, small amounts of ^{240}Pu and ^{241}Pu will also form, and some ^{237}Np will form by decay from ^{241}Pu. The end of the ^{237}Np decay chain is stable ^{209}Bi and in the zones where Np is proposed relatively high concentrations of Bi as well as Th are noted [i.e., from ^{235}U(N,γ)^{236}U that decays to ^{232}Th]. Frejacques *et al.* (1976) suggest that there may be a slight depletion of Bi in these zones, which, if real, can be due possibly to either loss of some Np species or to Bi loss.

Neptunium species are very complex, but the solid species NpO_2 and $NpO_2(OH)$ occupy much of the water stability field (Brookins, 1978b). The work of Bondietti and Francis (1979) further shows the importance of the field of NpO_2 for retention of Np under natural conditions. Since the boundary between NpO_2:NpO_2^+ falls well above the probable range in Eh and pH for Oklo ore formation, then Np retention is suggested for at least the 25°C diagram. Radiolytic effects, carbonate complexes, speciation identification in laboratory experiments, and other factors make predictions at higher temperatures very risky; hence no attempt is made here to present a diagram for 200°C. The writer again emphasizes, however, that if Np behavior at higher temperature is closer to that for Pu than for U, then retention as NpO_2: UO_2 solid solution is likely [i.e., the ionic radius for Np(IV) is 1.06 Å, between that for Pu(IV) and U(IV)].

Plutonium

Various Eh–pH diagrams for plutonium species (Fig. 11-6) have been presented by Polzer (1971), Apps et al. (1977), Allard et al. (1977), Rai and Serne (1978), and Brookins (1978f). Except for the diagrams of Polzer (1971) and Brookins (1978f), the others depict only the soluble species of

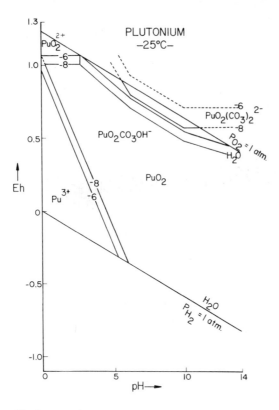

Fig. 11-6. Eh–pH diagram for part of the system Pu–C–O–H at 25°C, 1 Bar pressure.

Pu as Pu(III), (IV), (V), (VI). Problems with speciation are obvious with any attempt to deal with Pu forms, especially in solution. For this reason, the solid species PuO_2 is included in the 25° diagram. By so doing, the diagram may be somewhat oversimplified for many purposes, but is applicable for Oklo. At Oklo, the investigations of Frejacques et al. (1976) and others (IAEA, 1975; 1977) strongly indicate retention of Pu(IV) in host pitchblende. Since the ionic radii of Pu(IV) of 1.04 Å (eight-fold coordination) is virtually identical to that for U(IV) of 1.08 Å (eight-fold coordination), then preferential retention of Pu(IV) relative to U(IV) in MO_2 compounds is to be expected. Brookins (1978f) has discussed the reasons for Pu retention at Oklo even when some U (as U(VI)) may be remobilized, one apparent parameter apparently being the lower oxidation potential for U(IV):U(VI) relative to Pu(IV):Pu(V,VI). Further, as shown in Figs. 11-1 and 11-3, uranium carbonate complexes are important as transporting agents very close to the sulfide–sulfate boundary, whereas for 25°C, plutonium carbonate complexes are not important until very close to the upper stability limit of water. In short, since independent data (IAEA, 1978) suggest some remobilization of U at Oklo, but retention of Pu (Frejacques et al., 1975), the writer proposes that the hypothetical 200°C Eh–pH diagram would also show a large stability field for PuO_2 overlapping the uranium carbonate fields (Figs. 11-2, 11-3) at higher temperatures.

It should be emphasized that the 200°C diagram is not attempted here because of lack of reliable data for species known at 25°C, disproportionation of Pu^{4+}, radiolytic effects affecting many species, uncertainties in identication of species during sorption–desorption experiments, and other factors.

Concluding Statements

In Table 11-6 are the conclusions and inferences that can be made concerning retention or migration of various elements at Oklo; Eh–pH predictions are made for both 25°C and 200°C. The observations column is based on measurements reported by Naudet (1974), IAEA (1975), Brookins et al. (1976), and Gancarz et al. (1980). While both fissile and nonfissile elements are addressed, the point is made that the Eh–pH diagram predictions agree with the observations for most elements.

Oklo Studies

The Oklo Natural Reactor, Gabon, has been studied rather extensively for the last 12 years (see summaries in IAEA (1976, 1978). Conditions necessary for reactor start-up, burn-up, duration, and other physical–chemical aspects of the reactor are discussed in these compilations (IAEA, 1976, 1978). The geology of the Oklo ores, including formation of low-grade ore and high grade ore, has been given in Brookins (1980c) and by Gauthier-

Table 11–6. Important elements at Oklo.

Element	Comments	Eh–pH Predictions 25°C	Eh–pH Predictions 200°C
Krypton	Most migrated	Not applicable	Not applicable
Rubidium	Probable local redistribution	Not applicable	Not applicable
Strontium	Probable local redistribution	Not applicable	Not applicable
Yttrium	Most retained	Retention	Retention
Zirconium	Most retained; some local redistribution	Retention	Some migration
Niobium	Most retained	Retention	Retention
Molybdenum	Most migrated	Migration	Migration
Technetium	Local redistribution	Retention	Migration (?)
Ruthenium	Local redistribution	Retention	Minor migration
Rhodium	Most retained	Retention	Retention
Palladium	Most retained	Retention	Retention
Silver	Most retained	Retention	Retention
Cadmium	Most migrated	Migration	Migration
Indium	Most retained	Retention	Retention
Tin	Not yet studied	Retention	Retention
Antimony	Not yet studied	Possible migration	Some migration
Tellurium	Not yet studied	Retention	Retention
Iodine	Most migrated	Not applicable	Not applicable
Xenon	Most migrated	Not applicable	Not applicable
Cesium	Most migrated (?; perhaps locally)	Not applicable	Not applicable
Barium	Local redistribution	Not applicable	Not applicable
REE	Most retained	Retention	Retention
Lead	Variable migration	Retention or local redistribution	Some migration
Bismuth	Most retained	Retention or local redistribution	Some migration
Polonium	Most retained	Retention	Retention
Thorium	Most retained	Retention	Retention
Uranium	Some local redistribution	Retention or local redistribution	Retention
Neptunium	Most retained	Retention	Retention
Plutonium	Most retained	Retention	Retention
Americium	Not measureable	Retention	Retention

Lafaye (1977). The brief discussion below is from these sources and from Weber (1969). The Precambrian rocks of Gabon consist of granitic massifs with some metamorphic and volcanic parts, all of which are greater than 2 billion years old. Appearing as windows in older Precambrian rocks are folded and faulted sedimentary rocks of the Francevillian Series that have been dated as 1.8 to 2.0 billion years in age (Weber and Bonhomme, 1975; Bonhomme et al. 1978). The Oklo ores consist of syngenetic to early epigenetic sandstone types of deposits formed about 2 billion years before the present (BYBP) (Gancarz, 1978). In many respects the sandstone deposits are similar to deposits in the Colorado Plateau in the United States (Brookins, 1980b). Some of the early-formed ore was remobilized during folding and faulting relatively soon after sedimentation–mineralization. This remobilized ore was transported into zones of locally reducing conditions, where the uranium was again precipitated, in places reaching uranium contents of 50–70%. The uranium remobilized in this fashion was effectively segregated from most of the elements associated with it in the low-grade ore. These other elements, including V, Se, Mo, and most Fe, had they been remobilized with the uranium, would have prevented any nuclear reactions from taking place as they would have served as neutron poisons. The remobilized uranium was free from such poisons, however, except for some Fe (Brookins, 1980b).

The Oklo ores that sustained the fission reactions consist of pitchblende with gangue chlorite and illite. Quartz is rare in the reactor ores, but common in between these zones. Carbonates are in many places later than the reactor ores. Other accessory and gangue minerals consist of barite, chalcopyrite, pyrite, hematite, and others (see Geofroy, 1976). It is important to note that both pyrite and hematite are present, as this indicates that Eh–pH conditions during the reactor's operative lifetime were along the pyrite–hematite boundary; thus Eh–pH diagrams can, with caution, be used to attempt to determine which elements should be retained or migrated at Oklo (see Brookins, 1978a, b, f).

At about 2 billion years ago the ^{235}U content of normal uranium was about 3.2%, in the range to which today's uranium, with 0.7% ^{235}U, is enriched for use in man-made nuclear reactors. Criticality was achieved at Oklo ores because of the presence of a favorable water:uranium ratio and lack of neutron poisons. The water content of the rocks in which the remobilized uranium was precipitated had to have been on the order of 12 to 15%; a lower water content would not be efficient as a medium for neutrons to interact, and at much higher water contents the neutrons would be too diffuse and thus would not interact. Criticality was achieved during or soon after the precipitation of the remobilized uranium, at about 2 BYBP and continued for at least 500,000 years, during which time some 15,000 MWe-years of energy were generated (see discussion by Naudet, 1978). About 800 MT of uranium took part in the reactions at Oklo, and some 10 MT of ^{235}U underwent fission at about 2 BYBP. Smaller amounts of ^{239}Pu and ^{238}U also underwent fission, but most of these isotopes (the ^{239}Pu was produced from uranium by a series of neutron captures on initial uranium) decayed by alpha emission to ^{235}U and ^{230}Th, respectively. The fission reactions presumably ceased due to water being driven off because of the locally high (to 400°C) temperatures attained, and to the buildup of fission products, many of which acted as neutron poisons. Yet in excess of 10 MT of fission products were produced, which are identical to the fission products from man-made nuclear reactors. Furthermore, it has been documented that many of the fissiogenic (i.e., fission-produced) elements produced at Oklo either stayed in their host pitchblende or else migrated locally. In addition, the Oklo ores allow investigation of the actinides (U, Th, Np, Pu) and radiogenic lead as well.

The temperatures attained in Oklo reactor zones probably exceeded 400°C (Holliger et al., 1978; Openshaw et al., 1978) based on rare-earth nucleonics and fluid inclusion work, while temperatures in the low-grade (less than 1% U) ore and in the host rock probably did not exceed 250°C. The point is made, however, that even if temperatures greater than 400°C were attained in the reactor ore, temperatures very close to the reactor ore were not affected by a prolonged high-heat source. Thus in the low-grade ore fossil algae, and illite, which yields $t = 1.8 \pm 0.1$ Gyr by the K–Ar method (Bonhomme et al. 1965), are preserved, and an apparent abrupt temperature gradient as opposed to a mild gradient has been noted in traverses from low-grade rocks to the reactor zones (Openshaw et al., 1978). The depth to which the Oklo deposit was buried is not precisely known; Holliger et al. (1978) estimate burial to approximately 3500 m ($P = 1$ kbar). The geothermal gradient at about the time of reactor ore formation and subsequent diagenetic–epigenetic reactions can only be approximated at about 40–50°C/km.

A word on the age of the Oklo uranium ore and overlying rocks is in order. Bonhomme et al. (1965) have measured a Rb–Sr whole-rock isochron of 1.85 ± 0.1 Gyr for sedimentary rocks that overlie the uranium deposits in Couche 1; these same samples yield the above-mentioned K–Ar date of 1.8 ± 0.1 Gyr. Gancarz (1978) has discussed the significance of U–Pb and ages obtained by REE and Ru studies. A range of dates from 1.8 to 2.1 Gyr has been reported; Gancarz's determination of 2.09 Gyr from samples 2 to 10 m removed from the reactor ore is consistent with an interpretation that the age of the Oklo ore is closer to 2 Gyr rather than 1.8 Gyr. What is significant about the work of Bonhomme et al. (1965) is that the dates were obtained by the K–Ar and Rb–Sr methods on 1Md illites. The K–Ar method is especially sensitive to the type of mineral dated; 1Md illites in particular do not retain ^{*40}Ar at temperatures much above 125°C. Further, 1Md illites will be transformed to higher illite polytypes at relatively low temperatures (150–250°C) for prolonged periods of time or more rapidly to the 2Ml polytype at higher temperatures. Collectively the data from Oklo suggest that burial of the 1Md illite cannot have been greater than a few hundred meters and that temperatures must not have exceeded some 200°C or so. If these conditions were not met, then higher P, T would result in significant $*^{40}$Ar loss and thus $t_{Rb} > t_{K-Ar}$. Again, this is not the case. It is the purpose of this section to discuss elements from $Z = 34$ (selenium) to $Z = 71$ (lutecium) at Oklo and for U, Th, Np, Pu, and Am (actinides), and for Pb and Bi (formed by actinide radioactive decay) as well. The behavior of these elements at the Oklo reactor can then be used to discuss some aspects of geomedia for their possible suitability for nuclear reactor generated radioactive waste.

Behavior of Specific Elements at Oklo

The Oklo reactors can be considered as a moderate temperature hydrothermal system, although a direct comparison of Oklo elemental migration to conventional hydrothermal systems is difficult. At Oklo fissiogenic elements were formed within host pitchblende grains by combined fission of ^{235}U, ^{239}Pu, and ^{238}U, with ^{235}U being the most important. The ion imagery and other studies (IAEA, 1976) document that even now, some 2 billion years after reactor shutdown, fissiogenic isotopes of different elements are present in amounts close to 100% of original (REE, Th) to less than 5% (Rb, Sr). For those elements in which migration has occurred, the mechanisms for migration are at best poorly known. Elements not dependent on sulfide (including complexes of the type S_x^{z-}) are best described in terms of a groundwater transport medium; this group includes Rb, Sr, Zr, Nb, Sn(?), Cd, Ag(?), Cs, Ba, REE, and other, nonchalcophile elements. The Oklo conditions for chalcophile elements, even under sulfide-stable conditions, are not precisely known. In theory, at temperatures near 400°C (pressure variable; from very low to 300 bars), a typical hydrothermal solution (Barton and Skinner, 1979) would contain large quantities of total dissolved Cl, CO_3, S(2−), NH_3, and OH, while at lower temperatures $S_2O_3^{2-}$, SO_3^{2-}, CN^-, SCN^{3-}, and polysulfides would be important species. Oklo rocks have not been characterized as to any N-bearing species; sulfite and thiosulfite ions are short-lived and would be disproportionate to sulfide and sulfate ions. Further, flurocarbonate complexes are common to many hydrothermal systems but are not suspect at Oklo.

The Oklo system depends on the following conditions for elemental transport to take place: (1) transfer of the element from the interior of the host pitchblende to the grain surface, or transfer from the interior of the grain via a fracture; (2) desorption or other processes for removal of elements concentrated at the grain surface; (3) transport of the elements away from the grain surface as unassociated ions or with complexing ligands; (4) duration and length of transport will depend on total rock porosity, permeability, tortuosity, saturated vs. undersaturated conditions, kinetics of nucleation, resistance to secondary dissolution/desorption (i.e., dependent on fixation processes), and other factors. Thus the Oklo system is only partly like a conventional hydrothermal system. It is similar in that locally hot-water-rich solutions were present in the zones of the nuclear reactions, and that convective cooling of these fissiogenic-element bearing solutions resulted. The Oklo system is very unlike most hydrothermal solutions in that the fissiogenic elements of interest were generated *in situ* and not brought in from some outside source. In a sense the Oklo system is like a very small, finite reservoir for some elements transported under conditions of rapidly cooling aqueous solutions—very different from very large, essentially infinite reservoirs for conventional hydrothermal solutions.

The mechanisms for transfer of fissiogenic elements to the surface of, or removed via fractures from, the host pitchblende is probably due to a combination of diffusion and dissolution. A strong argument for a diffusion-controlling mechanism can be made based on the fact that those elements with similar crystal chemical properties (ionic charge, radius, electronegativity), such as the REE, Bi, Te, and Nb, are retained in host pitchblende in amounts of, or close to, 100%. The diffusion data for these elements in silicate–oxide matrices indicate diffusion processes to be too slow to allow any significant migration of these elements in the pitchblende crystal structure. Further, even where some dissolution (due, perhaps, to fracturing concomitant with introduction of solutions into the grains) has occurred, the solubility product constants for the REE, etc., oxides and hydroxides are so small that any release material would precipitate at the point of release from the pitchblende. Elements metastable in the pitchblende structure (Mo, Rb, Sr, Cd, Ag, Pd, Sb, Sn, Cs, I, Ba, etc.) may or may not be removed from the grain depending on the solubility of new phases that may form from release of the ions coupled with ions in solution. For example, any released Ba as Ba^{2+} would presumably react with SO_4^{2-} to form barite. Alternately, if the SO_4^{2-} content of infiltrating fluids is so low that the ion activity product of $(Ba^{2+})(SO_4^{2-})$ is much less than the K_{so} for barite, the Ba^{2+} will be free to move from the pitchblende grain. If so, it will either be incorporated onto clay mineral surfaces, replace Ca in carbonates, or be affixed on gangue barite. All these possibilities exist at Oklo.

For every other fissiogenic known or suspected to have migrated from host pitchblende a completely independent series of release, transport, and fixation mechanisms may be involved.

Fissiogenic Elements at Oklo: Elements from Z-34 to Z-71

M_f = fissiogenic isotopic composition,

M_n = normal isotopic composition,

M_t = total amount of element.

Selenium. This element has not yet been studied at Oklo, although behavior similar to sulfur would be expected. Sulfur was present in the system during the reactor lifetime and subsequent to its shutdown, hence compounds of the type $M(S,Se)$ newly formed in fractures may contain Se_f.

Bromine. This element has not yet been studied. Because Br^- is incompatible in the pitchblende structure, migration should occur. Fixation of Br_f may be at virtually any distance from point of origin, depending largely on anion exchange capacity (AEC) of clay minerals in the gangue. This element should be studied because the ratio 79:81 in Br_f is approximately 1:3 as opposed to 1:1 in Br_n, and thus indirect information on I_f can be obtained as well, since much of the chemical behavior of halides is identical.

Rubidium. Migration of Rb_f is well known (IAEA, 1976). At the same time, it has been argued that the migration of Rb_f may in fact be quite local as opposed to far removed from the reactor zones (Brookins, 1981). This is based on the fact that Rb_f is found in, at the edge of, and withing a few meters of reactor zones. Since the material removed from the reactor zone is mixed with Rb_n, if the total amount of Rb_n is considered relative to the amount of Rb_f lost from the pitchblende, then the slight increments of Rb_f detected in these samples suggest that up to 60% of the Rb_f may be fixed within a few meters of point of origin (Brookins, 1981a).

Strontium. Migration of Sr_f is also well known. As is the case of Rb, Sr_f has been found in the reactor ore, at the edges of ore zones, and as adsorbed material on some samples removed from the reactor zones (Brookins, 1981a). In brief, in the reactor zones the remaining Sr_f is structurally bonded in the pitchblende, and Sr_f is found in the insoluble residue of leached samples. At the edges of reactor zones, more Sr_f is found in the leachate fraction, suggesting fixation by adsorption on clay minerals or possibly in $(Ca, Sr)CO_3$ phases. Only small amounts of Sr_f have been found in samples away from the reactors, but, again, when mixes of Sr_n (infinite reservoir) and Sr_f (small, finite reservoir) are considered, the preliminary data argue for loss of Sr_f from reactor ore but fixation closeby (Brookins, 1981a). Transport for Rb and Sr is likely as Rb^+ and Sr^{2+} at concentrations of 10^{-8} (or less) for Oklo samples.

Yttrium. No evidence for migration of Y_f has been noted and, from crystal, chemical and aqueous chemical considerations, this is consistent with theory. Y^{3+} substitutes easily in the structure of pitchblende for U^{4+} and, if leaching occurs, $Y(OH)_3$ (which will age to Y_2O_3) will precipitate. Further, $Y_2(CO_3)_3$ mixed with other carbonates in gangue would prevent migration even if this occurred.

Zirconium. Zr_f migration on a very local scale may have occurred (Frejacques et al., 1976) although Brookins (1978a, f) has shown that most Zr_f has been retained in the Oklo samples. Zr^{4+} is metastable in pitchblende, but due to its charge, is commonly diadochic with U^{4+} despite a 20% smaller ionic radius. Migration as $Zr(OH)^{2+}$ at low pH and as ZrO_3^{2-} at alkaline pH may result in limited migration, but the K_{so} for $Zr(OH)_4$, ZrO_2 and even $ZrSiO_4$ are so small as to indicate fixation near any hypothetical point of release.

Niobium. No evidence for Nb_f migration has been reported. Since Nb^{5+} is extremely stable in nature, this is expected. Further, any Nb_f released from destruction of pitchblende will either precipitate as a Nb-oxyhydroxide or be incorporated into an Fe-bearing phase.

Molybdenum. Mo_f migration is expected (Brookins, 1978f). Mo^{6+} occurs as either MoO_4^{2-} or $HmoO_4^-$ under slightly oxidizing to strongly oxidizing conditions at T = 200°C and above. The small field of Mo_3O_8 stable in Eh–pH space at lower temperatures has disappeared by 125–150°C. Under more reducing conditions, Mo^{4+} may be stable if incorporated into a sulfide phase, but some data suggest transport of Mo^{4+} complexed with polysulfide. Mo_f is found removed from the reactor ore at Oklo (D. Curtis, personal communication), but not removed and disseminated so much that it cannot be detected. This argues for rather local migration.

Technetium. Tc_f migration is well known. ^{99}Tc decays to ^{99}Ru and zones of ^{99}Ru-enrichment have been identified. Transport of Tc^{4+} is problematic, however, due in large part to lack of critical data. Transport as TcO_4^-, while of importance under surface, high dissolved $O_2(g)$ conditions, is unlikely at Oklo, as TcO_2 is stable well into the hematite Eh–pH field (Brookins, 1978f) under sulfate-stable conditions, and TcS_2 should be stable under sulfide-stable conditions. The Eh–pH diagram for Re species supports this predicted behavior of Tc. If high Eh conditions were common at Oklo, then elements retained in the Oklo ore, such as Pd, Te, In, and Pu, should have migrated, but this is not the case. Overall reducing conditions may allow an argument for transport as a polysulfide complex, but this is admittedly speculative. Most important, though, is that Tc_f transport appears to be confined to within several meters of the reactor ore (Gancarz et al., 1980).

Ruthenium. Ru_f has locally migrated at Oklo. It is fixed in clay seams that cut the reactor ore and in sulfides within 10 m of the reactor ore (Gancarz et al., 1980). These samples also contain Tc_f, indicating similar transport mechanism. Ru_f transport as an oxyion at Oklo is questionable since too high an Eh is required. Since RuS_2 is found in nature, and Ru is found in high concentrations in other sulfides, transport as a polysulfide is possible. Regardless, the extent of migration away from the reactor zones is small.

Rhodium. Rh_f has not migrated at Oklo. Rh^{3+} is compatible with U^{4+} in pitchblende and $Rh°$ is stable under the Oklo Eh–pH conditions indicated (Brookins, 1978f).

Palladium. Pd_f has not migrated at Oklo. Recent studies (DeLaeter et al., 1980) indicate retention of Pd_f even where other elements are known to have migrated. This is of importance, because if polysulfide complexes of a simple type are responsible for transport of Tc_f, Mo_f, and Ru_f why, then, should Pd_f also not be transported in a similar fashion? The answers to this question may lie in greater stability of $Tc-S_x$, $Mo-S_x$, and $Ru-S_x$ complexes than for

Pd–S_x, but there are virtually no data for any of these species although there are limited data for Mo.

Silver. Ag_f has migrated on a local scale in some Oklo samples (DeLaeter *et al.*, 1980). This migration, although limited, supports the concept of polysulfide complexes for migration, because the Eh–pH diagram indicates a large stability field for $Ag°$ with smaller fields for Ag-sulfides. Since Ag^+ is a very large ion, it is difficult to incorporate into simple, newly formed sulfide phases and would be expected to be excluded from such phases. The extreme differences in equilibration times of formation versus temperature (Barton and Skinner, 1979) for Mo and Ag argue against similar behavior in simple sulfide systems, and that Ag_2S will equilibrate much more rapidly than MoS_2. At Oklo this may explain the absence of Ag_f in Mo_f zones away from the reactor. The Ag_f should be closer to the point of release, but no data are available to test this hypothesis. Only part of the Ag_f has migrated, however (DeLaeter *et al.*, 1980), thus also suggesting retention in pitchblende as $Ag°$ and $Ag_2S(?)$.

Cadmium. Cd_f migration is well known at Oklo (IAEA, 1976). The easiest explanation is based on Eh–pH arguments (Brookins, 1978f) for transport as either Cd^{2+} or CdO_2^{2-} under the slightly oxidizing conditions proposed. The small field of $Cd(OH)_2$ stable at near 25°C disappears well below 200°C. Yet transport as polysulfide under more reducing conditions is also possible. In such a case, segregation of Cd_f while transported as Cd^{2+} (or CdO_2^{2-}?) under sulfate-stable conditions from Tc_f, Ru_f, and possibly Ag_f may result. If true, then Cd_f may occur with Mo_f or, due to the dissimilarities between Cd(II) and Mo(IV, VI) transport (i.e., as a simple M^{2+} species for Cd_f but as oxyion for Mo_f) they may, too, be segregated. Further, it is not known whether Cd_f is concentrated more in Mo_f-bearing or Ru_f, Tc_f-bearing zones. Unfortunately no studies of barren samples for Cd_f have yet been undertaken. Finally, in support of at least partial transport as a polysulfide data are the recognition and data for $Cd(HS)_2$, $Cd(HS)_3^-$, and $Cd(HS)_4^{2-}$ (Barton and Skinner, 1979).

Indium. In_f has been retained at Oklo. While this is consistent with Eh–pH predictions (Brookins, 1978f), again the problem of polysulfide complexes for transport of some chalcophile elements and not for others is raised. Presumably In_f is more stable than Mo_f and Cd_f in Oklo ores because In^{3+} is not only diadochic with U^{4+} but, even if released, either In_2O_3 or In_2S_3 will form, depending on Eh. Like other M^{3+} and M^{4+} chalocophile elements, polysulfide complexes are not well understood.

Tin. Some local migration of Sn_f has been reported (DeLaeter *et al.*, 1980). Sn is chalcophile (as SnS_2) although SnO_2 is the dominant Sn-bearing phase in nature. The Eh–pH diagrams show an extremely large stability field for SnO_2 that probably encompasses the Oklo U-ore field, yet some migration has been noted. The ionic radius of Sn^{4+} is considerably smaller than U^{4+}, hence retention of Sn_f in host pitchblende is difficult. The fact that 40–60% of Sn_f has been lost from some Oklo samples (DeLaeter *et al.*, 1980) is thus not surprising, and it is equally probable that Sn_f may be located very close to the host pitchblende, possibly fixed as an oxyhydroxide. Alternately, transport with other chalcophile elements is possible but must await further study.

Antimony. From Eh–pH considerations, some migration of Sb_f may occur. The field of Sb_2O_3 above Sb_2S_3 is fairly small (Brookins, 1978f), and transport as SbO^+, $HSbO_2$ are possible under sulfate-stable conditions, and as a polysulfide under sulfide-stable conditions. The similar behavior of Hg and Sb is well known and, in the absence of halides, transport as polysulfides is probable. Despite the importance of Sb_f from man-made reactors, Oklo ores have not been studied for distribution of Sb_f.

Tellurium. Data for complexing ligands for Te are poor (Barton and Skinner, 1979). The available data for Oklo samples, however, shows that Te_f has been retained in ores, presumably in the pitchblende (DeLaeter *et al.*, 1980). This is in agreement with both Eh–pH and crystal chemical arguments (Brookins, 1978f).

Iodine. Migration of I_f at Oklo is well known (IAEA, 1976), but the path and fate of the I_f has not yet been established. The sulfides containing Ru_f and/or Mo_f should be studied for the presence of Br_f and/or high concentrations of I_t. Work on sulfides (Allard *et al.*, 1980) has shown that I_f can be incorporated into sulfides and, if I_f is transported with M_f, then both may be incorporated in the same zones. Alternately, any clay mineral with a proper AEC may incorporate some I_f, possibly where M_f elements are fixed under sulfate-stable conditions. The clay seams known to contain M_f should be tested for the presence of I_f. Anywhere a high I_t content is noted, if I_f is in part responsible, it can be tested by searching for Br_f.

Xenon. The few data for xenon suggest extensive loss from the Oklo ores (Drozd *et al.*, 1974). This is not surprising, since any noble gas would not be expected to be retained in such high-permeability, high-porosity sedimentary rocks such as those at Oklo. On the other hand, it has not yet been unequivocally demonstrated that xenon, and krypton as well, are not fixed in part in zones of secondary mineral formation (i.e., Christensen *et al.*, (1981) have shown fixation of Kr in zeolites and other geologic materials).

Cesium. Migration of Cs_f is also well known. The Rb_f studies (Brookins, 1981) and data for Ba_f suggest retention of Cs_f until after reactor shutdown, with subsequent loss of $^{133}Cs_f$ after perhaps 25–30 MY. If so, then Cs_f may be found in zones enriched in Rb_f, but the problem is not as easy because $Cs_f = Cs_n$ for this monoisotopic system. High Cs_t concentrations in which Rb_f is found may indicate presence of Cs_f *if* reasonable background values for Cs_n can be established. The various clay minerals with high CEC should be examined, especially where other M_f have been identified.

Barium. Probable retention of Ba_f at Oklo has been proposed by several investigators (IAEA, 1976). Although only questionable amounts of $^{138}Ba_f$ have been reported at Oklo, $^{135}Ba_f$ (from $^{135}Cs_f$) and $^{137}Ba_f$ (from $^{137}Cs_f$) have been detected (Brookins et al., 1976; Brookins, 1981a). The high background of Ba_n makes detection of Ba_f extremely difficult, because barite and $(M,Ba)CO_3$ as well as Ba-bearing clay minerals have also been mentioned. Ba_f will, upon release from host pitchblende, travel as Ba^{2+}. This ion will be fixed in sulfate-bearing waters assumed to be important under slightly oxidizing conditions at Oklo, and Eh–pH considerations argue for coexistence of sulfide and sulfate under some of these conditions as well. If any Ba^{2+} should escape inclusion in barite, then capture in carbonates or, in the absence of carbonate, adsorption on clay minerals, is likely.

Rare earths. The REE have been essentially 100% retained at Oklo (IAEA, 1976, 1978). This is important as not only is their immobility under Oklo conditions noted, but some limits on species in solution are implied as well. If a phosphate complex were present as a transporting agent, then complexes of the type $M_2P_2O_7^{2+}$ should effectively transport the REE (and Th). Instead, the retention of the REE is thought to be due to their ability to substitute for U^{4+} in host pitchblende, with the LREE more easily fixed in the structure than the HREE. Any released M^{3+} will, for total dissolved phosphate at activities greater than 10^{-3}, precipitate as M_2O_3 or $M(OH)_3$. If excess amounts of total dissolved carbonate are present, than $M_2(CO_3)_3$ will form. Transport as a carbonate complex is *not* considered likely. The recent literature (McLennon and Taylor, 1979) arguing for the importance of carbonate complexes to transport REE, Th, and U is not supported by reinvestigation of the original sources (Brookins, unpublished).

Nonfissiogenic Elements at Oklo

Lead. Very local migration (to 3 m) of *Pb has been noted at Oklo (Gancarz et al., 1980), and veins of *Pb-enriched galena cutting primary pitchblende and gangue have also been reported (IAEA, 1976). The first study indicates migration of *Pb in a path different from that of Ru_f (and Tc_f), but this is not surprising since Ru_f, Tc_f migrated during the reactor's operative lifetime, but *Pb formed during that same interval of approximately $5–10 \times 10^5$ years would be impossible to measure. Volume diffusion of *Pb from host pitchblende, probably as Pb^{2+}, is metastable in the UO_2 structure. The fixation of this *Pb in and near the host ore indicates only limited migration, although no comment on transporting mechanism can be made without further study. Samples of barite could be tested for the presence of *Pb to see if long-duration migration of Pb has occurred. Further, the *Pb-bearing galena should also be tested for other chalcophile elements to see if any M_f can be found: this, in turn, would imply migration well after reactor shutdown.

Bismuth. Retention of *Bi is indicated by local high concentrations of Bi in those parts of reactor zones where ^{237}Np formed (IAEA, 1976). The *Bi correlates with *Th and proposed ^{239}Pu-enrichment as well (IAEA, 1976). Retention of Bi is due to crystal chemical considerations, stability fields of Bi_2O_3 and Bi_2S_3, and the Eh–pH observation that aqueous species such as BiO^+ and $Bi(OH)_2^+$ occur at pH values outside the probable Oklo field (Brookins, 1978b).

Thorium. Th_n and *Th are compatible in the UO_2 structure (i.e., ThO_2–PuO_2–UO_2 all show extensive solid solution and possess identical structures with equal space groups). A low total dissolved phosphate content is indicated, because it has been shown that Th(aq) can be transported as a Th-phosphate complex. The extreme stability of Th_t in the Oklo ores is also attested to by retention of both Th_n and *Th (from ^{236}U and minor amounts of ^{240}Pu).

Uranium. Uranium migration of both reactor and low-grade, nonreactor ore is known (IAEA, 1976). Post-900 MYBP, some U_t migration occurred as southerly trending solutions remobilized some uranium that was concentrated on the north side of a 900-MYBP dolerite dike that cuts normal to the south–southeasterly trending ore lineaments. This remobilized U_t zone has not been studied for the presence of other M_f. The low-grade U_n could easily be transported as a carbonate complex, because the abundance of late- to post-reactor carbonate mineral formation is well established (IAEA, 1978). How much reactor U, if any, was transported is unknown. This is also an important point, because if transport of reactor U has been established, then other, carbonate complexed, M_f migration should be possible.

Transuranics. Retention of Np, Pu is documented both during the reactors' lifetimes and for at least 25 MY after reactor shutdown (IAEA, 1976). In addition, Eh–pH and crystal chemical considerations argue for retention of Am as well (Brookins, 1978b), and thus hypothetically formed Cm would also have been retained. The suite Np–Pu–Am–Cm as oxides (IV) are all isostructural with UO_2 and none form carbonate complexes as does U. The reader is referred to the discussion of these elements as the beginning of this chapter.

Comments on Oklo Studies:
Relevance to the U.S. Radwaste Program

The U.S. radwaste program requires knowledge of the behavior of radio-nuclides in various geomedia. Laboratory studies, despite the often high quality of scientific control, suffer in part due to the necessarily short duration of the experiment (i.e., hours to years when isolation requirements are often on the order of 10^4–10^5 yrs) and in part to the lack of characterization of the starting materials (i.e., different samples of a rock unit may be pulverized and subjected to quantitative leaching without first establishing whether or not the samples are truly identical, separate aliquots of the rock unit; thus different sorption data for supposed aliquots of the same rock unit may actually be due to different mineralogies and/or chemistry and, if corrected, yield equal sorption data). Further, geochronologic studies at the Oklo site provide useful data on closed or open systems for U, Th, Pb, Rb, Sr but not for other elements except by inference.

Oklo remains the only naturally occurring fossil fission reactor documented to date. The reactor was operative at least at part in water-saturated conditions, and no "canister or engineered backfill" equivalents were present for much of the reactor zones. Indeed, the reactor ore is analogous to SURF + HLW + TRU and some LLW. The host rocks are conglomeratic in part, arenaceous in part, amplitic in part, and monomineralic in part. The porosity and permeability of the overall Oklo geologic site are widely variable, and the rocks have been folded, faulted, and jointed. In short, these rocks would not be prime candidates for any realistic radwaste program under consideration today. Yet the fission products for the most part remained in or close to sites of formation. Some local migration of M_f occurred during the reactors' lifetimes, and some M_f and $*M$ migrated locally after reactor shutdown. Yet much of this migration is very local. While transport mechanisms and agents are not known, limits can be placed on them. The limits suggested would apply to *any* rock under consideration for radwaste storage and should be treated accordingly. The temperature distribution about the Oklo ore zones can be taken as approximately 400°C for the average high, and 200°C at a distance of about 1 m; ambient temperature is unknown but presumably is on the order of 50°–100°C close to the reactor zones. Pressure estimates vary from 0.3 to 3 kbars, but the effect of lithostatic pressure on solid and aqueous phases is not as important as temperature variation and chemical speciation for most elemental systems. With these assumptions in mind, the following statements can be made.

(1) The role of phosphate and carbonate complexes was of minor, if any, significance, but some U transport as UDC is possible.

(2) Transport of some species as simple unassociated ions or as oxyions under slightly oxidizing conditions is probable, but fixation occurred nearby either as more reducing conditions and lower temperatures were encountered and sorption or precipitation occurred in favorable sites.

(3) Transport of some of the chalcophile elements as polysulfides under lower (i.e., sulfide stable) conditions is likely, with precipitation caused by dilution.

(4) Clay minerals may have served as effective getters for some M_f either by ion exchange or sorption.

(5) I_f and Br_f behavior is unknown, but has not been adequately investigated.

(6) The Oklo hydrothermal system was low in chloride, ammonia, and CN, and relatively low in sulfur; thus migration was limited (i.e., Cl

complexes are more effective transporting agents for some hypothetical M_f than polysulfides, etc.).

(7) With continued support and interest, quantitative limits can be established for M_f distribution, pathways for migration, conditions for M_f concentration–fixation at new sites, and inferences for rates and species for essentially *all* radionuclides of interest to the U.S. radwaste program.

Concluding Statements on Oklo

Radionuclide migration based on laboratory studies cannot, in most instances, be extrapolated to the 10^4–10^5 years required without introducing as yet untested assumptions. Studies of elements of normal isotopic composition in terms of fixation, remobilization, etc., are extremely useful; there are too few of these and, further, threshold values are often so poorly defined that unequivocal retention vs. migration cannot be determined. Only Oklo provides rocks in which the fate of fissiogenic, radiogenic, radioactive, and radioactive transuranics can be evaluated. While admittedly there is much to do to fully document the Oklo site, the observations presented herein clearly show that the Oklo rocks have served remarkably well to prevent loss of fissiogenic elements, actinides, and lead and bismuth, for the time interval from 2 BYBP to the present. Further, where migration of elements has been noted, the migration is either known or thought to be extremely local; and some of this migration (i.e., Cs) did not take place until some 25 million years after reactor shutdown. In short, rocks with lower permeability and porosity and an overall chemically reducing environment should be even better suited for radioactive waste retention than those at Oklo.

Studies of the Eldora–Bryan Stock: Idaho Springs Formation, Colorado

The contact zone of the Eldora–Bryan Stock and the intruded rocks of the Idaho Springs Formation have also been tested as an analog for radwaste storage in crystalline rocks. Samples from the stock, and from the stock into the intruded rocks to a distance of 5000 m have been studied. Petrography, Rb–Sr and K–Ar geochronology, neutron activation analysis, oxygen isotopic analysis, and fission track studies have been conducted in the area. The reader is referred to Hart *et al.*, (1968) for discussion of the early studies of this well known area, and to Brookins *et al.* (1982b), for the sample descriptions and locations and chemical data. The radiometric data have been explained in terms of a model that assumes the stock is a 8000 m by 3000 m magma brick with an initial temperature of 780°C; contact effects are noted to approximately 8000 m. Their estimate of cooling time for the intrusive body is 10^5 to 10^6 yrs., which covers the 2.5×10^5 retention span for Pu-bearing repositories. Since the stock is very different chemically from the intruded rocks, study of the whole rocks of the Idaho Springs Formation allows testing of elemental migration as a function of heat–distance from the contact without benefit of canister of engineered backfill.

Rb–Sr Geochronology

A 1.4 BYBP age for five feldspars 2 to 7100 m from the contact has been reported (Hart *et al.*, 1968). The sample from 7 m falls above the isochron (Hart *et al.*, 1968, Fig. 8) and may represent a reset mineral point. New data (Brookins *et al.*, 1982b) show a 1.5 ± 0.1 BYBP age based on five carefully chosen whole rocks. Five mineral-dominated samples as well as four samples from the stock are plotted. The mineral-dominated samples exhibit wide scatter as expected, and even the stock samples show considerable spread in the $^{87}Sr/^{86}Sr$. For the former such behavior is due to redistribution of $^{*87}Sr$ during the contact metamorphism; for the latter no ready explanation is offered except to point out that the same observations have been made by others (Simmons and Hedge, 1978), who report a range in $^{87}Sr/^{86}Sr$ from 0.7054 to 0.7088. U–Pb dates of 1.5–1.6 BYBP for zircons from the Idaho Springs Formation and 1.2–1.4 BYBP dates for secretion pegmatites have also been reported (Hart *et al.*, 1968). The latter presumably formed during the strong regional metamorphism that affected the Idaho Springs Formation. Newer data (Brookins *et al.*, 1982b) are more important to stress the point that the Rb–Sr whole rock systematics have not been affected by the intrusion of the stock. Had either Rb-fixation or $^{*87}Sr$-loss from whole rocks occurred, then the whole rock data would behave like mineral systems and, as usual, fall below the isochron. That this is not the case reaffirms the isotopic and chemical closed system integrity of the whole rock samples.

REE Distribution Patterns

Brookins *et al.* (1982b) have shown the REE/chondrite normalized REE (lanthanide) distribution patterns for the Eldora–Bryan stock samples from the 0–3-m contact zone (including some stock samples), samples from the Idaho Springs Formation, and samples from two of the secretion pegmatites cutting the Idaho Springs Formation. Data for two samples of the Eldora–Bryan stock have been given (Simmons and Hedge, 1978). The data show rather normal REE/chondrite distribution patterns with pronounced enrichment of the light REE (LREE) and a slightly negative Eu anomaly; some samples show a very slight enrichment in the heavy REE (HREE). In near-contact samples, the stock samples show normal enrichment of the LREE and a somewhat more pronounced Eu negative anomaly; the embayed Idaho Springs Formation sample shows no infiltration of the REE from the stock into it (Brookins *et al.*, 1982b). The REE patterns for embayed Idaho Springs samples, however, show enrichment of the LREE and depletion of the LREE in the mafic and felsic parts, respectively. This same pattern is observed at all distances from the contact and is possibly due to REE partitioning during regional metamorphism at 1.2–1.4 BYBP, and not due to intrusion by the Eldora–Bryan Stock. This is further substantiated by the REE distribution patterns for samples from the Idaho Springs Formation further from the contact (Brookins et al., 1982b, Fig. 5), where both negative and positive Eu anomalies as well as local Ce anomalies are noted. Further, adjacent pegmatites show extremely different REE patterns; some samples are greatly enriched in the REE and show a major negative Eu anomaly while some samples show both positive Ce and Eu anomalies although the overall REE enrichment is far less than that for REE enriched samples. These data (Brookins et al., 1982b) argue for internal REE patterns in the Idaho Springs Formations to be controlled by prestock segregation, probably caused by the regional metamorphism at 1.2–1.4 BYBP. Only the samples actually infiltrated by stock material show stock-like REE patterns consistently.

Other Studies

The stable O isotopic data (Brookins et al., 1982b) show a range in $del^{18}O$ from 7.6 to 10.0 o/oo for the Idaho Springs Formation and a much more narrow range of 9.0 ± 0.3 o/oo for the igneous rocks of the Eldora–Bryan stock. There is thus no evidence for whole rock O isotopic exchange in the intruded rocks even in the immediate contact zone (Brookins, et al., 1982b). This, in turn, is supportive of cooling by conduction rather than by convection (Brookins et al., 1982b).

Chemical data reported (Brookins et al., 1982b) show variations in the Idaho Springs Formation felsic vs. mafic units at all distances (to 5000 m) from the contact. This is attributed to the regional metamorphism at 1.2 to 1.4 BYBP. No evidence for movement of Th, Cr, Co, Sc, Rb, Sr, U, and Fe from the stock into the intruded rocks is noted, however. Further, Hart et al. (1968) have pointed out that radiogenic Pb from the stock is not detected in intruded rocks more than 3 m from the contact.

Concluding Statements on the Eldora–Idaho Springs Problem

Our conclusions are: (1) The Idaho Springs Formation was not penetrated by hydrothermal fluids from the Eldora–Bryan magma except possibly on a local scale within 4 m of the contact (and possibly 3 m for Pb). (2) The light lanthanides may be locally redistributed in the immediate contact zone, but without additions from the magma. (3) The oxygen isotopic data argue for lack of hydrothermal fluids from the magma penetrating the intruded rocks, even in the highest-temperature contact zones (except where apophyses of the stock cut the host rocks). (4) Previous and new geochronologic studies argue for loss of radiogenic Ar from some minerals, and redistribution of radiogenic Sr and Pb in others, both due to diffusion, which is in turn due to heat from the magma. Whole rock data argue for closed-system conditions for Rb, Sr, Th, U, and Pb even where mineral ages have been lowered. (5) Data for Co, Cr, Sc, Fe, and Cs also indicate retention in whole rocks systems and no exchange with the magma. (6) The Idaho Springs Formation has been strongly metamorphosed, yet apparently readily accessible conduits for fluid transport (i.e., cleavage traces, zones of secretion petmatites, cross fractures) were not invaded by fluids from the magma. This is significant when it is noted that the contact was without benefit of canister or engineered backfill, and that even thermally induced mineral recrystallization was insufficient to allow fluid penetration. (7) The combined chemical, isotopic, petrographic, and theoretical data and calculations indicate suitability of rocks of the Idaho Springs Formation, and thus for many types of crystalline rocks as well, for their possible use for the storage of radioactive waste.

Alamosa River Stock-Platoro Complex Studies

The Alamosa River Stock is an Oligocene monzonite (29.1 ± 1.2 MYBP) intrusive into the La Jara Tuff and Summitville Andesite (both also earlier Oligocene) of the Platoro Complex, San Juan Mountains, Colorado. Lipman (1975) has described the area in general, and Williams (1980) has reported on the oxygen isotopic systematics of the Alamosa River Stock and its host rocks. Williams (1980) has shown that the contact zone rocks, well out into the intruded rocks, have been subjected to hydrothermal alteration involving, in part, widespread oxygen isotopic exchange between circulating waters and constituent minerals of the rocks. In his study of 260 minerals and whole rocks, he showed that over a 100-km^2 area the igneous rocks are depleted in del^{18}O, and that this depletion is due to exchange with even lower del^{18}O-bearing circulating fluids. The water:rock (stock mass) ratio calculated by Williams (1980) is 0.2 for the area. Convective cooling of the system is shown by Williams (1980). The thermal events proposed by Williams involve temperatures in the range of 250–370°C, and were active perhaps over 1–2 million years post-stock emplacement. Recently, the author and colleagues (see Brookins et al., 1983) have studied several traverses made from the stock into the altered tuffaceous and andesitic rocks for possible elemental exchange due to the convective system. Samples, analyzed for their chemistry by neutron activation analysis, show favorable chemical gradient between the rocks for several elements, U, Th, K, Cr, Co, Ni, Zr, Ti, Sr, Ba, V, and others, and near-equal abundances of the REE in the stock and in the tuff. Despite the favorable chemical gradients, no elemental migration between stock and tuff has occurred, except in a 1-m zone at the contact, where some evidence for physical infiltration of the tuff by stock is observed. Brookins et al. (1983) calculate elemental migration rates for Cs, Co, and Cr of no more than 10^{-5} moles/cm^{2}y, even with circulating waters. Further, the REE are unaffected by the intrusion and the subsequent hydrothermal activity, even in the 1-m affected zone, and Th and U show no redistribution as well. The Sr isotopic systematics have also been examined in the rocks (unpublished data), with no evidence found for Sr migration or for isotopic perturbations. Collectively, the data show that the host rocks to the intrusive monzonite, and the monzonite itself, behaved as closed systems to most elements despite the oxygen isotopic exchange noted.

Other Analogs

Several other analogs have been studied in conjunction with radwaste investigations, and other studies provide indirect information of value to the program for radwaste isolation. The intrusion of a lamprophyre dike into the evaporitic rocks of the WIPP area has been discussed in Chapter 7. In brief, although the dike was emplaced at a high temperature, and without benefit of any protective rim, there was no transfer of elements from dike into the evaporite (see Brookins, 1981c). Further, the evaporitic rocks were largely unaffected by the intrusion. Fluid inclusions were locally affected, but there was no large-scale migration of fluids toward the dike (i.e., no fluid transport toward the heat source). Recrystallization of clay minerals and salts occurred within 1–2 m of the contact, but without elemental transport from the dike. This is significant, as the evaporites are highly depleted in the REE, U, and Th and yet these elements were retained in the dike.

The Notch Peak pluton, a quartz monzonite intrusive into a carbonate sequence mixed with shales, has been studied as an analog by Laul and Papike (1981). While the investigators suggest some evidence for migration of the REE from the stock into the carbonates, there are no supporting isotopic data to enforce their interpretation. Lithologic variations would account for the patterns noted (Brookins, unpublished).

Shea (1982) has studied the uraninite vein deposits of the Marysvale Mine, Utah, and concluded that some U, within a few millimeters of the contact of the vein with metamorphic rocks, has diffused outward from the vein. The time over which this movement occurred is unknown, and could range from a few tens of years to perhaps thousands. Regardless, the distance covered by the U is on the scale of millimeters only.

In the 1960s and part of the 1970s, a number of studies of igneous contact zones were made, primarily to test for conditions of equilibrium and geochronologic variations as a function of distance from the igneous contacts. Brookins and Dennen (1964) concluded from a study of several such contacts (intrusive-host rock), basalt–rhyolite, basalt–granite, granite–arkose, basalt–shale, basalt–diabase, that major and trace element variations were a function of recrystallization in the contact zone rocks, but that there were no data to support elemental transport across the contacts except perhaps on a millimeter–centimeter scale. These studies, too, are not supported by any isotopic work, but the results testify to the chemical inertness of the overall systems.

Hart et al. (1968) and Hanson and Gast (1967) have studied variations in U–Pb, K–Ar, and Rb–Sr ages in contact zones, also as a function of distance from the contacts. As mentioned earlier in this chapter, Hart et al. (1968) studied these effects in the Idaho Springs Formation (Precambrian) intruded by the Eldora–Bryan Stock (Laramide). Hanson and Gast (1967) studied the intrusive effects on the Snowbank stock, Minnesota, by the Duluth gabbro. In both cases, age lowering as a function of distance from the contact is explained by simple diffusion models. Whole-rock chemistry and isotopic balance, except for ^{40}Ar, is maintained. Their studies also argue for no chemical exchange between intrusive and intruded rocks.

Active and paleogeothermal systems may also serve as analogs for some radwaste scenarios. If it is assumed that any fluid in possible contact with the waste package is a hydrothermal fluid, then the reactions involving the waste form, canister, ± backfill, and repository rock are analogous to such reactions in hydrothermal systems. Studies of such deposits are fairly common, but not for an exact chemical, isotopic, and mineralogic balance. Still, the work by the author (Brookins and Laughlin, 1982, in press) on the hydrothermally altered core from the HDR drill core, Fenton Hill, near Los Alamos, New Mexico shows millimeter-scale redistribution of ^{*87}Sr and some Rb and Sr in fractured core. The bulk samples have basically retained their internally closed system conditions for U, Pb, Rb, and Sr.

Other analog systems are under study. The author and colleagues are actively investigating basaltic rocks intruded by rhyolites in the Columbia River basalts, a peridotite dike intrusive into halite in New York, and dikes intrusive into evaporites from the WIPP site. They continue to examine contact effects and vein relationships, and are starting work on active and paleohydrothermally altered rocks as well. These studies, when complete, should provide even more meaningful data for radwaste application.

The author (see discussion in Chapter 13) has also considered uranium ore bodies in sedimentary rocks as analogs both for buried spent fuel and for engineered backfill, and, for some of the argillaceous sedimentary rocks in the Grants Mineral Belt, as additional analogs for engineered backfill. Within limits, the use of such materials as analogs is very informative (see Chapter 13).

More recently, my students and I have investigated various Columbia River basalts. In one area, near Hood, Oregon, a dacite is intrusive into Wanapum basalt, and there is a pronounced chemical gradient between the two rocks. Yet detailed chemical and isotopic studies (unpublished data) again show no elemental transfer between the rocks. In addition, in a series of drill core from near the BWIP site, fracture filling minerals yield Sr isotopic compositions consistent with derivation from basalt and not from infiltrating meteoric waters (unpublished data). Although not yet complete, these studies further show the importance of natural analogs for radwaste studies.

Conclusions

The use of Eh–pH diagrams in dealing with various aspects of radwaste disposal is now well documented. The geochemical predictions based on the use of these diagrams is in excellent agreement with observation (Oklo).

Studies at the Oklo Natural Reactor site show a remarkable ability of the Oklo rocks to retain fission products, actinides, and actinide-daughter products. In brief, except for Xe and Kr, which have not been rigorously studied outside of reactor zones, all elements may be accounted for within a few meters of their sites of origin. This attests to the ability of the host rocks to prevent radionuclide for at least 5×10^5 years, and perhaps on the order of at least 3.5×10^7 years.

Studies in the contact zones of the Eldora–Bryan and Alamosa River Stocks, both in Colorado, show only very limited chemical exchange between the intrusive and the host rock—no more than a few meters, if that. These two examples are typical for cases for conductive and for convective cooling.

Other analogs, such as igneous dike rocks intruding evaporites, pitch-blende veins in various rocks, and many igneous–igneous contacts, show very limited, if any, migration of elements between the rocks. Collectively, the analogs indicate a high degree of retention of elements of interest in a very large number of samples. The data are of direct interest to buried radwaste packages, and also show that the candidate materials for engineered backfill are favorable as well.

CHAPTER 12
Waste Forms

Introduction

Waste forms under consideration internationally for the first barrier storage of radioactive wastes include various glasses, cement, cermet, calcine, supercalcine, phosphate, composite forms, synroc, and others. Intensive work on all of these has been carried out over the last several years, and many of these forms exhibit extreme resistance to leaching under assumed repository conditions.

Glasses

The chemically treated radioactive wastes from reactors may be stored for encapsulation in any of a number of waste forms. Glasses have been intensely studied in the United States, Germany, the United Kingdom, France, India, Belgium, Canada, Japan, and the Soviet Union. High-level liquid wastes are vitrified with additives to the liquid as desired. Low-melting and high-temperature glasses are both under consideration.

Low-melting glasses are those that can be processed at temperatures under 1200°C. These are commonly borosilicate glasses, although considerable work on phosphate glasses (see Mendel, 1978; Kupfer, 1979) has been carried out. The borosilicate glasses commonly are about one-third high-level waste oxides, the balance being a chemically inert glass due to additives during the vitrification. The glasses do contain fractures on a small scale and they are metastable, although the point is well made that their metastability does not necessarily mean they will fail with age. Naturally occurring glassy material of virtually all ages exist, including reliably dated terrestrial glasses 40 million years in age, tektites (high SiO_2, very low H_2O bearing glasses formed in extraterrestrial environments) of up to 70 million years in age, and lunar impact glasses probably hundreds of millions of years in age. The lunar glasses are preserved because of the chemically inactive lunar atmosphere, and most tektites are found in very dry climates on earth (see discussion in Ewing, 1979). Hence it is obvious that glasses can exist in the terrestrial environment for very long periods of geologic time. The glasses containing HLW must be cooled very rapidly so that excessive devitrification does not occur. For conditions of geologic disposal the temperatures will be so low that thermal devitrification should not pose a problem. Where some devitrification is observed, the effect is to alter the leach rates somewhat (DOE, 1980), possibly by a factor of three increase. Further, it has been shown that these glasses can withstand the equivalent of 500,000 years alpha radiation (DOE, 1980). The borosilicate glasses exhibit good durability although they will react with water, with reaction rates in the range 10^{-7} to 10^{-5} g glass/cm^2/day (25°C; few weeks duration). At higher temperatures the leach rate is higher.

The manufactured glasses are designed to be cast in cylindrical steel canisters. These canisters will be about 3 m long and the diameter will be 0.3 m for commercial HLW and 0.6 for lower heat-generating defense HLW.

Numerous glasses have been investigated for use as waste form. In Table 12-1 are shown the analyses of several of the borosilicate glasses under consideration. Leaching of glasses, and their other properties, are discussed later in this chapter.

High-temperature glasses are those that melt at temperatures of 1200°C and higher. These are more silica and alumina rich than the low-temperature glasses. The Canadian Government produced a glass of the approximate composition of nepheline syenite and buried it, and after several years it was monitored for leachibility. The leach rate for ^{90}Sr decreased as a function of time until about 5×10^{-11} g glass/cm^2/day was reached (see Merritt, 1977). In the United States, a high-temperature, silica-armored, stuffed glass has been tested (see Simmons et al., 1979). This product contains the HLW as oxides interstitially, and has the advantages of durability and silica on its surface that acts as a barrier to prevent radionuclide loss.

A series of specially formulated glass ceramics are also available. These have been especially studied in Germany (see Guber et al., 1979). These are basically glasses containing HLW that are devitrified (i.e., crystallized) under carefully monitored and controlled conditions, with about 50% each of glass and crystalline products. They are very durable with high thermal stability and physical ruggedness, although their leach rates are about equal to the low-level waste glasses (see DOE, 1980).

Crystalline Waste Forms

There are many crystalline waste forms under investigation in the United States today. For purposes of simply distinguishing these from glasses, virtually everything else can be considered as a crystalline waste form, including cement, calcine, and synthetic minerals.

Table 12–1. Nominal compositions of typical waste glasses.

Component	Glass Composition (wt%)		
	76-68 PW-8a-1	77-107 PW-9-1	77-260 PW-7c-1
Glass Frit			
Al_2O_3			2.0
B_2O_3	9.5	13.0	9.0
CaO	2.0	2.0	1.0
CuO			3.0
K_2O		4.0	2.0
Na_2O	7.5	2.0	8.0
SiO_2	40.0	38.0	36.0
TiO_2	3.0	3.0	6.0
ZnO	5.0	5.0	
	67.0	67.0	67.0
Waste			
Ag_2O	0.03	0.05	0.02
BaO	0.56	1.13	0.55
CdO	0.03	0.08	0.03
CoO	0.11	0.20	0.10
Cr_2O_3	0.41	0.21	0.02
Cs_2O	1.02	1.98	0.78
Fe_2O_3	9.68	0.90	1.19
Cd_2O_3			10.14
MnO_2			0.11
MoO_3	2.27	4.41	1.92
Na_2O	5.00	3.47	3.14
NiO	0.52	0.16	0.31
P_2O_5	0.48	0.40	2.36
Rb_2O	0.13	0.25	0.11
RE^a	4.76	9.24	3.84
RuO_2	1.06	2.20	0.85
SrO	0.38	0.72	0.29
TeO_2	0.25	0.50	0.22
$U_3O_8{}^b$	4.54	3.80	5.40
ZrO_2	1.76	3.30	1.68
	33.00	33.00	33.00

[a]A commercial rare earth mixture having the nominal composition in wt% 0.2, H_2O_3; 24.0, La_2O_3; 48.0, CeO_2; 5.0, Pr_6O_{11}; 17.0, Nd_2O_3; 3.0, Sm_2O_3; 0.8, Eu_2O_3; and 2.0 Gd_2O_3.
[b]U_3O_8 prepared from depleted uranium.

Much of the TRU and LLW materials are encapsulated in cement. Slurry or liquid wastes are mixed with specific amounts of dry solids. Portland cement is commonly used as solid, as are cement mixed with fly-ash and clay minerals (grout). Even specially prepared high alumina cements are used (see Lokken, 1979; Stone, 1977). Cements are porous, and susceptible to radiation damage and long-term thermal exposure due to the presence of hydrated minerals. Cements have been considered for both defense and commercial radioactive waste storage. For cement processed at 250°C at 600 psi (Moore et al., 1981) the porosity is reduced as is the water content, after which this material may be used for HLW storage.

Cement

Use of cement for the impoundment of LLW and TRU waste is feasible for many reasons. Cement is readily available, inexpensive, and, under shallow burial, very stable in the near-surface earth environment. For greatest stability, a low-humidity environment is desirable. It is also resistant to radiation (see Kibbey and Godbee, 1979) and possesses good mechanical strength. Cement is better for some organics than others, but problematic organics can either be pretreated or special additives to cement can reduce the potential deleterious effects of the admixed organics. Leach rates are, as expected, quite variable, ranging from unacceptably high values of near 10^{-1} g/cm^2/day to excellent leach-resistant values of 10^{-9} g/cm^2/day. Addition of barium salts to cement helps scavenge ^{90}Sr and, at some time in excess of 10^3 years, also retain radium that will build up in the encapsulated waste. Further, Morgan et al. (1979) have shown that ^{129}I can successfully be retained as $Ba(IO_3)_2$ in cement-stored wastes. Leach rates for ^{90}Sr and ^{137}Cs have been studied extensively. Rudolph et al. (1981) have shown that Portland cement (which is made from limestone plus shale or clay) and bentonite-bearing (as additive) cements are able to retard Sr and Cs leaching even when the cements are attacked by salt brines. The leaching curves they report are essentially identical regardless of leachant or specimen composition.

Cermet

Cermet is the name for a uniform dispersion of waste oxides in a metal matrix (see Quinby, 1978). The HLW and additives for the ceramic oxide phases are dissolved with the metal matrix alloy in molten area, followed by precipitation, calcination, and compaction into desired shapes. Reducible metal oxides such as Cr, Co, Fe, or Ni are used, and these are reduced in a CO or H$_2$ atmosphere to form alloys that surround the nonreduced ceramic oxides. An attractive feature of cermet is that the reducing conditions lessen the problems with volatility during the processing.

The composition of cermet must vary widely in order to accomodate varying compositions of HLW. As metal content increases, the waste loading can be increased since the metal forms part of the matrix (see Aaron *et al.*, 1979). When the sodium content of the waste increases, the integrity of the cermet requires more aluminosilicates to be added during processing. The metal used in the alloy matrix is designed to be a mixture of iron and nickel due to the abundance of these elements in the waste. The metal alloy matrix gives the cermet high thermal conductivity and good mechanical strength while the ceramic material provides the sites for many of the HLW constituents.

Leaching studies of cermet are difficult to properly evaluate because the experimental domains in the crushed cermet and other factors are especially problematic (see Aaron *et al.*, 1979; Blanco and Lotts, 1979a). In the work by Aaron *et al.* (1979) a Soxhlet leach test (72 hours) gives a leach rate for Cs of 7.1×10^{-6} g/cm^2/day, and these authors indicate a leach resistance equal to (or possible greater than) that for alternative waste forms. The cermet program is relatively new, and much more experimental study is needed to document their chemical and physical characteristics.

Calcine

Calcine has been produced from defense HLW at the Idaho Chemical Processing Plant (ICPP) since the early 1960s. In their process the HLW is converted to dry salts and oxides. The calcine waste is low-volume and essentially noncorrosive.

Composite Waste Forms

Composite waste forms include those in which the HLW is contained in one material surrounded by different nonradioactive materials. The nonradio-active material and the HLW material are chosen so that the properties of the composite are more favorable than the HLW form by itself. The HLW form may be fragments of crystalline waste forms or glasses, or specially produced HLW-bearing spheres or particles. Surrounding materials include ceramics or metals that, when employed, increase thermal conductivity, are more resistant to fracturing, and provide an additional barrier for possible radionuclide release from the surrounded HLW form.

Metal matrices provide increased ductibility as well as improved thermal conductivity and reducing potential for fracturing. A joint Germany–Belgium project (PAMELA; see Salander and Zuhlke, 1979) in which HLW glass beads are mixed with a lead matrix is planned for the 1980's. Most of the attention on metal matrices has involved low-temperature metals such as lead or aluminum or alloys of the two, although copper and steel can be used by special sintering techniques (see DOE, 1980). Leaching of these wastes is still under study. Water will attack the HLW form in the metal, and if this happens in the laboratory, the bond with the metal matrix is lessened and the attack will be promoted. Ways to prevent this from happening include surrounding the waste–metal matrix with even more metal so as to prevent active attack on any of the waste by this extra layer of metal. One drawback to the lead metal matrix is that the lead itself may possibly, with time, pose an undesirable hazard (see chapter 14).

Coated pellets consist of ceramic pellets coated with graphite and silicon carbide embedded in a graphite matrix. The HLW core material is designed to be synthetic mineral calcine, although any HLW form, in theory, can be used. The coated particles are essentially under investigation for fuels for the high-temperature gas-cooled reactors (HTGR) in which particles will be used as the fuel element. The coated particles are remarkably stable, and can be coated with an outer layer of alumina for greater durability. The alumina may be subject to mechanical weakness, however, so a metal sleeve (i.e., like the canister or canister sleeve) must be used to protect the waste form core. Coated particles have been described in detail by Rusin *et al.* (1979).

Synthetic minerals are designed to chemically incorporate radioactive waste constituents into crystalline mineral species. This has the advantage that, in most cases, the naturally occurring mineral species can be studied as analog to the artificial, radioactive waste-bearing mineral. When there is marked difference between the two, then extensive laboratory work is necessary to evaluate the stability of the artificial mineral as in the case of the effect of alpha-radiation damage of the HLW-bearing species.

So-called *supercalcine* is an outgrowth of the calcine process independently arrived at by Battelle Pacific Northwest Laboratories and by Pennsylvania State University (see McCarthy, 1979; McCarthy and Davidson, 1975). The HLW liquid waste is not calcined directly, but instead is mixed with additives of specific quantities of Ca, Al, Si, and other elements so that after calcination an assemblage of predetermined mineralogy results. A disadvantage is that the HLW is variable in chemistry and the mineral assemblage is hard to characterize and predict. More recently, this process has been used with additives to remove just certain elements from the HLW. Fluorite and monazite are, for example, used for removing and storing the actinides, since the actinides are easily captured in these structures and both minerals are remarkably stable in the natural environment.

The phases identified in supercalcine are varied, but some of the more common ones include a complex alkaline earth–rare earth silicate, rare earth phosphate, pollucite-equivalent, scheelite, actinide–rare earth–zirconium oxides, spinels, corundum, and rutiles. The approximate distribution of various elements into these phases is given in Table 12-2. From crystal chemical reasoning, the elements listed in Table 12-2 should be incorporated into the phases as indicated. There will be some partitioning of elements between phases. The heavy rare earth elements (HREE) will be partially incorporated into the $(Zr, U..)O_{2+x}$ phases as well as the apatite and monazite structures while the light rare earth elements (LREE) may be partially camouflaged in Ca^{2+}-sites in the apatite and scheelite. All REE will be accounted for in these phases, however. The fission products and

Table 12–2. HLW elemental distribution in supercalcine phases.

Elements	Approximate Composition of Synthetic Mineral	Structure
Sr, Ln	$(Ca,Sr)_2Ln_8(SiO_4)_6O_{12}$	Apatite
Ln, PO_4	$LnPO_4$	Monazite
Cs, Rb, Na	$(Cs,Rb,Na)AlSi_2O_6$	Pollucite
Ba, Sr	$(Ca,Sr,Ba)MoO_4$	Scheelite
U, Ce, Zr	$(U,Ce,Zr)O_{2+x}$	Fluorite
Zr, Ce, U	$(Zr,Ce,U)O_{2+x}$	Tetragonal
Fe, Ni, Cr	$(Ni,Fe)(Fe,Cr)_2O_4$	Spinel
Fe, Ni, Cr	$(Fe,Cr)_2O_3$	Corundum
Ru	RuO_2	Rutile

Note: The additive ions are Al, Si, and Ca. Ln = the rare earth elements, La, Pr, Nd, Sm, Eu, Gd, Y. See text for details.

actinides, not shown in Table 12-2, will include Te, Pd, Rh, Tc, Np, Pu, Am, and Cm. The actinides (Np, Pu, Am, and Cm) will exhibit behavior similar to uranium and be concentrated in the two $(M)O_{2+x}$ phases. Palladium and Rh will partially be camouflaged in the RuO_2 phase and partially for Fe-sites in the spinel and corundum–Fe_2O_3 (ss) phases. The Tc produced will be preferentially incorporated into the RuO_2, and Te may substitute (as Te^{3+}) into Fe-sites in spinel or corundum-Fe_2O_3 (ss).

The importance of these phases produced in the supercalcine is their extreme stability in the crust of the earth. Apatite is an extremely durable structure, and while the phase produced is not a chemical apatite, its structure indicates that it should be as stable as naturally occurring apatite. In that case, it is important to point out that apatites have been successfully dated by the U–Pb methods, as well as used as petrogenetic indicators based on radiogenic [87]Sr content (see Faure, 1977). The fact that a large-ionic-radius, chalcophile ion such as Pb^{2+} is retained in the structure of apatite means that the equivalent supercalcine phase may contain fissiogenic Cd, Ag, Pd, and Sb and possibly other elements. Monazites are also commonly dated by the U–Pb and Th–Pb methods and their extreme crustal stability has been well documented. Thus some U, Th, and other actinides as well as the REE will be well preserved in this phase. The pollucite deserves special mention. It is a well-known phase from experimental work, yet its occurrence in nature is limited. This is due, however, to the relative scarcity of Cs in nature coupled with the camouflaging of Cs by K in most minerals (Cerny, 1979). The stability of the artificially produced phase should be excellent based on the preservation of pollucites from very old pegmatites and other geologic occurrences. This phase will house fissiogenic Rb, Cs, and some Sr and Ba, The alkaline earth molbdate, the scheelite phase, will also prove to be stable. Scheelites are not incorporated into geochronologic work, but their phase relations are well known (Scheetz *et al.*, 1982). This phase is important not only for the incorporation of fissiogenic Mo but also for fissiogenic Sb, and possibly for Te. The fluorite structured Mo_{2+x} phase will, along with the other MO_{2+x} phase (tetragonal fluorite structure), contain most of the uranium and other actinides. In addition to Zr and the REE, some Ru, Tc, and Te may be incorporated into these phases. More important, UO_2 is one of the minerals used extensively for U–Pb dating, attesting to its extreme geologic stability, especially under chemically reducing conditions. Spinel and corundum-Fe_2O_3 (ss) phases in nature are also among the most stable, as attested to by their abundance in placer deposits ranging from very young to very old, as well as in such rocks as kimberlites. These phases are among the most stable phases known.

Sodalite $(AE_2(NaAlSiO_4)_6(MoO_4)_2)$ is a minor phase of supercalcine. It has the potential to incorporate Cs and Sr into its structure as well as many other elements (i.e., in the AE and Mo sites). While little radiometric dating has been attempted on sodalite, the reported cases (see Dalrymple and Lanphere, 1969) show sodalite to contain excess argon, a fact that potentially makes this phase very important, as, in theory, it should incorporate Kr and Xe into its structure.

Synroc

Synroc is the acronym (by A. E. Ringwood) for synthetic rock. In synroc the HLW radionuclides are held in solid solution in three distinct minerals: perovskite, hollandite, and zirconolite (Ringwood *et al.*, 1979). An attractive feature of syncroc is that a low waste loading is maintained, hence the known stability of the minerals is not disturbed. To make synroc, calcined HLW is mixed with synroc additives and hot pressed at 1200 to 1300°C.

The composition of the HLW sludge and slurries is variable, hence the exact amount of the three different minerals listed above, plus smaller amounts of others, is variable. The sludge or slurries must be treated for sodium removal, and the dried material then mixed with oxides of CaO, BaO, TiO_2, Al_2O_3, and ZrO_2. For an alumina-rich HLW starting material, the resultant synroc mineral assemblage is hollandite $(BaAl_2Ti_6O_{16})$, perovskite $(CaTiO_3)$, zirconolite $(CaZrTi_2O_7)$, and hercynite $(FeAl_2O_4)$. For an iron-rick HLW starting material, the assemblage will contain an ulvospinel (Fe_2TiO_4) instead of hercynite. If some sodium remains, it will be incorporated into nepheline $(NaAlSiO_4)$. These minerals are known in nature, some being more common than others. Further, occurrences of these minerals show that they have been stable in the crust of the earth for at least two billion years. Hollandite is a rare mineral originally reported from manganese deposits, and no work on its use for radiometric age determinations has been reported. It possesses a structure of very high strength and should, based on the well preserved Mn deposits in which it is found, be stable for very long periods of time in the crustal environment. Perovskite, also, is not used for radiometric dating (note: in theory this mineral could be dated by the U–Pb and Th–Pb techniques) but it occurs in rocks of very great ages and is a common constituent of placers, attesting to its stability in the crust. Zirconolite is known from several places in the world, commonly in placers. The mineral has been successfully dated by the U–Pb and Th–Pb methods, with reasonable results. In natural zirconolites, uranium and thorium substitute for zirconium. In some places the alpha-radiation damage is suspected as a contributing factor to the chemical alteration of this phase (see Haaker and Ewing, 1981), such that the leachability of the altered zirconolite is greater than expected based on laboratory leaching of the synthetic mineral. A major problem is that zirconolite is very rare, and to test the effect of radiation damage versus nonradioactive chemical alteration due to oxygen-charged waters, zirconolite from under reducing conditions would have to be studied, and such material is not available. Even in cases of alteration, though, the U–Pb dates are not entirely perturbed, thus suggesting that if the synthetic material, thoroughly mixed with the other synroc phases, is kept under low oxidation potential conditions, it should be stable.

Ringwood et al. (1980b) have convincingly shown that, even for metamict Sri Lanka zirconolite, the radiometric $^{238}U-^{206}Pb$ and $^{235}U-^{207}Pb$ ages are in excellent agreement, even superior to zircon dates probably due to the fact the Pb is held better in the structure of zirconolite than zircon. It is also important to point out that the Sri Lanka zirconolite studied by Ringwood et al. (1980b) was subjected to a stronger alpha-radiation dose than will be present in the synthetic mineral, and therefore the artificial minerals should be even more stable than the natural species.

Radiometric ages are also not available for minerals such as spinels, although the spinels are remarkably stable in the earth's crust, occurring both as remnant minerals in otherwise chemically altered rocks and in abundance in placers. The nepheline formed in natural occurrences has been dated, though, and successful Rb–Sr and K–Ar dates have resulted. The mineral will break down with active chemical weathering (under conditions of tropical weathering, for instance) but, in reducing environments, it is stable.

The synroc minerals provide sites for the radioactive constituents of HLW. The actinides (Th, U, Np, Pu, Am, Cm) will be incorporated into the zirconolite phase, as will many of the REE and Y, Ru, Tc, Rh, Pd, and possibly some Sb, Te, In, and Sn. Any of these elements that escape capture in zirconolite, and have ionic radii closer to that of Ca^{2+} than Zr^{4+}, will be incorporated into perovskite. Higher-ionic-radii cations will be incorporated into hollandite.

Cesium and barium and much of the Sr will be preferentially incorporated into the hollandite, with some Sr for Ca in zirconolite and perovskite. Any Rb present will also be in hollandite. Iodine may pose special problems, as it is not easily incorporated into the minerals discussed above. Iodine in the HLW package, if released, is present as I^- ion, and which can be sorbed onto various minerals (see Allard et al., 1980; Bird and Lopata, 1980).

Properties of Waste-Form Materials

Six major groupings have been proposed by Mendel *et al.* (1981a) to address important material properties of waste forms.

(1) Physical properties, including density and thermal conductivity, which have only a peripheral effect on ultimate risk.

(2) Mechanical properties, which are important because the waste forms, while not structural materials, must be consolidated and resistant to dispersion. Thus mechanical properties measured on waste forms are aimed at addressing their coherence and resistance to impacts.

(3) Chemical durability, the most important waste form property. Normally the chemical durability is addressed by consideration of leach rates, although an understanding of factors such as solubility constraints are also necessary.

(4) Vaporization of radionuclides from waste forms of potentially great concern, and the theoretical dispersion pathways that could arise from different accident scenarios must be addressed. The various radio-nuclides that might vaporize need to be identified and characterized.

(5) Radiation effects, especially in the near field, are important to the waste package, and how these effects will vary with the long times for radioactive decay to take place must be known.

(6) Phase stability, especially in terms of response to radionuclide in-corporation into the original waste package phases, is of extreme importance. The newly formed phases accompanying devitrification of glass, for example, must be adequate for retaining radionuclides. Similarly, if there are phase changes in the crystalline waste forms after some residence time in the repository, then these phases, too, must also be able to retain radionuclides.

The physical properties of waste forms depend on chemical composition, manufacturing processes, and the conditions for storage. The chemical composition is controlled by the inert constituents used to promote coherence and in other ways to control the overall structure of the nonwaste part of the package and the nuclear waste constituents. The amount of nuclear waste constituents, by weight percent, is referred to as waste loading, and varies from a few percent to 100%. Most waste forms use a loading of 33%. The density, thermal conductivity, and coefficients of thermal expansion of various waste forms is given by Mendel *et al.* (1981b), Guber *et al.* (1979), Wald and Westsik (1979), and Blanco and Lotts (1979b).

The mechanical properties determine whether or not the waste forms will remain cohesive and monolithic, and both the impact and static strengths of the waste form are important to know. Various studies of impact strength, tensile strength and compressive strength are summarized by Merz (1981).

The chemical durability of waste forms has been summarized by Mendel *et al.* (1981a). Specifically the resistance of waste forms to reactions with water have been evaluated by these workers. The importance of chemical durability is that radionuclides released by reactions with water can be transported in the water either by diffusion (static conditions) or fluid flow (moving water). The overall reaction involving waste-water reaction is leaching, and experimentally determined leach rates allow the chemical durability of waste forms to be quantitatively assessed. There are many variables that affect leach rates, including the different techniques and equipment and even measuring methods used by different laboratories. Leaching experiments are typically carried out in order to study waste-form development and characterization, the analysis of the safety of waste management systems alternatives, and the quality assurance in a waste-solidification facility. Leach tests may be carried out under static and/or dynamic conditions. For static conditions, the leachant is not replenished, and the composition changes as more material is leached from the sample. With time, the composition of the leachant is only slightly affected by additional leaching, i.e., so-called steady-state conditions are realized. For dynamic leaching, the leachant is replenished either continuously or periodically. Most leaching tests done to date are of the dynamic variety. Of the two, static leaching tests are perhaps the most important for conditions to be met in a repository, since the flow rate of underground waters that might come into contact with any of the repository's multiple barriers will, by definition, be close to static conditions. Chemical durability is also very important and can be investigated for waste–rock interaction tests. For many experiments, the leachant is composed to simulate the actual site-specific waters and is pre-equilibrated with crushed rock from the repository site rocks (or as close to it as possible). Dynamic conditions are used during the experimental work. Flow rates are used that are assumed to represent those typical of the proposed repository.

An interesting test is under way at Chalk River, Canada. There radioactive glass blocks were buried in 1960 and have been monitored since. Five years were required for steady-state conditions to be met, and testing is still going on. This is the longest-duration experiment of man-produced radioactive waste.

Leaching

The data reported for leaching tests is, unfortunately, not uniform, although there is an attempt underway to standardize reporting. A normalized leach rate is the most common way of expressing chemical durability by the expression:

$$(\text{LR})_i = \left(\frac{A_i}{A_o}\right)\left(\frac{W_o}{\text{SA} \cdot t}\right) \qquad (12\text{-}1)$$

where

$(LR)_i$ = leach rate in $g/cm^2/day$ (normalized to the behavior of some
      component, i),
  A_i = amount of component, i, leached during a time interval, t,
  A_o = initial amount of component, i, in waste form,
  W_o = original weight of the waste-form sample,
  SA = surface area of the waste form sample, in cm^2,
    t = the time interval of leaching, in days.   (12-2)

It is convenient to convert the normalized leach rate to a penetration rate,
PR_i, by dividing $(LR)_i$ by the density, ρ so that:

$$PR_i = \left(\frac{A_i \cdot V}{A_o \cdot SA \cdot t} \right) \tag{12-3}$$

where

V is the volume of the waste form sample, in cm^3.

Another way of expressing the leaching date is by consideration of the
cumulative fraction released or cumulative penetration, so that:

$$C_{p_i} = \frac{A_i}{A_o} \frac{V}{S} \tag{12-4}$$

and

$$C_{F_i} = \frac{A_i}{A_o}, \tag{12-5}$$

where

C_{p_i} = the cumulative penetration, from behavior of component, i, as a
    function of depth in the surface of the waste form,
C_{F_i} = the cumulative fraction of i released,
  A_i = the cumulative amount of component i that has been leached from the
    waste form sample.

Additional ways of expressing leaching have been created by reporting the
concentrations of the species produced by leaching, or by weight loss of the
waste form sample. Table 12-3 is the tabulation of Mendel et al. (1981a) for
leach rates (see Mendel et al., 1981a; Jardine and Steindler, 1978).

Knowledge of how radionuclides are released from waste forms is
necessary in order to fully assess leaching data. Incongruent dissolution, for
example, usually results in differential release of ions to solution, which, in
turn, will be quite different from material released by simple dissolution or
diffusion. Study of the surface of waste-form samples used in numerous

Table 12–3. Leach rates for various waste forms.

Waste Form	Leach Rate $(g/cm^2/day)$	Water Temperature $(^{\circ}C)$
Pot calcine	10^{-1}	25
Fluid bed calcine	10^{-1}	25
Pot calcine	10^{-2}	25
Fluid bed calcine	10^{-2}	25
Cement	$10^{-2} - 10^{-3}$	25
Al metal matrix-sintered	3×10^{-3}	25
Concrete	10^{-3}	25
Al metal matrix-cast	$10^{-3} - 10^{-4}$	25
Aqueous silicates (clay)	$10^{-4} - 10^{-5}$	25
Bottle glass[a]	5×10^{-5}	100
Borosilicate glass	3×10^{-5}	100
Lead matrix[a]	2×10^{-5}	100
Vitromet (63 vol% phosphate/glass 37 vol% lead)	2×10^{-5}	100
Phosphate glass	1×10^{-5}	100
Lead matrix[a]	7×10^{-6}	20
Glass (devitrified)	5×10^{-6}	25
Asphalt	4×10^{-6}	25
Zn borosilicate glass (devitrified)	2×10^{-6}	25
Borosilicate glass (in-can melted)	1.3×10^{-6}	25
Borosilicate glass	$10^{-5} - 10^{-7}$	25
Phosphate glass	$10^{-6} - 10^{-7}$	25
Titanates	5×10^{-7}	25
Alumina phosphate glass	5×10^{-7}	25
Phosphate glass	5×10^{-7}	20
Vitromet (63 vol% phosphate glass/ 37 vol% lead)	5×10^{-7}	20
Zn borosilicate glass (as formed)	3×10^{-7}	25
Glass (as formed)	2×10^{-7}	25
Borosilicate glass	$10^{-4} - 10^{-7}$	25
Silicate glass (Canadian)	$10^{-6} - 10^{-7}$	25
Sintered glass ceramics	$10^{-5} - 10^{-8}$	25
Industrial glass[a]	$10^{-6} - 10^{-7}$	25
Industrial glass[a]	$10^{-5} - 10^{-8}$	25
Silicate melts (fired clay)	$10^{-6} - 10^{-8}$	25
Silicate glass (U.S.)	2×10^{-7}	25
	6×10^{-7}	

Source: Adapted from Jardine and Steindler (1978).

[a]These materials are listed for comparison purposes and contain no simulated waste.

Table 12–4. Leach test results to illustrate incongruent dissolution in deionized water.

Waste Form	Leaching Order	Reference
Glass	Ca Co Ag Zn Ba Zr Ce, Eu	Flynn *et al.* (1979)
Glass	Cs Sr, Pu	Plodinec and Wiley (1979)
Glass	Na Si B Cs Sr Zr Ce Tb	Johnson and Marples (1979)
Glass	Sr Cs actinides	Rankin and Kelley (1978)
Supercalcine	Na Mo Rb Cs, Ba, Sr, Ca	McCarthy *et al.* (1978a)
Concrete	Sr Cs Pu	Moore *et al.* (1979)
Spend fuel	Sm, Cs, Pu, Ce Sb, Eu, Sr U Ru	Katayama *et al.* (1980)

Source: From Mendel *et al.* (1981a).

experiments shows that incongruent dissolution can have widely varying effects on experimental results. Mendel (1978) have tabulated the results in Table 12-5 to illustrate incongruent leaching (in deionized water).

Some interesting observations from leaching experiments have been summarized by Mendel *et al.* (1981a).

(1) Westsik and Harvey (1981) have shown that release of the alkaline earths from a waste glass actually decreased as the temperature increased from 150 to 250°C. This decrease in release, apparently associated with solubility effects and secondary phase formation, was sufficient to make calcium, strontium, and barium releases comparable to those of the actinides.

(2) Both McCarthy *et al.* (1978a) and Westsik and Turcotte (1978) have reported that dissolution of silicon from glass and from supercalcine was less in salt brine than in deionized water, while releases of the other components to salt brine were higher.

Table 12–5. Leach test results to illustrate incongruent dissolution of actinides as a function of leachant and time.

Element	Observed Ranking of Element Release from Highest to Lowest Element Release	Ratio of Highest to Lowest Element Release ($t = 1$ day)	do. (40–600 days)
WIPP brine	Np, Am, U, Pu, Cm	36	20
CaCl$_2$	Am, Cm, Pu, Np, U	7	5
NaCl	Np, Am, Cm, Pu, U	18	10
NaHCO$_3$	Np, U, Am, Pu, Cm	12	3
Deionized water	Np, Cm, Pu, Am, U	37	11

Source: Modified from Mendel *et al.* (1981a).

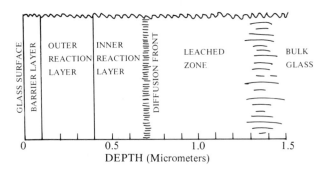

Fig. 12-1.(a) Leached layer of waste glass, schematic. Modified from Mendel *et al.* (1981a).

(3) Hench (1977) has suggested that congruent dissolution of glasses may be caused by a high solution pH.

(4) Waste-form developers use chemical additives to immobilize specific ions in the solidified waste. The resultant phases may then be less leachable than other parts of the waste form, thus leading to incongruent leaching. For example, Stone (1977) noted that the presence of MnO_2 in a particular sludge composition reduced the leachability of strontium, and Ringwood *et al.* (1979) modified their SYNROC B in response to the discovery that cesium was segregating into a highly leachable phase.

For most leaching experiments, two orders of magnitude are common between the most and least leachable results. There is a tendency toward congruency during leaching experiments; the closer to steady state the longer the experiment, but it is a slow trend. Knowledge of the leaching behavior of individual radionuclides is necessary , as this will determine the factors to be weighed in risk assessment. Barney and Wood (1980) report the 10 most important radionuclides in a basalt repository to be, after ^{90}Sr has largely decayed away, ^{99}Tc, ^{129}I, ^{237}Np, ^{226}Ra, ^{107}Pd, ^{230}Th, ^{210}Pb, ^{126}Sn, ^{79}Se, and ^{242}Pu. The leaching of ^{99}Tc is anomalous, since it is strongly leached early but less and less is leached until less ^{99}Tc than actinides is leached. This behavior is possibly due to the formation of Tc-bearing globules in the glass (Bradley *et al.*, 1979).

Examination of leached surfaces by various techniques shows that the surfaces are usually quite irregular due to the incongruent dissolution taking place on the surface. The leaching process is complicated by structural alterations, solubility, and reprecipitation. During the leaching process, a reaction layer forms (see Fig. 12-1), which increases in thickness as a function of temperature or duration of the leaching experiment or both. Commonly the alkali and alkaline earth elements are leached from these zones and silicon, the rare earths, and other transition metals enriched in the reaction zones. The composition of the leachant is commonly thought to be the controlling factor for surface reactions of this type, according to Westsik

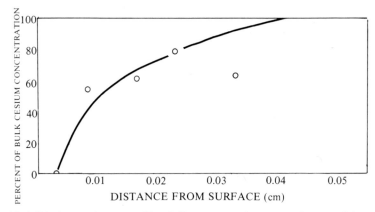

Fig. 12-1.(b) Approximate profile of Cs over reaction zone of supercalcine pellet post-salt-brine leaching. *Source:* Modified from Mendel *et al.* (1981a).

and Turcotte (1978). Interestingly, as devitrification takes place in glasses, the crystalline phases formed are more resistant to leaching than the residual glass matrix (see Rankin and Kelley, 1978; Wald and Westsik, 1979).

Temperature variations have a pronounced effect on leaching. As temperature increases, the amount of material leached also usually increases. In a repository located at some depth (~800 m), since the pressure at depth may increase to some 2300 psi, a wide range in temperature at various pressures needs to be considered in order to cover all aspects of leaching of the waste package. The actual self-heating of the solid radwaste form will be a function of such factors as the age of the waste, size and shape of the waste form, waste loading, constituents in the waste, and thermal properties both of the waste form and the surrounding materials. Fortunately, the heat generation rates for HLW decrease rapidly as a function of time (e.g., it is reported in ERDA-76-43, 1976 that HLW with 18.5 kW/MTU at 160 days out of reactor decreases to 0.86 kW/MTU after 10 years). Hence interim surface storage will allow the HLW to cool sufficiently so that thermal effects in the repository are not severe; i.e., the maximum allowable temperature in a repository due to the waste form may be only 100 to 150°C, although this is still under study.

Studies of temperature effects on leaching of different materials are summarized in Mendel *et al.* (1981a), who report that Lanza *et al.* (1980), in their study of water leaching of borosilicate glass, found that weight loss versus time of leaching experiments could be expressed by:

$$\text{Weight loss} = a\, t^{1/2} + b\, t, \tag{12-5}$$

where a and b are constants and t is time. The rate expression for this equation is given by:

$$\text{Rate} = \tfrac{1}{2} a\, t^{-1/2} + b \tag{12-6}$$

Table 12–6. Approximate activation energies for leaching of borosilicate glass by water as a function of exposure times at 50 and 80°C.

Exposure Time (hrs)	E_a (kcal/M)
0	6.92
100	8.01
1000	8.48
8000	13.93

for times up to about 7800 hours. For 50°C, $a = 10^{-4}$ and $b = 5.7 \times 10^{-7}$ while at 80°C, $a = 2.5 \times 10^{-4}$ and $b = 3.6 \times 10^{-6}$. From these data activation energies can be obtained from standard Arrhenius relationships. These (from Mendel *et al.*, 1981b) are summarized in Table 12-6.

Activation energies reported (see Mendel *et al.*, 1981a) commonly range from about 8 to 20 kcal/mole, and wide variations from experiment to experiment are commonly noted. This is due to the different reactions that may be taking place during the leaching, i.e., if all sites for release of elements leached were known, then $E_a = E_1 + E_2 + \ldots + E_{n-1} + E_n$. Unfortunately, not all such sites can be identified, hence the wide variation in values for E_a.

Temperature effects indicates a factor of 10 to 100 increase in leaching rates from ambient temperature to 100°C, and stresses again the importance of temperature monitoring during leaching experiments. The effect of pH on leaching is a lowering of pH values causing an increase in the leach rates (see Adams, 1979; Boult *et al.*, 1979). The pH effect at different temperatures is shown in Fig. 12-2. (Modified from Boult *et al.*, 1978.)

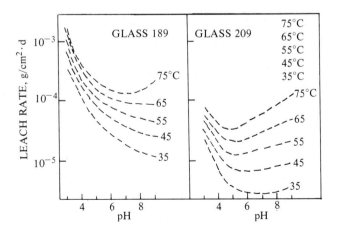

Fig. 12-2. Effect of pH on leach rate of glasses at different temperatures. *Source:* Modified from Boult *et al.* (1978).

Leaching is commonly carried out using one of the following tests: (a) modified IAEA test, (b) static test, (c) single-pass, continuous-flow test. The International Atomic Energy Agency (IAEA) test has the advantage that the solution used for leaching is renewed according to a preset schedule, with the sample being leached placed in a fresh volume of leachant and the experiment continued. Costs are higher for the IAEA test than for other leaching tests, and it requires more effort than other tests. Further, groundwater flow is not adequately modeled from the IAEA test because of the periodic replacement of fresh leachant, which in turn results in periodic changes in pH and solution concentrations.

The static test allows evaluation of buildup of concentrations of ratio-nuclides and matrix elements as well as reactions that retain or facilitate leaching. In the static test, the solution is not agitated and there is no solution flow or replacement. Tests are run in batches so that an aliquot after some time is separated from another aliquot at a different time and so on; this eliminates possible errors caused by resampling the same experiment. The tests are easy to run, require little laboratory space, and replicates are used to ensure proper evaluation of results.

The single-pass, continuous-flow test is used to provide information on the rate and nature of radionuclide release, it being recognized that the static test is severely limited in not providing for flow by a specific point where leaching may be taking place. The natural hydrologic gradient in an hypothetically flooded repository necessitates the continuous-flow testing. In the experiments, flow rate can be directly correlated with leach rate (and the volume of water required). A disadvantage of the tests is that they require more time and cost more than the static or IAEA tests.

Leaching of spent fuel has been undertaken by numerous investigators, including an excellent summary by McVay et al. (1981). In theory, leaching of spent fuel should be easier to interpret than that for borosilicate glass, but, in practice, the radiation history of the fuel has changed it sufficiently so that it, too, is difficult to interpret by simple leaching results. The inherent chemistry of spent fuel depends on such parameters as U/O ratio, amount of burnup, and its power rating (McVay et al., 1981). Irradiation effects include grain growth, cracking of fuel, porosity effects (i.e., open and closed due to formation of fission products, gas bubbles, etc.). One complication of experiments on spent fuel is that, due to their highly radioactive nature, they must be carried out in a hot cell.

There is a large temperature gradient from the center of spent fuel to the surrounding cladding, which results in differences in porosity, grain size, and other factors. Further, different elements are concentrated at different points in the fuel-cladding assembly. For best experimental results, parts of fuel-cladding assemblies must be used, as this ensures a good approximation of the stored material in a repository. As McVay et al. (1981) point out, there is no one best representative spent fuel sample, and it is difficult to compare experimental results. Leaching is observed to be relatively incongruent, however, with fairly rapid release of ^{137}Cs and ^{90}Sr in the short term.

Temperature effects on leaching are not well understood, primarily due to a lack of data. In addition, while the leach rate of Cs as a function of oxygen content of spent fuel-cladding is unchanged, other elements commonly show dissolution rates that decrease as oxygen content decreases. The leachant composition is presumed to have a pronounced effect on leach rate, although the experimental results to date are difficult to interpret (see McVay et al., 1981, for discussion). Solubility effects are not well known, although the effect of solubility on UO_2 under reducing conditions is so slight as to often prevent measurement of any material leached (i.e., due to extreme insolubility of UO_2 under reducing conditions). It is difficult to compare leaching of UO_2 with spent fuel, however, as the latter contains so many defects, etc., as to make a direct comparison very risky.

Some of the oxidation and dissolution reactions of UO_2 in water (i.e., carbonate-absent) are shown in Fig. 12.3.

Comparison of leaching of spent fuel and glass for elements common to both waste forms is shown in Fig. 12.4. There is about a two-orders-of-magnitude difference in leach rates for the elements shown. However, with sodium bicarbonate rather than distilled water as leachant (Fig. 12.5) the results are quite different, and there is no appreciable difference in leach rates. As McVay et al. (1981) point out, these two cases (Figs. 12.4 and 12.5) are the extremes, and most groundwaters would be expected to fall somewhere between the two. Further, the results shown in Figs. 12.4 and 12.5 are for air-saturated conditions, and under reducing conditions, very different results should be obtained.

Hydrothermal reactions will have a pronounced effect on leaching behavior. Hydrothermal reactions are used here to describe those reactions taking place in a repository at greater than 100°C with water under pressure. It must be emphasized, however, that for hydrothermal reactions to occur, several events must occur beforehand: (1) The temperature in the waste package area must be raised due to heat from the waste. (2) Water must have been introduced to the area of attack. (3) The repository system must be closed so that water does not escape as steam. (4) The overpack and canister barriers must fail almost instantly so that the heated water will retain its heat in order to attack the waste. The likelihood of all four of these events being realized in a repository is very small. The self heat from the waste package may locally raise the temperature in the surrounding canister–overpack and possibly the innermost part of the backfill, but an extreme temperature gradient is unlikely. How to introduce water and then seal the conduits off to prevent steam dissipation involves unlikely assumptions. The multiple barriers should minimize or prevent water from reaching the waste form, yet if such an event did take place, it is also unlikely that the repository would then magically seal to thus prevent steam loss. Item (4) is contrary to canister and overpack design. The intent is to use these as two of the multiple barriers

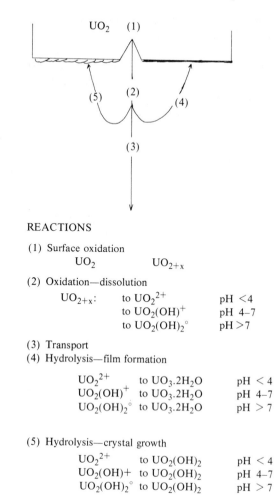

REACTIONS

(1) Surface oxidation

$$UO_2 \qquad\qquad UO_{2+x}$$

(2) Oxidation—dissolution

UO_{2+x}:	to UO_2^{2+}	pH <4
	to $UO_2(OH)^+$	pH 4–7
	to $UO_2(OH)_2{}^\circ$	pH >7

(3) Transport
(4) Hydrolysis—film formation

UO_2^{2+}	to $UO_3.2H_2O$	pH < 4
$UO_2(OH)^+$	to $UO_3.2H_2O$	pH 4–7
$UO_2(OH)_2{}^\circ$	to $UO_3.2H_2O$	pH > 7

(5) Hydrolysis—crystal growth

UO_2^{2+}	to $UO_2(OH)_2$	pH < 4
$UO_2(OH)+$	to $UO_2(OH)_2$	pH 4–7
$UO_2(OH)_2{}^\circ$	to $UO_2(OH)_2$	pH > 7

Fig. 12-3. Oxidation and dissolution mechanisms for UO_2 in deionized water. From McVay *et al.* (1981).

and under no sensible scenario for radwaste attack do these barriers get penetrated "instantly." Thus, in theory, the possibility for hydrothermal events of the type described here is very small. It is noteworthy that should some combination of these occur, however, the products formed by waste–rock–water and waste–backfill–water reactions should retain the HLW radionuclides (see McCarthy, 1977, 1979a; Braithwaite, 1979). At temperatures in the range of 250 to 350°C, however, most solid materials will react unfavorably (the amount and rates of material lost to solution during hydrothermal attack are variable and often large), thus temperatures in this range should be avoided. The suggested limits of T 100–150°C or so will not support active hydrothermal attack.

The leaching of samples of SYNROC, supercalcine, and glass (PNL-72-68) are shown in Table 12-7. It is clear that the 150°C column for each waste type is significantly lower than the corresponding values for higher temperatures, again emphasizing the need for temperatures of 150° or lower for the composite waste package.

Leaching is a function of composition of leachant, and the compositions of leachants may vary widely, from relatively clean water in the interim storage area to high-concentration brines in the vicinity of buried waste in

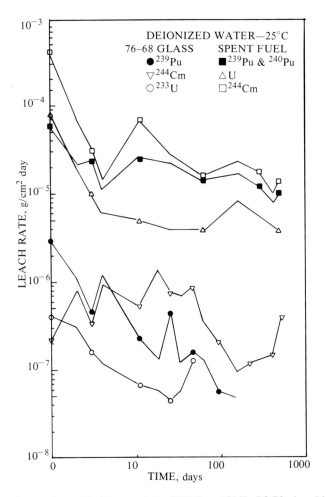

Fig. 12-4. Comparison of leach rates from SURF and PNL-76-78 glass (demineralized water, 25°C) From McVay *et al.* (1981).

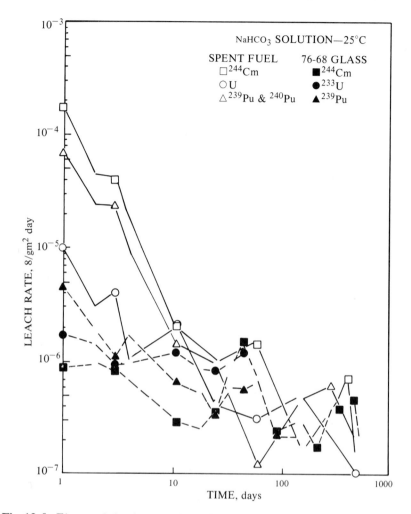

Fig. 12-5. Elemental leach rates from SURF and PNL-76-78 glass, 0.03 M NaHCO₃, 25°C. From McVay *et al.* (1981).

repositories. Site-specific tests are necessary to address the effect of different leachants on waste forms at the different repository sites.

Most natural waters yield pH values in the range of 2 to 10, most groundwaters fall in the range of 5.5 to 8, and sea water varies from 8.4 to 8.1 or 8.2. As the temperature of any water-bearing system increases, the pH is decreased due to increased ionization of water. The pH can also be lowered in radiation fields (Rai *et al.*, 1980). Some pH values in potential repository are anomalously low or high. A pH of 3 has been reported for hot salt brines and a pH of 10 for fluids in basalts. Yet pH values between 5.5

Table 12-7. Comparison of preliminary leach data for SYNROC, supercalcine SPC-4, and PNL 72-68 glass in deionized water.

	Fraction Leached/cm² × 10⁴ at 2000 psi								
	Synroc B			SPC-4			PNL-72-68		
Element	150°C	250°C	350°C	150°C	250°C	350°C	150°C	250°C	350°C
Cs	220	350	430	7.3	15	9.8	8	29	450
Na	126	215	224	8.2	13	12	10	58	1500
Mo	50	50	70	7	13	12	4	52	1100
Ba	0.53	0.18	0.18	—	—	—	1.6	0.99	1.2
Sr	0.35	0.14	0.12	0.85	0.88	0.58	2.3	3.5	3.6
La	0.05	0.08	0.32	—	—	—	—	—	—
Al	7.2	15.9	20.3	6.6	19	12	—	—	—
Zn	13.1	4.7	0.6	—	—	—	—	—	0.14
Ni	6.5	12.6	0.1	—	—	—	—	—	—
Si	—	—	—	24	38	52	7.5	22	150
Fe	2.8	3.7	0.7	—	—	—	0.16	0.56	1.9

Source: Adapted from Cornman (1980).

and 6.5 are common for evaporitic waters at the WIPP site and most pH values are under 9.3 at the BWIP site.

It is important to note that while most waste forms are more readily leached at very low and high pH values, minimum leach rates are common in the pH range of 6–8; this range ($\pm$ 1 unit from neutrality) is typical of most repository waters. In Fig. 12-6 is shown the leaching behavior as a function of pH for several waste-glass compositions. What is evident by inspection of Fig. 12-6 is that minimum leaching occurs at about pH $= 7$. Further, from Fig. 12.6 it is seen that unlike waste glass, where there is more active attack at low pH, the commercial glass is more readily leached at high pH. This is due to the fact that the glasses for radwaste are low SiO_2 glasses (about 50%), whereas the commercial glasses are high SiO_2 (70%). Acid attack on many rock-forming minerals indicates the same pattern of more active attack on at low pH, very little attack near neutral pH conditions, and increased attack with increasing basicity (see Brookins, 1978b; White and Claasen, 1979). For glasses, the low-SiO_2 radwaste glass contains elements such as Cs and Sr, which are more readily attacked at low pH. Commercial, high-SiO_2 glass is not as readily attacked under low pH due to the fact that the matrix remains relatively intact. At high pH, however, the high-SiO_2 glass is more favorably attacked due to the increase in dissociation products from H_4SiO_4; vis. $H_3SiO_3^-$. The radwaste glass not only contains less SiO_2, but many of its constituents of the waste load are not affected by high-pH conditions, hence the leach rates are low at high pH. Elements such as Zr, Fe, Ti, and some actinides can actually be used as additives to various glasses for alkaline-condition durability.

The Eh of the repository environment is much lower than that of the surface. At the surface waters are in direct communication with the oxygen in the atmosphere, and at shallow depths dissolved oxygen is readily available. At depth, however, the conditions are much more anoxic, and the overall environment is chemically reducing. Thus species commonly soluble as oxyions (i.e., TcO_4^-, RuO_4^{2-}, UO_2^{2+}, etc.) are not found under reducing conditions due to formation of phases containing reduced ions (i.e., TcO_2, RuO_2, UO_2, etc.). Several well-known Eh-sensitive boundaries are known in nature. The boundary between reduced S (as S^{2-}) and oxidized S (as SO_4^{2-}) will, with time, control many other reactions (note: species such as $S_2O_3^{2-}$ and SO_3^{2-} while locally of interest, age to SO_4^{2-} in relatively short times). Another couple is Fe^{2+}–Fe^{3+}, as well as C(O); C(IV). It is especially important to note that under the proposed repository conditions, the Eh will be very low—values about -0.1 to -0.5 volts are likely, as compared to much higher Eh values of $+0.4$ to $+0.8$ for most air-saturated experiments (near pH $= 7$). Under these reducing conditions, the reduced, usually insoluble forms of elements with more than one valence are found. Hence Tc, Ru, Pd, Rh, Cd, Mo, In, Sb, Ag, U, Np, Pu, Am, and Cm would, in theory, be expected to be highly insoluble even if waste forms were destroyed. Further, although the Eh in the near field may increase, the effect of the

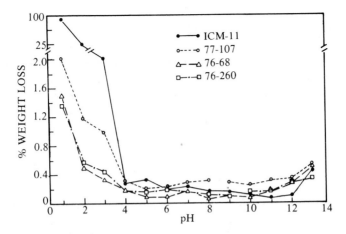

Fig. 12-6.(a) The effect of pH on leaching characteristics of borosilicate glasses. *Source*: McVay *et al.* (1981).

canister and engineered backfill will be to lower Eh thus preventing escape of any initially formed oxyion as some reduced-element precipitate. Brookins (1978b,f) has pointed out the importance of low-Eh conditions at the Oklo Natural Reactors for the retention of fissiogenic elements, and actinides, Pb and Bi (see Chapter 11).

Leachant chemistry is complex, but only a few major cations and anions dominate any particular system. Table 12-12 presents data for the chemistry of natural waters. Only Na^+, K^+, Ca^{2+}, Mg^{2+}, Si (as H_4SiO_4), Cl^-, SO_4^{2-}, and HCO_3^- (pH 6–10) are major species in most waters. Under hydrothermal solution conditions, high amounts of Fe, Mn, Al, and other metals may be present, along with Br, BO_3^- and some F and I. The chemistry of natural waters is complex (see Drever, 1982), but fortunately, the species listed in Table 12-8 are not complexed and can be readily identified and measured. Organics can, if present, pose special problems. Many cations

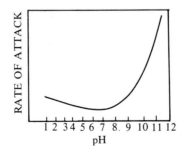

Fig. 12-6.(b) pH leaching effects on commerical container glass. *Source*: McVay *et al.* (1981).

Table 12–8. Representative concentrations (ppm) of ions and silica in natural waters.

	Rainwater	Terrestrial Water	Hydrothermal Solution
Na	1.1	5.8	51,000
K	0.26	2.1	25,000
Ca	0.97	20	40,000
Mg	0.36	3.4	700
Si	0.83	8.1	>110
NH_4^+	—	—	500
Sr	—	—	700
Li	—	—	300
Ba	—	—	200
Rb	—	—	170
Cs	—	—	20
As	—	—	15
Fe	—	—	3200
Mn	—	—	2000
Al	—	—	300
Zn	—	—	970
Pb	—	—	100
Cu	—	—	10
Ag	—	—	1
Cl^-	1.1	5.7	185,000
SO_4^{-2}	4.2	12	60
$HCO_3^- + CO_3^{-2}$	1.2	35	—
Br^-	—	—	150
F^-	—	—	18
I^-	—	—	22
BO_3^-	—	—	520

Source: Adapted from Wedepohl (1967).

form chelates with organics (especially humic and fulvic acid derivatives), and organics are present in most natural surface waters. Bacteria and other organisms can also affect water chemistry. Fortunately, many waters at depth in basalt, granite, and evaporites are depleted in organic material.

The effects of substantial radiation doses on candidate waste glasses and supercalcine have been investigated extensively (see summary in Mendel *et al.*, 1981a). Their findings show that, by and large, most candidate waste glasses and supercalcines are unaffected by large doses of radiation, even when equivalent to the dose that might be received over a 100-year period for stored waste. Alpha doses to borosilicate glasses have been studied by Boult *et al.* (1978) and Bibler and Kelley (1978), who find that even large amounts of alpha irradiation has little effect on leach rates for Pu and Cm. Bibler and

Kelley (1978) have shown the leach rates for both ^{244}Cm and ^{238}Pu are extremely small, on the order of 1.4×10^{-7} to 9.8×10^{-7} g/cm^2 for ^{244}Cm and 1.6×10^{-8} to 7.3×10^{-8} for ^{238}Pu. The leach rates for the heavily alpha-irradiated ^{238}Pu-doped glass are higher, but the leach rate remains fairly constant after about 1–2 years. This means that while the leach rate increases initially, it tapers off so that additional leaching is minimal (McVay, 1979; Bibler and Kelley, 1978).

Long-term leaching is of great importance to the storage of radioactive wastes in repositories, and the short-duration leaching experimental data must be extrapolated to determine what the long-term leaching will be. The expression:

$$F_i(t) = at^{1/2}$$

is generally followed, where $F_i(t)$ is the fractional release of component i at time t, and a is a constant. For simple diffusion, from a semi-infinite solid, Godbee and Joy (1974) show that:

$$F_i(t) = 2(S/V)(Dt/\ )^{1/2},$$

where $S =$ surface area, $V =$ volume of the solid, $D =$ effective diffusivity of species i within the glassy solid, $t =$ duration of leaching experiment. However, if the corrosion is rate determined, then

$$F_i(t) = bt,$$

where b is a constant.

The equation that normally applies to glassy solids, however, is

$$F_i(t) = at^{1/2} + bt,$$

which takes into account simultaneous dependency on diffusion and corrosion; and that for the observed fractional release rate is

$$R_i(t) = at^{-1/2} + b.$$

While more complex expressions are necessary for late stage leaching, these expressions (above) are sufficient to approximate the leaching process. Since the leach data are variable, it is a common practice to use the results of a short-duration, high-flow-rate experiment to serve as a conservative figure for modeling purposes. This is not good practice because the flow rates around the waste forms will be much lower, temperature will be less in the surroundings than in the self-heated waste-form area, and the leachant chemistry will be modified by the backfill and other barrier materials. Of interest, though, is the comparison of the leachability of natural minerals and rocks with HLW devitrified glass (Table 12-8). Here it is seen that the glass has chemical durability equal to or better than many naturally occurring materials. This is of importance because it is an established fact that the rocks and minerals are indeed stable in the earth's crust, and any material with a comparable leachability must also be stable.

The leach rates in Table 12-8 are fairly high, and reflect the results of laboratory experimental conditions. To make a simple extrapolation of these leach rates to nature would suggest that all land masses should have been dissolved away long ago. Their existence firmly shows that the true leach rates over long-term conditions are much smaller than the laboratory, short-term leach rates, and that the latter must not, without due caution advised, be applied to the long-term leaching of any material. Rate-inhibiting processes in nature prevent widespread dissolution, and while not precisely known, it is clear that the short-term leaching experiments do not adequately duplicate natural conditions. Yet the laboratory data for rather extreme conditions must be attempted as more realistic, low-temperature, near-neutral-pH, low-ionic-strength solutions are so ineffective in leaching many materials that experiments would be infinitely long.

The chemical durability of different waste forms can be compared by considering "steady-state" release rates postulated to occur under specific conditions. The NRC's repository release criterion is, for HLW, proposed as 10^{-5}/yr of the radionuclide inventory present after 1000 years. In Fig. 12-7

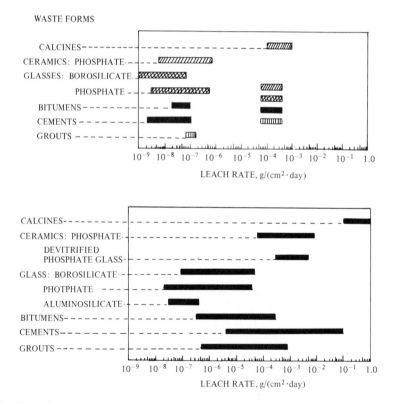

Fig. 12-7. Leach rates of alkali and alkaline earth elements and REE and actinides from different waste materials. From McVay *et al.* (1981).

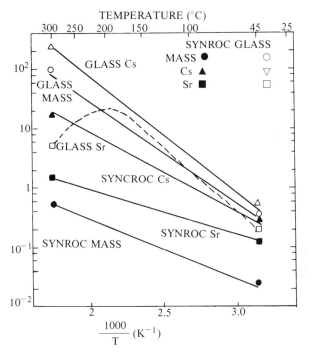

Fig. 12-8. Temperature effect on initial leach rate of borosilicate glass and synroc (with 10% simulated radwaste). From McVay *et al.* (1981).

is shown a tabulation of leach rates for alkali and alkaline earth elements and for the rare earths and actinides. The leach rates are calculated for a 1000-year situation, and the net rate of between 10^{-7} to 10^{-8} g/cm^2/day corresponds to 10^{-5}/yr. In fact, the actinide data (Table 12-5) will control the leach rates at about 1000 years.

Distinctions between waste forms is greatest between 0 and 1000 years, and it is not yet possible to make waste-specific tests to fully postulate the reactions that may take place in this period. Yet one must remember that the leach rates reported are in the absence of canister, sleeve, overpack, and engineered backfill, and it must be demonstrated that these barriers can indeed be penetrated in just a 1000-year period before even the reported leach rates become meaningful.

Mendel *et al.* (1981a) report that modifications of glass compositions appear feasible for borosilicate glasses. These modifications would result in glasses with lower leach rates by about a factor of 10 or so. For crystalline materials, and especially SYNROC, they possess lower leach rates than borosilicate glasses. In fact the decrease in leach rates with time is also more pronounced for SYNROC than for borosilicate glass. The comparison of leach rates as a function of temperature is shown in Fig. 12-8. In the case of

Table 12–9. Comparison of the chemical durability (Soxhlet Test) of waste glass and common minerals.

Minerals	Wt% Leached
Quartz crystals	0.41
Milky quartz	0.50
Dolomite	0.55
HLW glass	0.70
Garnet	0.73
Corundum	0.77
Orthoclase	0.90
Granite	1.10
Quartzite	1.20
Felsite	2.10
HLW glass (devitrified)	2.50
Marble (dolomite)	2.90
Calcite	5.80
Basalt	6.10

Source: Adapted from Ross (1975).

other crystalline materials, the Savannah River HLW contains a large amount of Al, and high-Al ceramics are appropriate for it. Harker *et al.* (1980) report leach rates of 10^{-5} g/cm^2/day for this high-Al ceramic as opposed to 10^{-4} to 10^{-3} g/cm^2/day for borosilicate glass.

Factors that may affect waste form stability are shown in Table 12-9. As expected, acidic pH enhances the leaching of various waste forms, while increased temperature is variable in its effect on leaching. An order of magnitude in leaching results can be caused by compositional variation within the fluid phases, while radiation effects generally cause leaching increase, but at a low rate. Solubility effects are difficult to evaluate due to the very long times required for solids to reach equilibrium with dissolved ions. When chemically reducing fluids are used, the overall effect is to decrease leachability of the waste form. The mechanics for leaching are still under investigation. In the short term, diffusion appears to be the important mechanism, but for the long-term, surface controlled mechanics, e.g., sorption, and precipitation, dominate.

Conclusions

Various waste forms under intensive study both in the United States and abroad show that most exhibit strong resistance to leaching under assumed repository conditions, and that most have favorable mechanical and thermal properties as well. The decision to use glass as a waste form has already been made in France, and glass will probably be used at least for defense waste in

Table 12-10. Factors affecting waste-form stability.

Variable	Borosilicate Glasses	Natural Silicate Glasses	Spent Fuel	UO_2, Single Crystals
pH (for oxidizing conditions)	Increased leach rate* at acid (<6) and basic (>10) pH's	Increased leach rate[a] at basic (>8) pHs		Increased leach rate** at basic pHs (>9) and acidic pHs; presence of HCO_3^-
Temperature	Leach rate element dependent; $\geq$Si rate; most experimental data $<100°C$	Increased dissolution and recrystallization with increased temperature	Lower leach rate with increased T, except for ^{137}Cs; measurements to 150°C	Decreased leach rate[b] with increased T to 150°C
Fluid composition	One order of magnitude variation in leach rate for different ground waters	Presence of Cl^- increases leach rate of certain elements; Na^+ and K^+ facilitate ion exchange with glass	Decrease in groundwater leach rates compared to H_2O; U results conflicting	Leach rate strongly dependent on fluid composition ($HCO_3^- \gg H_2O > Cl^-$:75°C: $HCO_3^- \gg H_2O \simeq Cl^-$:150°C)
Radiation effects	Magnitude of radiation damage to glass uncertain with respect to leach rate; radiolysis effects in N-bearing fluid produces HNO_3, increasing leach rate	Increased alteration rate	Probable effect is to increase leach rates; under study presently	Radiolysis increases leach rate; greatly enhanced if N present for formation HNO_3
Solubility	Solubility limit for amorphous silica not reached after 1 year in H_2O at $<100°C$	Dissolved SiO_2 concentration determined by amorphous SiO_2 solubility	Not investigated	Supersaturated fluid with respect to UO_2 (?)

Table 12-10 (*continued*)

Variable	Borosilicate Glasses	Natural Silicate Glasses	Spent Fuel	UO$_2$, Single Crystals
Oxidation state of fluid	Decreased leach rate of Pu, U, Nd with reducing fluid		Decreased leach rate with reducing fluid except for ^{137}Cs, no effect	Decreased leach rate of U with reducing fluid
Static *vs.* dynamic	Increased flow rates; increased leach rate			
Mechanism	Diffusion-controlled in short term; long-term surface, steady state, and/or precipitation/ sorption	Diffusion-controlled in short term; long-term surface-controlled	Incongruent leaching ^{137}Cs, ^{90}Sr in short term; long-term steady state concentration for other elements	Oxidation/dissolution, precipitation of UO$_3$-hydrates

Source: Modified from Moody (1982).
[a]Determined by Si weight loss and Na extraction.
[b]Determined by U concentration change in reactive fluid.
T = temperature.

the United States. Synthetic waste forms are varied, and most also exhibit favorable resistance to leaching and favorable thermal properties, although not all have favorable mechanical properties. Synroc is, in theory and backed by numerous experiments and natural analogs, remarkably resistant to attack by leachants under crustal conditions.

Naturally occurring glasses are known that are tens of millions to hundreds of millions of years old, thus attesting to their resistance under surface oxidizing conditions. Under reducing conditions, these glasses would be even more resistant to chemical attack. Minerals as analogs to phases present in supercalcine and synroc also are known to be extremely stable in the crustal environment, as attested to by radiometric age determination and other chemical and isotopic information.

Some laboratory leach rates are clearly too high for materials tested; if these rates were valid, the continental land masses should have been levelled and inundated eons ago. More recent work shows that leach rates of less than 10^{-5}/yr (NRC) are to be expected. Finally, it must be remembered that in the multibarrier system, the waste form is only the first barrier, and canister, backfill, and repository rocks proper will pose further barriers to radionuclide migration.

CHAPTER 13

Engineered Backfill and Canisters

Introduction

The waste form is the initial barrier in the multibarrier system, after which is the canister encasing the waste form. The canister(s) is, in turn, surrounded by clay minerals, with or without additives, that constitute the engineered backfill. Clay minerals, because of their high sorptive capacity for the radionuclides of interest, are prime candidates for backfill materials. Naturally occurring clay minerals, including shales, are not only good for retaining the radionuclides of interest, but also are impervious to fluid flow. The choice of canister materials has narrowed to Ti-alloys that have been shown to be resistant to leaching under simulated repository conditions.

Engineered Backfill

Part of the multibarrier system is the blanket or continuous layer of material surrounding the waste canister. This engineered backfill may consist of any of a number of materials, all of which are chosen because of their potential to sorb or otherwise retain radionuclides should they escape from a breached radwaste canister. These materials commonly are clays, shales, or mixtures of clay minerals or shales and sand or other materials. According to Klingsberg and Duguid (1980), some of the purposes of backfill are:

(1) To absorb the limited amount of water present in a dry repository (i.e., fluid inclusions in salt).

(2) To impede flow of any groundwater to or from the waste package.

(3) To remove radioisotopes by sorption from water that has reached the waste in the event of breaching of the canister.

(4) To change the chemistry and the composition of groundwater so as to minimize leaching of the waste form and corrosion of the waste package.

(5) To provide mechanical relief for emplacement hole closing as well as stresses on the waste package due to rock movement.

(6) To serve as a heat transfer medium.

For example, Nowak (1980b) has estimated that a one-foot thickness of backfill with a linear distribution coefficient, $K_d = 2000$ ml/g, can delay the release of the transuranic elements past the backfill by some 10^4 to 10^5 years, and fission products with $K_d = 200$ ml/g (^{90}Sr, ^{137}Cs, and others) can be delayed for 10^3 years. This time is sufficient for decay of ^{90}Sr and ^{137}Cs to negligible amounts. While the use of K_d values for this and other purposes is very approximate due to the difficult task of demonstrating equilibrium, speciation, matrix characterization, and other factors, the values are nevertheless useful working approximations to attempt to predict the fates of radionuclides in the waste package–backfill system. The importance of the K_d values mentioned above is that, while they may vary by a factor of ± 1–3, they do indicate the probability of retardation of the radionuclides from a breached radwaste canister.

The earlier work on engineered backfill (or equivalent) has been summarized by Robinson (1962), Akatsu et al. (1965), and Tamura (1972), among others. Nowak (1980a,b) has presented literature reviews, including site-specific studies for shale and bedded salt geomedia, as well as a discussion of backfill materials. The studies of Dosch and Lynch (1980), for example, demonstrated the importance of the addition of activated charcoal to the engineered backfill in order to act as a getter for TcO_4^-. Brookins (1980a), Bird and Lopata (1980), and Bondietti and Francis (1979) have argued that small amounts of magnetite (or possibly Fe^{2+}-bearing clay minerals) may also be effective getters for TcO_4^- and other species. At many of the radwaste disposal sites, the waste package will be initially stored at depth under a chemically reducing environment. Under such conditions, local oxidizing conditions are expected in the near field due to radiolytic effects, release of oxygen from fluid–waste package interaction, and other factors. Such oxidizing conditions are not expected to exist past the near field, however, as the surrounding backfill and rock will first buffer the fluid choked with radwaste and reduction will be concomitant with the buffering or following immediately afterward so that soluble species such as TcO_4^-, RuO_4^-, and U^{6+}-carbonates will be reduced to insoluble TcO_2, RuO_2, and UO_2, respectively. Bondietti and Francis (1979) have shown the ability of simple shale backfill materials to effectively reduce and remove from solution dissolved Tc, Np, and U.

The use of clay minerals in engineered backfill has been recognized for a long time (see discussion in Nowak, 1980a,b). Clay minerals with an intermediate cation exchange capacity (CEC) and radionuclide-specific exchange sites are under investigation by numerous investigators. Nowak (1980b) has presented the results of detailed chemical studies of materials, including clays, soils, zeolites, and charcoal. Although his studies are aimed more or less specifically at the WIPP site, the results can be applied to other systems as well. In fact, at the WIPP site any fluid that might possibly reach the waste canister will be, by definition, a brine, and the brine–backfill reactions are quite possibly more complex than simple water–backfill reactions. Nowak (1980b) used Brines A and B (see Chapter 7 for discussion) with the different materials. The pH of the various experiments was regulated to approximate that of bedded salts (about 4.5). Nowak (1980b) used radioactive species for his experiments: ^{152}Eu(III), ^{238}Pu(IV), ^{243}Am(III), ^{137}Cs(I), ^{85}Sr(II), and ^{99}Tc(VII). The radiotracers were added to brines A and B and then reacted with bentonite, hectorite, clinoptilolite, mordenite, activated charcoal, tuff, caliche, sand, Dewey Lake Redbed (a sedimentary unit consisting of interlayered siltstone, gypsum, clays from near the WIPP site). All experimental results were given in terms of a semiempirical distribution coefficient, K_d, which is defined as:

$$K_d = \frac{C_s}{C_1}, \qquad (13\text{-}1)$$

where C_s is the concentration of the sorbing species on or in the solid phase for a specific experiment, in M/g, and C_1 is the concentration of the sorbing species in the fluid phase, again for a specific experiment, in M/ml. I again stress that the K_d values are just empirical ratios of concentrations for any specific experiment. If slightly different starting materials or reaction fluids are used, the K_d values are not quantitative, but are useful working guides for use in attempting to predict radionuclide behavior in various systems. The calculation of K_d values can be done in numerous ways. One convenient expression is

$$K_d = \frac{V}{W} \cdot \frac{R_f - R_s}{R_s - R_b}, \qquad (13\text{-}2)$$

where V is the volume of fluid added to the sample, W is the mass (in grams) of the solid during the experiment, R_s is the average count rate of filtered brine, R_f is the average count rate per unit volume of feed brine, and R_b is the average background count rate (normalized to a unit volume of counted brine). For the ^{152}Eu, widely variable K_d values were obtained, with values ranging from 11 to 26,000. Low values (i.e., less than 100) were common for mordenite, caliche, and kaolin-solid experiments. In all other experiments the K_d values ranged from several hundred to tens of thousands. Zeolite behavior is complex (see Neretnieks, 1977), yet the K_d values are still significantly greater than 0, thus suggesting that all materials examined are favorable for removing Eu from solution. The Eu experiments are of interest not only for WIPP-specific experiments, but because at high pH values, removal of any Eu(III) as $Eu(OH)_3$ or possibly as $Eu_2(CO_3)_3$ is likely based on Eh–pH considerations (Brookins, unpublished).

Montmorillonitic clay minerals have been examined for use as backfill by numerous investigators, primarily because they are capable of sorbing radionuclides, possess low permeabilities for fluids, and for other reasons, Bentonite, a montmorillonite-rich clay derived from volcanic ash, is especially useful because it swells on the addition of water or other fluids and thus flows plastically into cracks and holes, and is relatively impermeable to infiltration by waters. A mixture of bentonite and sand has favorable thermal conductivity and support strength for radwaste applications (see Pusch, 1977). Hectorite also has many of these same properties (Nowak, 1980b), and K_d values indicating retention of Eu, Am, and Pu in the solid phase for the different experiments were obtained.

In two separate studies, Brookins (1980a, 1982a,b) has investigated naturally occurring rocks as analogs for backfill. In the first study (1980), he argued that the uranium deposits of the well-known Grants Mineral Belt, New Mexico could be treated as analogs for buried spent fuel rods without benefit of canister and overpack, and that the rocks surrounding the uranium ore could be treated as a natural backfill. In this treatment, elements commonly concentrated in the ore zone are, in many instances, the normal crustal counterparts of fission products, and how these elements behave in the natural rocks can be taken to study fissiogenic element behavior. For example, in the typical uranium deposits, the elements U–Mo–Se–V–As are carried together as metal oxyion complexes and removed from solution as a result of reduction. As the ore forms, a series of complex reactions occurs, which forms a new generation of clay minerals that is intimately mixed with the uranium ore. These clay minerals are commonly chlorite or montmoril-lonite or illite or mixed-layer varieties of these (see Brookins, 1979b, 1982a). Brookins (1976, 1981d) has shown that these clay minerals can be used, with the Rb–Sr method, to date the age of uranium mineralization. Since the ages demand closed-system conditions for Rb and Sr, the very fact that these ages have been undisturbed since age of formation indicates no migration of Rb or Sr since this age. Further, Cs is more effectively retained in rocks and minerals than Rb (see Kharaka and Berry, 1973; Brookins, 1982a), hence the same rocks and minerals are also closed for Cs. In addition to Mo, U, Rb, Sr, and Cs, elements of interest that are retained in the ore–clay mineral–rock system include Sb, Ba, and the lanthanides. While data are not available for many other elements, the overall chemically reducing environ-ment, known mineralogy, and Eh–pH and activity diagrams argue for favorable conditions for retention of Zr, Tc, Ru, Pd, Rh, Ag, Cd, In, and Sn (see Brookins, 1979b).

In some parts of the Grants Mineral Belt, a series of tectonic events caused destruction of some early-formed uranium ore deposits with remobilization and reprecipitation of new deposits close to the original sites (see Brookins, 1981d). This, too, is important, as the new generation of clay minerals can also be dated by the Rb–Sr method (Brookins, 1981d), and the same trace elements are again concentrated with the uranium ore. It must be emphasized that the destructive events that affected the early uranium ore are those *not* expected to occur in stable tectonic areas where geologic sites are under consideration, hence events of similar magnitude are not to be expected. However, the destructive events probably occurred some 15–20 million years after initial uranium ore mineralization and some 5–10 million years were required for destruction and remobilization and reprecipitation; the new ore deposits were formed within a few hundred meters of the probable original site. This information, qualitative as it is, suggests that even in the event of a major tectonic disruptive event, the mobilization of uranium and the other elements was not extreme. Had the uranium ore, for example, been located some 800–1000 below the surface, it would have been reprecipitated within 0–200 m of the original site, still well below the surface. In addition, if the series of disruptive events required 5–10 million years to take place, that would be well beyond the 0.25 million years "isolation period" (i.e., 10 times the half-life of ^{239}Pu) commonly cited.

In the second analog (Brookins, 1982a), study was made of the Dakota Formation of the San Juan Basin in New Mexico and Colorado. This formation consists predominantly of sandstone mixed with bentonite in many places. The rocks chosen for study approximated a 1:9 mix of bentonite and quartz (with minor feldspar and other minerals), in the range of the bentonite–sand mix used by the Swedish Government (Brookins, 1982a,b). These rocks have been extensively studied because of their proximity to the uranium deposits in the San Juan Basin and for other purposes (Brookins, 1982a,b). Brookins (1982b) showed that the Dakota Formation has served as an effective getter for uranium (and associated elements) during destruction of uranium ore from nearby rocks, and that joints, bedding plane fractures, and argillaceous parts within the formation were favorable sites for deposition of uranium and other elements. Rb–Sr age of primary clay mineral assemblages again yielded the age of formation, and the secondary ore possessed a new suite of clay minerals that could be dated by the Rb–Sr method. The trace element chemical studies showed that the newly formed ore possessed chemical signatures similar to those of suspected source rocks, and that the chemical integrity of the Dakota Formation was unaffected by the younger events. In short, the original rock and constituent minerals retained its chemical and isotopic integrity, and newly formed deposits of secondary minerals occurred primarily by void filling and some sorption. The Dakota Formation thus served as a repository and as backfill. The reader is again referred to Brookins (1979b, 1982b) for more details on this matter.

Nowak (1980b) reported K_d values from 350 to 5200 for 1:9 mixes of bentonite or hectorite and sand for ^{152}Eu, thus showing the effectiveness of the hectorite and bentonite to sorb the Eu. For ^{238}Pu with bentonite, hectorite, and clinoptilolite, both individually and mixed with sand, the K_d values ranged from 400 to 40,000, again showing that these materials are well suited for removal of Pu from solution. For Am, the K_d values are, as expected, very high, ranging from 4100 to 16,000. Since the actinides all behave more or less similarly, this is expected based on similar predicted behavior from Eh–pH considerations.

In the case of Cs and Sr, K_d values were much smaller, ranging from approximately 0 to 4.9. The concentration of these elements in the brines A and B is higher, however, than the amount added as radiotracer. Thus the concentration of Cs in brines A and B is 10^{-5} molar as opposed to 2×10^{-7} molar used as a tracer, and Sr is 6×10^{-5} molar (Brine A) and 2×10^{-4} molar (Brine B) while the Sr radiotracer was 5×10^{-11} molar. This means that the exchange of radiotracer Cs and Sr with the normal elements in the brines masked an elemental exchange with the solid samples. Cs K_d values ranged from 0.4 to 4.9, strongly suggesting that the solids were effective in scavenging radiocesium from solution. The 6 to 7 orders of magnitude between normal Sr and radiostrontium was too extreme, however, for similar results to be obtained for ^{85}Sr. The analog studies of Brookins (1981d, 1982b), indicating clays with closed system behavior of Rb and Sr even in the presence of brines, suggests the likelihood of such clay minerals scavenging Rb (and Cs) and Sr from solutions, but, admittedly, more follow-up work is needed to verify this.

The experiments for TcO_4^- cannot directly be used to comment on the suitability of the materials used (activated charcoal, hectorite, bentonite) under actual repository conditions, except for the activated charcoal. The TcO_4^- species was stable under the laboratory experiment conditions, but Tc(IV) is expected under repository conditions. Thus the K_d values of 0–1 for the hectorite and bentonite are a simple reflection of the point that these minerals are not efficient anion adsorbers. The K_d values of 290–380 for TcO_4^- on the activated charcoal are to be expected. The studies of Lopata and Bird (1980), Bondietti and Francis (1979), and Brookins (1980a) show that TcO_4^- will be removed as TcO_2 as a result of reducing conditions, especially if some magnetite or similar Fe^{2+}-bearing phase is present in small amounts. Addition of activated charcoal would, of course, remove even TcO_4^-, but it may not be necessary as abundant Fe^{2+} is already present. Nowak (1980b) was careful to point out that the K_d values determined cannot be rigorously applied to radwaste repositories because the reaction mechanisms for removal of Eu, Pu, and Am are not precisely known (i.e., complex versus single reactions are likely), and the high concentrations of Cs and Sr in brines prevent favorable K_d values for ^{137}Cs and ^{85}Sr from being obtained. The TcO_4^- should not be removed from the clay minerals under oxidizing conditions, and that is precisely what is observed.

The analog studies, coupled with the experimental data, especially for fluids other than brine or for diluted brines, support use of clay minerals such as montmorillonites and chlorites (i.e., hectorites) in engineered backfill. It is also theoretically possible to remove even I^- from solutions with the presence of sulfides (see Bird and Lopata, 1980), for example. The entire study of backfill may be used to argue in support of shales for repositories because, by definition, an infinite number of barriers in the multiple barrier concept, are present.

Nowak (1980c) has reported follow-up experimental results on engineered backfill materials specifically aimed at investigating brine–backfill reactions. His results are summaried as follows: For a mixture of 30% bentonite, 70% activated charcoal, the following retention times (in years) were obtained: Pu—5×10^4, Am—2×10^4, TcO_4^-—1×10^4, I^-—1×10^3. For 30% bentonite, 40% mordenite, 20% activated charcoal, and 10% sodium titanite the following retention times were obtained: Pu—5×10^4, Am—2×10^4, Cs—6×10^2, Sr—7×10^2, TcO_4^-—4×10^3, I^-—4×10^2. These results show that while individual repository needs (i.e., bedded salt versus crystalline rocks) may need to be tailored with mixtures of different materials, the tailored mixtures are remarkably efficient in retarding any potential radionuclide flow. Pusch and Bergstrom (1981) have also advocated use of bentonite as backfill, primarily because not only does it effectively retard radionuclide flow, but its swelling properties allow the clay to flow into minute openings in rocks and thus seal them. This, in turn, ensures direct contact between the backfill and the waste package and the surrounding rock, which is critical to prevention of lateral channeling at the interface between the waste package–backfill and backfill–repository host rock.

Numerous potential backfill materials have been studied by Komarneni and Roy (1980), including natural zeolites (erionite, mordenite, clinoptilolite, chabazite, phillipsite), artificial zeolites, clay minerals (montomorillonite, vermiculite), and oxides–hydroxides (commercial Fe-, Mn-, and Al species). Their approach was to investigate mixtures of zeolites and clay minerals, both with and without oxides–hydroxides present, based on the assumption that virtually all 40 or so fission-produced radionuclides can be sorbed by either zeolites or clay minerals. The zeolites and most clay minerals are structurally unaffected by modest increases in the partial pressure of water in a system, and remain unchanged even up to temperatures of 200°C. Studies were made (Komarneni and Roy, 1980) of the ability of the zeolites, clay minerals, and oxides–hydroxides to sorb Cs, Rb, Sr, Ba, La, Nd, and U. The rare earths La and Nd as well as U were totally sorbed (not unexpectedly). The natural zeolites were effective in removing Cs from solution, and were also good for Rb sorption. Removal of Sr by sorption (26–75%) was realized by clinoptilolite and mordenite, but the other zeolites did not sorb Sr. The synthetic zeolites were far more effective in sorbing Sr and Cs and fair for Rb sorption. The clay minerals were fair for Cs and Rb sorption, but poor for Sr, while oxide–hydroxide phase sorbed up to 94% Sr but were ineffective in sorbing Cs or Rb. Sorption of Ba was only attempted for the oxide–hydroxide mixture, and was most effectively removed (75%) by Mn–oxide–hydroxide. These studies show that individual minerals do not uniformly sorb Cs, Rb, and Sr from solutions emanating from a breached radwaste package. Komarneni and Roy (1980) then tested mixtures to assess their suitability. Their results show that a mixture of either clinoptilolite plus mordenite or clinoptilolite plus phillipsite are most effective in removing Cs (83–85%), Rb (63–84%), Sr (72–80%), and Ba (85–100%). These data, while very preliminary, according to Komarneni and Roy (1980), can nevertheless be used to comment on some aspects of engineered backfill materials as well as repository host rocks. First, rocks with fairly high zeolite contents (e.g., basalt, zeolite-bearing tuffs) will contain excellent natural barriers outside of the engineered system to prevent radionuclide migration through the rock. Second, rocks with zeolite–clay mineral mixes (e.g., tuffs, some basalt) will also prevent radionuclide migration through them. Third, rocks containing both montomorillonite and illite ($\pm$ chlorites) can, based on studies of naturally occurring shales and clay minerals (Komarneni and Roy, 1980), retard or retain radionuclide flow. Hence the mixes of zeolites with clay minerals such as hectorite (Nowak, 1980b) or montmorillonite–illite–chlorite (Brookins, 1980a) will serve very effectively as backfill.

Specific sites, as mentioned earlier, possess specific problems essentially unique to any one particular site. In the case of bedded salt, the engineered backfill must not only prevent water from reaching the waste package, and in the event this does happen, prevent radionuclides from escaping, but must also minimize the amount of water (or brine) present. To accomplish this additives to the proposed clay mineral–charcoal ($\pm$ other tailored ingredients; see Nowak, 1980c) may include desiccants such as CaO and MgO. These materials can be readily produced during the calcination of limestone and dolomite, and are very effective in removing water by forming hydroxides. Thus the reduction in brine volume will increase the likelihood of removal of any radionuclides entrapped on the other tailored ingredients of the backfill. Simpson (1980), who has discussed the use of such desiccants in detail, points out that not only will the volume of fluid be reduced by formation of $Ca(OH)_2$ and/or $Mg(OH)_2$, but any entrapped CO_2 will be removed by formation of $CaCO_3$ and/or $MgCO_3$.

In the case of basalt, a different suite of tailored ingredients may possibly be employed. Coons et al. (1980) have pointed out that two concentric layers of backfill, each with a distinct purpose, will be effective. The composite backfill will consist of bentonite, other clay minerals, quartz, zeolites, phosphates, basalt, basalt glass, sodium feldspar and bornite. The bornite is added to scavenge I^- and possibly some TcO_4^-. Under chemically reducing conditions the Tc(VII) will be reduced to Tc(IV) and removed in the basalt by reduction, perhaps induced by Fe^{2+}. The sodium feldspar will remove Cs and Sr (and Rb), while the phosphates will remove Ba, the rare earths, and the actinides. The clay minerals, including bentonite, will scavenge much of the alkali and alkaline earth elements from solution, and the zeolites from the remaining Cs, Rb, Sr, and Ba. Some materials may be sorbed directly on the basalt glass. An inner concentric layer will consist of crushed basalt plus additives, which will be surrounded by bentonite (plus clay) plus additives.

The outer layer is designed to prevent waters from reaching the waste package, while the inner layer will be designed to remove radionuclides.

Shales and clay minerals as used for engineered backfill have been investigated to test their ability to react with, and retain, radionuclides from spent fuel. Komarneni *et al.* (1979) have examined a number of common shales (Salona, Antrim, Brallier, Conasuaga, Catskill) with spent fuel analogs under hydrothermal conditions. In particular, they wished to test for Cs, Mo, and U behavior in such systems. The results of their experiments are of interest in that the illite–montmorillonite–chlorite–kaolinite ($\pm$ other phases) found in the shales is sufficient to fix Cs and other radionuclides. CsOH, CsI, Cs_2MoO_4, and b-$Cs_2U_2O_7$ were taken as probable Cs-bearing phases in spent fuel (O'Hare and Hoekstra, 1975). In all experiments, the pressure was 300 bars, and samples of shale were reacted with the Cs phases at $100°$, $200°$, and $300°C$ for 4, 2, and 1 months, respectively. In all experiments pollucite was formed. Once locked into this mineral's structure, Cs is very stable. Some Cs was removed by fixation in montmorillonite by ion exchange. The Mo in Cs_2MoO_4 reacted to form powellite ($CaMoO_4$) because of the excess of Ca in the shales; powellite is stable in the crustal environment.

The U from b-$Ca_2U_2O_7$ was removed by two means. In samples of oxidized shale, the Cs was removed into new phases (pollucite, montmorillonite) and the U^{6+} fixed in schoepite ($UO_3 \cdot 2H_2O$) or a weeksite analog (($Cs,Na)_2(UO_2)_2(Si_2O_5)_3 \cdot 4H_2O$). These new phases may be produced by even locally oxidizing conditions and, once formed, are remarkably stable in the earth's crust (see discussion in Brookins, 1982b). For samples of shale containing reduced phases such as organic matter and/or pyrite, the U^{6+} was reduced and removed by formation of UO_2. In a natural repository the overall chemical environment will be reducing. Hence if U^{6+} is formed in the near field, it may be fixed in a schoepite or weeksite phase, which, if somehow altered, will lose its U^{6+}, and in turn be removed by reduction in UO_2. This is important because if conditions that prevent U migration are found, the other actinides, which are less soluble than U-compounds, will be even more successfully retained in the backfill.

Nowak (1982) has further investigated Cs and Sr behavior in brine-saturated backfill for an hypothetical bedded salt repository. He first pointed out that there are many problems with simple volume/mass K_d data in that many aqueous species may be affected by complexing, pH, Eh, sorption, precipitation, and other factors. Reactions involving just-distilled water are not directly applicable to a bedded or dome salt medium. Nowak (1982) used mordenite and sodium bentonite for essential backfill materials in brine-saturated columns to which ^{137}Cs and ^{85}Sr had been added so that their behavior in the columns could be monitored by simple gamma counting. In other experiments, sodium titanate was added to bentonite, and sodium titanate plus activated charcoal was added to bentonite plus mordenite. Brine A, representative of brines likely to be found in WIPP rocks, was used. To model the results, Nowak (1982) used the following formula, which describes one-dimensional diffusion from an instantaneous planar source:

$$\frac{C}{M} = \frac{1}{(\pi Dt)^{1/2}} \exp\left(-x^2/4Dt\right), \qquad (13\text{-}3)$$

where

C = concentration of diffusing materials at distance x (M/cm^3),

M = total amount of diffusing substance in instantaneous source per unit area normal to direction of diffusion (M/cm^2),

$D = D^*/\alpha(1 + \varepsilon K_d/\varepsilon) = D^*/\alpha R,$

α = totuosity, = $\sim(2)^{1/2}$,

R = retardation factor = $(1 + \rho K_d/\varepsilon)$,

ρ = packing density (g/cm^3),

K_d = distribution coefficient for linear equilibrium sorption (ml/g),

ε = void fraction,

x = distance from source in direction of diffusion (cm),

t = time after introduction of source (sec).

Nowak (1982) then compared his K_d results based on diffusion for Cs and Sr with batch sorption measurements (from Winslow, 1981); the results for Cs diffusion are given in Table 13-1. The results show that the diffusion K_d agrees with the batch K_d within a factor of 2. For strontium, the results are given in Table 13-2. The results of the Sr studies show agreement within a factor of 4 between the diffusion and batch K_d data. The low-diffusion K_d for use of nonsulfate brine may be explained by precipitation of SrSO$_4$ in the sulfate-bearing experiments. The diffusion K_d values argue for strong retardation under Brine A (i.e., repository) conditions for the backfill mixtures shown. The addition of mordenite and sodium titanate (+ charcoal)

Table 13–1. Comparison of K_d results based on diffusion for Cs and Sr and batch sorption methods.

Backfill Mixture	Average K_d from Diffusion Data (ml/g)	Average K_d from Batch Sorption Data (ml/g)
Bentonite	9	5
Bentonite + mordenite	19	14
Bentonite + mordenite + sodium titanate + charcoal	22	12
Bentonite + mordenite (non-Cs, Brine A)	19	14

Data from Nowak (1982); Brine A used as fluid.

Table 13-2. Comparison of K_d results based on diffusion for Sr and batch sorption methods.

Backfill Mixture	Average K_d from Diffusion Data (ml/g)	Average K_d from Batch Sorption Data (ml/g)
Bentonite	5	0
Bentonite with non-sulfate Brine A	2	—
Bentonite + sodium titanate	40	13
Bentonite + mordenite + sodium titanate + charcoal	45	13

Data from Nowak (1982); Brine A used as fluid except where noted.

strongly increases the retardation of Cs and Sr. Uncertainties in actual repositories may include the following (from Nowak, 1982).

(1) Radiation exposure may change retardation properties of backfill materials.

(2) The factor R is likely to change for highly compacted materials (i.e., low-density data cannot be unequivocally extrapolated to high-density materials).

(3) Diffusing species other than Cs^+ and Sr^{2+} may be present in leachates.

(4) Hydrothermal solutions may alter the retardation of properties of bentonite.

(5) Diffusion of strontium in a temperature gradient may be complicated by temperature-dependent solubility of $SrSO_4$.

(6) A Soret effect in a temperature gradient may occur.

Despite these uncertainties, the studies of Nowak (1982) convincingly show that readily available backfill materials such as bentonite, with additives, can retard Cs and Sr, and that these mixtures have already been shown to successfully retard many other fission products and actinides.

The extensive work on backfill and canister materials in the waste package has been summarized by Moak (1981), and the material below is abridged from his report. Several potential backfill materials have been identified for additional study (Table 13-3a), including sodium bentonite and other swelling clays with additions of other materials for greater stability and efficiency in retarding and retaining radionuclides. The sodium bentonite chosen as the base material has the ability to swell and thus minimize migration of groundwater (which will also self-seal cracks and other openings or breaks), to sorb specific radionuclides, and to react with intruding waters to prevent attack on canisters and/or to remove from solution dissolved radionuclides. Some of the additives to the backfill now under investigation are listed in Table 13-3b.

Table 13–3a. Preliminary list of candidate backfill materials.

Clays
   Sodium bentonite
   Calcium bentonite
   Illite
   Treated sodium bentonite
Sand
   Quartz sand (10–230 mesh)
Zeolites
   Clinoptilolite
   13X
   Zelon-900
Metal Powders or Fibers
   Iron
   Aluminum
   Lead oxide
Minerals/Rocks
   Pyrite
   Ferrosand (glauconite)
   Basalt
   Tuff
   Serpentine
   Anhydrite
Charcoal
Desiccants
   MgO and CaO

Source: Modified from Moak (1981).

Table 13–3b. Selected additions to bentonite.

Material	Attributes
Quartz sand	Increases thermal conductivity and strength of the backfill
Zeolites	High ion-exchange capacity increases sorption of selected radionuclides
Metal powders (Fe, Al)	Increases thermal conductivity, lowers oxygen fugacity
Graphite/charcoal	Increases thermal conductivity and sorption of selected radionuclides, e.g., Tc, I
Host rock (basalt in basalt repository)	Lower oxygen fugacity, enhances sorption

Source: Modified from Moak (1981).

The screening results on corrosion-resistant canister metallic materials for both brines and basaltic groundwaters are pure titanium and the alloy TiCode-12, with backup by nickel-based alloys. Tables 13-4 and 13-5 give the results of screening tests in salt and basaltic environments, respectively. The information presented in these tables indicates the same conclusions for both salt and basalt conditions. Although the screening for backfill and canister materials is now complete, candidates other than those listed in Tables 13-4 and 13-5 may be included in the future.

Backfill materials have been rigorously assessed (see Moak, 1981). Among the materials that received attention are bentonites, quartz, crushed host rocks, zeolites, kaolinite, attapulgite, iron, aluminum, CaO, MgO, Al_2O_3, $CaCl_2$, PbO, tachyhydrite, anhydrite, gypsum, apatite, chalcopyrite,, bournonite, ultramarine, azurite, borates, charcoal, peat, and mixtures of the above. The backfill must provide a multifunctional role in the overall performance of the waste package. Water exclusion is best assured by using a mixture of bentonite and quartz sand, roughly about 2:8. Clay:sand ratios lower than this are less effective as sealing substances, although the amount

Table 13-4. Summary of corrosion resistance screening tests in a salt environment.

Material	Attributes
Selected for further study	
Ti-base[a]	Very low uniform corrosion rates; no
Titanium	evidence of nonuniform corrosion
TiCode-12	
Ni-base superalloys	Very low uniform corrosion rates; some
Inconel-625	evidence of nonuniform corrosion;
Incoloy-825	higher cost than titanium
Hastelloy-C276	
Cast iron/steel[b]	Expected high corrosion rates but may be applicable in very thick sections
Currently rejected from further study	
Low alloy and mild steel	High corrosion rates
Stainless steels	Good uniform corrosion resistance but known to be subject to stress corrosion cracking
Lead	High corrosion rates
Copper and Cu–Ni alloys	High corrosion rates under oxidizing conditions; poorly adherent, heavy scale develops
Zircoloy-2	Comparable corrosion behavior to titanium but costs considerably more

[a]Primary candidate for borehole design concept.
[b]Primary candidate for self-shielded design concept.

Table 13–5. Summary of corrosion resistance screening tests in a basalt environment.

Material	Attributes
Selected for further study	
Ti-base[a]	Very low uniform corrosion rates; no
Titanium	evidence of nonuniform corrosion
TiCode-12	
Ni-base superalloys	Very low uniform corrosion rates; higher
Inconel-625	cost than titanium
Incoloy-825	
Hastelloy-C276	
Cast iron/steel[b]	Expected high corrosion rates but may be
	applicable in very thick sections
Currently rejected from further study	
Stainless steels	Good uniform corrosion resistance but
	known to be subject to stress corrosion
	cracking
Copper and Cu–Ni alloys	High corrosion rates
Zircoloy-2	Comparable corrosion behavior to
	titanium but costs considerably more

Source: From Moak (1981).
[a] Primary candidate for borehole design concepts.
[b] Primary candidate for self-shielded design concept.

of piping and other openings is small (Moak, 1981). The sorption capability of the backfill is important, as it is the role of the material to remove radionuclides, presumably by sorption, from waters that enter the backfill environment. As pointed out earlier, Nowak (1979) has shown that a 0.3-m-thick backfill around a waste canister is sufficient to delay the breakthrough of Pu and Am for $10^4 10^5$ years, and Cs and Sr for 10^3–10^4 years, so that these radionuclides would have decayed to below background levels. Similarly, Swedish studies (Moak, 1981) show that a 0.2-m-thick clinoptilolite barrier could retard breakthrough of Cs and Sr for 10^4 years, and of Pu and Np, some 2×10^6 years. How diffusion in the interstitial waters relates to sorption and other processes or radionuclide removal is not too well known (see Moak, 1981), although the ability of the materials to effectively sorb specific radionuclides is well documented. In bentonite–sand mixtures and in clinoptilolites diffusivities on the order of 1 to $1.3 \times .0^{-10}$ m^2/sec have been observed, although diffusion data for compacted clays are badly needed.

The backfill materials also essentially buffer the groundwaters to ensure a stable, near-neutral pH, and also a fairly well controlled Eh. This is important, as for waste repositories, Eh conditions favoring mild- to stronger-reducing conditions mean that species of Tc, Ru, the actinides, and others are likely to occur in low-valence, highly insoluble forms. The self-sealing ability of backfill materials has also been studied. Basically, the bentonite–sand mixture, and other materials and mixtures as well, has ideal self-sealing properties (Pusch, 1977; Moak, 1981). These materials and mixtures are essentially insensitive to radiation damage (Moak, 1981).

The heat-transfer capability of the backfill is also important. In general, the thermal conductivity increases with increasing degree of water saturation. A 10% bentonite mixture will take up to 20–30% water if no swelling takes place. Hectorite and clinoptilolite also possess thermal conductivity favorable for consideration for backfill materials.

The mechanical properties of backfill can be affected by processes such as cementation; noncemented backfill, under stress, will permanently deform, whereas the cemented backfill could produce cracks and fissures. Sodium bentonites, in particular, reduce in volume under compression, much more so than kaolinite or illite.

The long-term compatibility and stability of backfill materials has been investigated at low temperatures (room temperature) and at elevated temperatures and moderate to high pressures. Sodium montmorillonites maintain their high cation exchange capacity even at elevated temperatures relative to Ca and Li forms.

From the extensive studies in the United States and Sweden, mixtures of bentonite and quartz sand appear to have many properties that make them ideal for backfill concentration. In Table 13-6 the properties of bentonite and quartz are given.

Canister and Other Corrosion-Resistant Structural Components

As part of the NWTS program, the materials being considered for canisters and other corrosion-resistant parts of the waste package have been investigated, primarily at Sandia National Laboratories (SNL) and Battelle Pacific Northwest Laboratories (PNL). Various materials have been tested in the presence of WIPP brines, seawater, basaltic groundwaters and de-

Table 13–6. Main advantages (+) and disadvantages (−) of bentonite/quartz material.

Feature	Quartz material	Bentonite
Long-time stability	+	+
Bearing capacity	+	−
Plasticity	−	+
Permeability	−	+
Swelling properties	−	+
Heat conductivity	+	−
Ion-exchange capacity	−	+

Source: Modified from Moak (1981).

ionized water. The major ions present in Brines A and B and in seawater are given in Chapter 6. Brine A represents brines likely to be found in a repository in bedded salt, while Brine B is more typical of dome salt. About 20 candidate alloys studied for corrosion have been listed by Moak (1981). The corrosion rates were measured by weight change after either 6 months or 28 days of testing. The testing was usually carried out at 250°C, which is the approximate temperature (maximum) on the outer surface of the overpack for commercially reprocessed HLW in a dry salt repository. This temperature is too high; a maximum temperature of 150°C is more credible for a repository flooded with brine. Thus the 250°C data (Moak, 1981) are "overtests" in that the temperature will be lower, and the corrosion rates reported are also higher than will be encountered should canister breaching take place.

When both corrosion testing and economic considerations are evaluated, it appears that TiCode-12, a titanium alloy (ASTM grade 12, titanium with 0.3% Mo, 0.8% Ni, 0.2% O_2, plus impurities of Fe, H, C) is the most suitable for canister considerations. Another form of titanium, Ti-50A (ASTM grade 2, chemically pure titanium) is being considered as back-up for temperature less than 100°C. Ti-50A is less expensive than TiCode-12 but the latter is more corrosion resistant at temperatures in excess of 100°C.

Alternative overpack materials to TiCode-12 were also studied, and include inconel-625, hastelloy-C-276, ferritic alloys, 20-4, Ni–Fe alloys, and Cr–Mo alloys. These are being studied in the event that additional testing indicates TiCode-12 is unsuitable.

In support of conceptual waste package designs (for salt repositories) conducted by the Westinghouse Advanced Energy Systems Division (see Moak, 1981) for the NWTS program, several low-cost cast-iron and stainless steel materials are being tested. Of primary interest are several nodular cast irons and 1018 mild steel for the so-called self-shielded concept (SS). Corrosion tests are under way in Brine A as a function of temperature (50, 70, 90, and 150°C) and time (10 days to about 6 months).

In addition to the Sandia work, a wide variety of alloys were also examined in the PNL screen studies (Westerman, 1980). Specimens of the materials listed in Table 13-7 were exposed in one or more of the tests.

Two water chemistries were used in the study: a Hanford basaltic ground water (HGW) and an NaCl–MgCl$_2$ brine considered possible under accident conditions in bedded salt. All tests were performed at 250°C under anoxic conditions (50 ppb to 1 ppm O_2).

The results of some of the testing of alloys in the presence of Hanford ground water are given by Moak (1981). The cast irons, copper, nickel, and copper–nickel alloys all showed considerably higher corrosion rates than the inconel, incoloy, hastelloy, titanium, and zircaloy-2. Zircaloy-2, however, has been removed from further consideration because it is very expensive.

Moak (1981) is careful to point out, however, that the leach test results for laboratory experiments may not be directly applicable to an actual

Table 13–7. Alloys studied in the Battelle Pacific Northwest Laboratories.

Stainless steels
  304 SS
  304 L SS
  316 SS
  321 SS
  405 SS
  410 SS
Cast irons
  No. 180-7 (ductile iron)
  No. 22-8 (ductile iron, with 0.73% Cu)
  No. 142-12 (gray iron, with 0.68% Cr, 1.34% Cu)
  No. 166-3 (gray iron)
  No. 136-4 (gray iron, with 3.05% Ni)
Zirconium-base alloy
  Zircaloy-2
Nickel-base alloys
  Inconel-600
  Inconel-625
  Incoloy-800
  Hastelloy-C-276
  Ni-200
Titanium-base alloys
  Titanium-Code 12
  Titanium-Grade 2
  Titanium-6Al-4V
Copper-base alloys
  Copper
  Copper-nickel-70-30

Source: From Moak (1981).

repository, where the nature of reactions may be quite different. These generic conclusions are directly applicable to waste-package studies, though.

In salt media a series of corrosion experiments have also been carried out. Variables for Brines A and B and seawater include temperature (25–250°C), time (28 days–6 months), pH (1–8), oxygen content (30 ppb–1750 ppm), moisture content (dry–flooded), and gamma irradiation (dose rates of 10^7 and 10^5 rad/hr, doses up to 2×10^{10} rad/hr). The effects of dissolved oxygen, temperature, and irradiation on corrosion are discussed by Moak (1981).

TiCode-12 and Ti-0.2% Pd are quite corrosion resistant in hot brines and in seawater, and Ti-50A is corrosion resistant up to temperatures of 100°C or so. Part of this resistance to corrosion is due to the formation of a highly adherent, oxide film; in fact, these materials may be more corrosion resistant under oxidizing than reducing conditions, although they are corrosion resistant under both. The formation of crevices and pits in Ti-50A was observed at 250°C (Braithwaite, 1980), although not in TiCode-12 or Ti-0.2% Pd. The results of the irradiation testing indicate that only for the overtest dose of 10^7 rad/hr did irradiation, especially with increased time, have any effect on the corrosion rate. Glass (1981) has shown that materials such as TiCode-12, over which an oxide film forms and increases resistance to corrosion, are even less susceptable to irradiation effects. How materials such as TiCode-12 may act in the presence of large amounts of hydrogen is unknown, but, since H_2 is likely to be present in the vicinity of a radwaste package breached as a result of radiolysis, it must be pointed out that the oxide layer will largely controls the rate at which H_2 may actively attack the metal. In nature, Ti-oxide films (usually leucoxene) are remarkably stable, even in the presence of hot spring waters.

It is also known (Moak, 1981) that stress corrosion cracking may affect TiCode-12 and Ti-50A. Studies by Abrego and Rack (1981) show that both alloys are less affected by stress corrosion cracking in the presence of brine than in the presence of air. Further aspects of mechanical testing of titanium alloys are given by Moak (1981).

Under conditions of basalt groundwaters (Pitman, 1980), Moak (1981) summarized the following conclusions:

(1) Titanium-grade 2 and titanium-grade 12 exhibited strain-rate-dependent degradation of ductility in slow-strain-rate tests at 250°C.

(2) Tests in air confirmed that the loss of ductility is not limited to the simulated repository environment, and may be caused by an internal strengthening or embrittlement mechanism, such as dynamic strain aging.

(3) The ductility degradation in grade 12 titanium was found to be highly orientation dependent. Only one orientation of grade 2 titanium was tested.

(4) No fractographic evidence of an environmental cracking mechanism was found in the slow-strain-rate tests. All specimens failed by microvoid coalescence.

(5) The fatigue-crack growth rate of grade 2 and 12 titanium was not affected, relative to air or high-purity water, by any of the environments used in this study, and no frequency dependence was observed. This tentatively indicates that no environmental cracking mechanism relative to a Hanford basalt repository is significantly operative under the conditions employed, i.e., high-oxygen fugacity and a temperature of 90°C.

(6) All of the ceramic materials tested suffered extensive attack in one or more of the leach tests. Overall, the two grades of alumina appeared to be the most leach resistant of the ceramic materials tested. Zirconium dioxide and TiO_2 were the best of the remaining materials.

(7) Localized attack or alteration reactions become a serious problem for most of the candidate materials at temperatures above 150°C. Only graphite, ZrO_2, and TiO_2 did not appear to suffer from localized attack or alteration reactions, although some of the TiO_2 specimens exhibited cracking that may have resulted from fabrication problems rather than from attack by the leach solutions.

(8) The leach rates increased with temperatures as expected, but in some cases the increases were not as great as one might anticipate. For example, in the tests with flowing demineralized water the leach rates for several materials, including graphite, TiO_2, and ZrO_2, increased tenfold or less as the leaching temperature was increased from 90 to 250°C. However, for other materials, such as mullite and Pyroceram 9617, the leach rates increased 500–1000 times over the same temperature range. In the static tests at 150 and 250°C the leach rates were generally highest in the demineralized water and lowest in the brine solution, with the Hanford groundwater giving intermediate results.

(9) In most of the tests the leach rates varied with time, although in some of the dynamic tests it appeared that steady-state conditions were attained. In most of the static tests it was apparent that saturation of dissolved species in the leach solutions served to limit dissolution of the ceramic materials.

The conclusions stated above are based on an overall evaluation of the test results, and may not be valid for a specific repository environment, where conditions could be quite different from those used in the leach tests. Although graphite and alumina show promise as chemically resistant barriers, no further work is anticipated, since current package design concepts do not require ceramic barriers.

Conclusions

Clay minerals, or shales, or mixtures of clay mineral-rich material and silicates such as quartz or sandstone are resistant to radionuclide flow under simulated repository conditions. Times of in excess of 10^5 years have been reported (Nowak, 1982), and, with additives such as activated charcoal, Fe-bearing minerals, and possibly sulfides, the composite engineered backfill will be able to retain or largely retard any product of the waste form. The canisters also have been shown to be highly resistant to leaching under assumed repository conditions. Estimated times of resistance to leaching are commonly in excess of 10^3 years. The combination of waste form, canister, and engineered backfill under assumed conditions of leaching indicate times in excess of 5×10^5 years, well in excess of suggested containment times of 10^3 or 10^4 years.

CHAPTER 14

Toxicity

Introduction

In this chapter an attempt will be made to assess the toxicity, i.e., the toxicity of naturally occurring materials and man-made materials, of radwaste in its various forms, as well as to briefly discuss some aspects of health as related to toxicity. The concept of a total hazard index (cf. Smith *et al.*, 1980) is discussed.

Geotoxicity Index

The basic problem of radioactive waste is that of its toxicity, and how this toxicity compares with that of other elements and materials in the crust of the earth. Smith *et al.* (1980) have recently proposed the following mathematical formula for a geohazard index (GHI):

$$GHI_{(i)} = TI_i \cdot P_i \cdot A_i \cdot C_i, \qquad (14\text{-}1)$$

where TI_i is a basic toxicity index, P_i is a modifying factor related to persistence of a radionuclide, other element, or other material, A_i is a modifying factor related to a species availability, and C_i is the buildup factor for a daughter element more toxic than its parent. Therefore,

$$TI = Q/MPC \text{ or } Q/DWS, \qquad (14\text{-}2)$$

where Q is the quantity of material (in curies or grams), MPC is the maximum permissible concentration for radionuclides in public drinking-water supplies (Ci/m^3), DWS is the drinking-water standard or maximum permissible concentration of stable toxic materials in public drinking-water supplies (g/m^3), and TI is expressed as units of volume (m^3). In theory (Smith *et al.*, 1980), TI is the volume of water necessary for dilution of a quantity of toxic material to drinking-water levels. It must be modified, however, by the parameters, P, A, C:

$$P = \frac{1 - \exp\left(-T^* \cdot \ln\left(2\right)/T_{1_2}\right)}{T^* \cdot \lambda} \qquad (14\text{-}3)$$

where T_{1_2} = half-life in years, λ = corresponding decay constant, and T^* is a reference period (300 yrs as used by Smith et al., 1980). In the case of stable materials, $P = 1.0$; for decaying material, $0 < P < 1.0$.

For the availability factor of material i, A_i, reference is made to the materials' availability in the crust of the earth, A_o. Therefore:

$$A_{oi} = \frac{\text{Ingestion rate } (i)/\text{crustal abundance } (i)}{\text{Ingestion rate } (^{226}\text{Ra})/\text{crustal abundance } (^{226}\text{Ra})} \qquad (14\text{-}4)$$

where natural availability of any element is referenced to naturally occurring ^{226}Ra (i.e., A_o for ^{226}Ra $= 1.0$). In the actual case, though, $A_i = A_o\, m$, where m is the modification of availability of any given waste or mineral component, taking into account chemical and physical form and burial conditions. For example, Smith et al. (1980) argue that increasing isolation would result in lower values of m, and $m = 1$ could be taken as a first approximation. A_o requires knowledge of the biogeochemistry of an element, and A_i is a refinement of A_o, in turn based on the precise understanding of a specific natural accumulation of a toxic material.

For materials that decay or are transformed into more toxic daughters than parents,

$$C = \frac{\text{TI}_{\max}}{\text{TI}_o}, \qquad (14\text{-}5)$$

where $\text{TI}_{\max}$ is the maximum total toxicity index of parent plus daughter and TI_o is the original toxicity index (Smith et al., 1980).

For TI, the intrinsic toxicity, data for radioactive species are referenced to MPC and those for stable toxic materials to DWS (cf. Tonnessen and Cohen, 1977). Smith et al. (1980) propose use of the 1962 IAEA (IAEA, 1962) MPC data for radionuclides, since they are in good agreement with drinking-water standards for stable materials.

For the persistence factor, P, $T^* = 300$ yrs is reasonable since this is about 10 times the half-lives of ^{90}Sr and ^{137}Cs, two of the most important radionuclides in radioactive waste. After 300 years, these two isotopes will have decayed to inocuous levels.

For values of A_i, Cohen and Jow (1978) have tabulated data for approximately 50 elements (see Table 14-1). Values of A will vary for different types of repositories, being greater for surface or near-surface disposal sites and very much lower for deep disposal sites.

For build-up correction factor, C, it is important to point out that this factor will not be important for the majority of radionuclides, as usually the

Table 14–1. Data for determination of A_o.

Element	Z	Intake (g/day)	Abundance in sediment (ppm)	Intake/ abundance ((g/day)/ppm)	Element	Z	Intake (g/day)	Abundance in sediment (ppm)	Intake/ abundance ((g/day)/ppm)
Li	3	2.0×10^{-3}	38	5.2×10^{-5}	Zn	30	13×10^{-3}	52	2.5×10^{-4}
Be	4	12×10^{-6}	1.9	6.3×10^{-6}	Ge	32	1.5×10^{-3}	1.4	1.1×10^{-3}
B	5	1.3×10^{-3}	68	1.9×10^{-5}	As	33	1×10^{-3}	1.8	2.4×10^{-4}
C	6	300		1.5	Se	34	1.5×10^{-4}	0.36	4.2×10^{-4}
N	7	16	600	2.7×10^{-2}	Br	35	7.5×10^{-3}	4.8	1.6×10^{-3}
O	8	2600	4560	5.8×10^{-1}	Rh	37	2.2×10^{-3}	127	1.7×10^{-5}
F	9	1.8×10^{-3}	410	4.45×10^{-6}	Sr	38	1.9×10^{-3}	375	5.1×10^{-6}
Na	11	4.4	4560	1.0×10^{-3}	Zr	40	4.2×10^{-3}	165	2.5×10^{-5}
Mg	12	0.3	18,100	1.7×10^{-5}	Nb	41	6.2×10^{-4}	11	5.6×10^{-5}
Al	13	45×10^{-3}	65,300	7.0×10^{-7}	Mo	42	3×10^{-4}	1.2	2.5×10^{-4}
Si	14	3.5×10^{-3}	226,200	1.6×10^{-8}	Ag	47	7×10^{-5}	0.08	8.8×10^{-4}
P	15	1.4	550	2.5×10^{-3}	Cd	48	1.5×10^{-4}	0.19	7.9×10^{-4}
S	16	0.85	1970	4.3×10^{-4}	Sn	50	4×10^{-3}	5.8	6.9×10^{-4}
Cl	17	5.2	130	4.3×10^{-2}	Sb	51	5×10^{-5}	1.2	4.2×10^{-5}
K	19	3.3	15,850	2.1×10^{-4}	I	53	2×10^{-4}	1.2	1.7×10^{-4}
Ca	20	1.1	81,000	1.4×10^{-5}	Cs	55	1×10^{-5}	7	1.4×10^{-6}
Ti	22	0.85×10^{-3}	2960	2.9×10^{-7}	Ba	56	0.75×10^{-3}	460	1.6×10^{-6}
V	23	2×10^{-3}	81	2.5×10^{-5}	Tl	81	1.5×10^{-6}	0.77	1.9×10^{-6}
Cr	24	1.5×10^{-4}	67	2.2×10^{-6}	Pb	82	0.44×10^{-3}	14.6	3.0×10^{-5}
Mn	25	3.7×10^{-3}	596	6.2×10^{-6}	Bi	83	20×10^{-6}	0.01	2.0×10^{-3}
Fe	26	14×10^{-3}	21,700	6.5×10^{-7}	^{210}Po	84	7.13×10^{-16}	1.67×10^{-10}	4.3×10^{-6}
Co	27	3×10^{-4}	11	2.7×10^{-5}	^{226}Ra	88	2.35×10^{-12}	7.59×10^{-7}	3.1×10^{-6}
Ni	28	4×10^{-4}	58	6.9×10^{-6}	Th	90	3×10^{-6}	6.9	4.3×10^{-7}
Cu	29	3.5×10^{-3}	34	1.0×10^{-4}	U	92	1.9×10^{-6}	2.3	8.3×10^{-7}

Source: Data taken from Cohen and Jow (1978).

Table 14–2. Reference containment-facility parameters.

Capacity	$6 \times 10^5/m^3$	(Sufficient volume to contain the wastes generated by 1000 GW(e)/yr of power production)
Site plan area	$2 \times 10^6/m^2$	
Size of trenches	$200 \text{ m} \times 10 \text{ m} \times 8 \text{ m}$	
Number of pits	100	
Distance to site boundary	160 m	

Source: Data from Ford *et al.* (1977).

parent radionuclide is more hazardous than its daughter(s), so correction is unnecessary (i.e., $C = 1.0$). For ^{238}U, however, both ^{226}Ra and ^{206}Pb are intermediate and final daughters, respectively, and these two nuclides are considerably more toxic than their parent. Hg is a good example of a nonradioactive species. Should monomethyl mercury form by an organic reaction involving native mercury, C would be high, since monomethyl mercury is more hazardous than native Hg (see discussion in Smith *et al.*, 1980).

Smith *et al.* (1980) have applied their geotoxicity hazard index to a reference low-level radioactive waste, a high-level radioactive waste repository, a reference uranium-ore body, and two ore deposits containing stable metals. The assumed parameters for these cases are given in Tables 14-2 to 14-7.

In the case of the low-level radioactive site (Table 14-3), the waste is assumed to have accumulated for 1000 GWe years of power production. The total GHI (Smith *et al.*, 1980) is 2.7×10^{11}, for which ^{237}Np is of major importance due to its initial toxicity and build-up correction factor.

In the case of the reference high-level waste repository (Table 14-4), Smith *et al.* (1980) used the data of Cohen *et al.* (1977) for radioactive waste

Table 14–3. Low-level waste inventories and GHI.

	Reference inventory (Ci)	GHI_i
Strontium-90	24,000	5.54×10^{10}
Iodine-129	0.73	9.87×10^7
Cesium-137	35,000	1.14×10^8
Neptunium-237	12,000	1.98×10^{11}
Plutonium-239	65,000	8.45×10^9
Americium-241	65,000	8.22×10^9
Total GHI for the facility		2.7×10^{11}

Source: Smith *et al.* (1980).

Table 14–4. Fuel-cycle parameters for high-level waste.

Reactor power level	30 MW per metric ton of uranium
Burnup	33,000 MW days per metric ton of uranium
Neutron flux	$2.92 \times 10^{13}/\text{cm}^{-2}/\text{sec}^{-1}$
Thermal-electrical efficiency	35.4%
Total energy produced	1000 GW(e)/yr
Reprocessing time	150 days
Holding period	10 years

Source: Data from NUREG (1979); Cohen *et al.* (1977).

inventory from light water reactors; again assuming waste accumulated after 1000 GWe years of power production and a 10-year above-ground interim storage period. The GHI calculated is 4.40×10^{13}. The value of m, the modification factor for A_o, is taken as 0.01 because of the deep isolation of the buried waste. The GHI is too high, since the inventory assumes all contained isotopes to be present in amounts of 100% at the time of start of interim storage, and, in actuality, it would take a long time (years) for generation of 1000 GWe years power production, during which time radioactive decay will be occurring. The initial inventory of ^{90}Sr is relatively large, as is its specific toxicity, and it dwarfs the other species shown in Table 14-5.

For a reference uranium-ore body of the given dimensions of one km^2 area by 10 m thickness, and a grade of 0.2% uranium, the inventory of important elements is given in Table 14-6. Smith *et al.* (1980) consider only species in

Table 14–5. High-level waste inventories and GHI.

	Reference inventory (Ci)	GHI$_i$
Strontium-90	1.9×10^9	4.39×10^{13}
Technetium-99	4.4×10^5	9.46×10^8
Tin-126	1.7×10^4	1.89×10^9
Cesium-134	2.6×10^8	1.29×10^9
Cesium-137	2.7×10^9	8.51×10^{10}
Europium-154	1.4×10^8	8.82×10^9
Plutonium-238	3.2×10^6	1.59×10^9
Americium-241	5.5×10^6	7.06×10^9
Curium-244	5.4×10^7	4.36×10^9
Total GHI		4.40×10^{13}

Source: Data from Smith *et al.* (1980).

Table 14–6. Reference uranium ore body.

	Inventory	GHI_i
Uranium-238	1.68×10^4 Ci	5.2×10^{11}
Thorium-234	1.68×10^4 Ci	3.8×10^4
Uranium-234	1.68×10^4 Ci	2.5×10^{11}
Thorium-230	1.68×10^4 Ci	2.5×10^{11}
Radium-226	1.68×10^4 Ci	1.8×10^{12}
Lead-210	1.68×10^4 Ci	1.8×10^{11}
Polonium-210	1.68×10^4 Ci	6.0×10^7
Lead (Stable)	8.19×10^6 kg	1.6×10^{10}
Total GHI		3.0×10^{12}

Source: From Smith *et al.* (1980).

the ^{238}U decay chain, excluding ^{222}Rn. The reference ore body is given an $m = 1$ value because it is further assumed that the ore body is considered as equivalent to a shallow-buried mass of material. By inspection of Table 14-6, it is seen that the most important component is ^{226}Ra, although the uranium and lead isotopes are also of major importance. The uranium-ore body has a higher GHI (3×10^{12}) than the low-level waste repository (2.7×10^{11}) and is lower than that for the high-level waste repository (4.4×10^{13}).

Two heavy-metal ore deposits were considered by Smith *et al.* (1980): lead and mercury. Both elements are toxic. Lead production in the United States is about 5 million tons and mercury production about 1000 tons. For the reference cases, each ore deposit is assumed to contain 10% of annual production, although this is not a good assumption for mercury. The modification factor, m, is taken as unity, and since the deposits are of stable metals, P is assigned a maximum value of 1.0. The data are presented in Table 14-7. The GHI for mercury is 1.3×10^{13} and the GHI for lead is 9.7×10^{12}. The high value for mercury results from its high mobility and specific toxicity, and the nearly as high value of lead results from its high inventory and relatively high mobility and toxicity. Basically these two metal deposits are as hazardous as the uranium ore body and the high-level waste repository. Further, detailed analysis of the two metal ore bodies may result in higher GHI values. Lead deposits contain many other elements: 100% Bi, 40% Sb, 25% Zn, 8.7% Ag, and variable amounts of Te, Co, S, Au, and W are produced as by-products of lead operations, and in final toxicity consideration, the total GHI must account for these. In the case of mercury, high concentrations of Cd, Sb, As, and other potentially hazardous elements occur, and these, too, must be incorporated into a more refined GHI.

One purpose of the assessment of Smith *et al.* (1980) of various GHI values is to point out that naturally occurring accumulations of elements such as uranium, lead, and mercury are more hazardous than LLW and about as harzardous as HLW. Further, although data are not available to document it, there are low-level nonradioactive dumps possibly possessing very high GHI values that are indiscriminantly scattered about without proper inventory, disposal plan, or contingency plan.

Finally, it must be pointed out that in the case of the LLW and HLW their GHI decreased with time; this is especially true for the HLW due to the extreme importance of ^{90}Sr, which has a short half-life. But the GHI of the stable metal deposits does not diminish with time and, in fact, may increase for mercury as the chance of reaction with organic matter to form monomethyl mercury may increase with time.

Data for GHI Calculations

This section contains the data used to compute the geotoxicity hazard index for the sample cases discussed in Table 14-8. For each material, values are given for the appropriate water standard (DWS or MPC), half-life, P (persistence), A_o (availability, as in nature), and C (build-up correction factor).

The TWS values are from the National Interim Primary Drinking Water Regulations (EPA, 1975), and the MPC data are from IAEA Safety Series 9 (IAEA, 1962), except where otherwise noted.

Half-life values are from the Chart of the Nuclides (GE, 1977), and P is calculated from these values according to the formula given in Section 3.0 of GE (1977).

The A_o values are calculated from data in Cohen and Jow (1978) by dividing the radio of intake to abundance for a particular element by the corresponding ratio for radium. For elements not included in this reference, A_o is computed using the authors' recommended values according to their chemical grouping on the periodic table.

Data from Ford $et\ al.$ (1977) are used to simulate a large, low-level waste facility (LLW). The reference site includes the significant radionuclide inventories resulting from 1000 GW(e) years of power production. Table 14-2 lists the assumed parameters related to the facility, and Table 14-3 lists the radionuclide inventories present in the facility, together with the computed nuclide geotoxicity hazard indices and the total index. Mathematically, the GHI is the simple product of these parameters and, for any material (i),

$$\text{GHI}_i = \text{TI}_i \cdot P_i \cdot A_i \cdot C_i. \qquad (14\text{-}6)$$

Table 14–7. Calculational results for stable metal ore bodies.

Ore body	Inventory (tons)	DWC (mg/L)	A_o	GHI
Mercury	10^2	0.002	2.55×10^2	1.3×10^{13}
Lead	5×10^4	0.05	9.68	9.7×10^{12}

$Source:$ From Smith $et\ al.$ (1980).

Table 14–8. Data used to compute the geotoxicity hazard index.

Material	DWS or MPC (Ci/m³)	Half-life (yr)	Persistance factor (P)	Availability factor (A_o)	Correction factor (C)
90Sr	1×10^{-7}	28.1	1.4×10^{-1}	1.7	1.02
99Tc	3×10^{-4}	2.1×10^{5}	1.0	64.5	1.0
126Sn[a]	2×10^{-5}	10^{5}	1.0	222.6	1.0
129I	4×10^{-7}	1.7×10^{7}	1.0	54.8	1.0
134Cs	9×10^{-6}	2.05	9.0×10^{-3}	0.45	1.0
137Cs	2×10^{-5}	30.2	1.4×10^{-1}	0.45	1.0
154Eu	2×10^{-5}	8.2	3.9×10^{-2}	3.23	1.0
210Pb	1×10^{-7}	21	0.10	9.7	1.15
210Po	7×10^{-7}	37.8×10^{-2}	1.8×10^{-3}	1.4	1.0
226Ra	1×10^{-8}	1600	0.94	1.0	1.1
230Th	2×10^{-6}	8×10^{4}	1.0	0.14	206
234Th	2×10^{-5}	6.6×10^{-2}	3.2×10^{-4}	0.14	1.0
234U	3×10^{-5}	2.5×10^{5}	1.0	0.27	1682
238U	4×10^{-5}	4.5×10^{9}	1.0	0.27	4476
237Np	3×10^{-6}	2×10^{6}	1.0	0.65	79
238Pu	5×10^{-6}	86	0.38	0.65	1.0
239Pu	5×10^{-6}	2.4×10^{4}	1.0	0.65	1.0
241Am	4×10^{-6}	458	0.79	0.65	1.0
244Cm	7×10^{-6}	18.1	8.8×10^{-2}	0.65	1.0
Hg[b]	2×10^{-3c}		1.0	255	1.0
Pb	5×10^{-2c}		1.0	9.7	1.0

Source: From Smith *et al.* (1980).

[a]MPC value taken for 125Sn.

[b]For Hg, buildup of the toxic methylated form is ignored and C is assigned the value 1.0. The value for A_o is taken to be the same as that of Cd, which is also a Group-2b transition element.

[c]In mg/L.

For mixtures of various materials,

$$\text{GHI} = \Sigma \text{GHI}_i. \qquad (14\text{-}7)$$

The GHI can be applied to assessments of radioactive and stable toxic material buried in the earth either by man or by nature. Sample calculations applied to various "reference" sites indicate the values shown in Table 14-9.

One of the key terms of the GHI is the *availability factor*, which in turn depends on the *m* factor (Smith *et al.*, 1980), where

$$m = A_i/A_{i0}, \qquad (14\text{-}8)$$

where

A_i = actual availability of material i, in a specific burial setting, A_{i0} = average availability of analog material i.

Wachter and Kresan (1982) have investigated the availability factor as used by Smith *et al.* (1980) in detail, and separate natural parameters from engineered parameters (both are included by Smith *et al.*, 1980). In the treatment by Smith *et al.* (1980), the A_{i0} term is listed simply as a reference availability, while Wachter and Kresan (1982) define it (above) as availability of an analog for the ith element. More specifically, they (Wachter and Kresan, 1982, p. 6) consider the *m* factor to reflect the influence of a certain condition of the mobility/solubility for selected elements. Thus a positive $(+)$ *m* factor indicates an increase in mobility of a certain element, a negative $(-)$ *m* factor decreased mobility, and a value of zero (0) *m* factor is chosen as the standard of essentially immobile uranium species. Further, Wachter and Kresan (1982) report their *m* values on a logarithmic scale, which is different from the Smith *et al.* (1980) scheme. Wachter and Kresan (1982) further assume elemental immobility when the molar concentration of certain element is at or less than 10^{-11}. To a first approximation, *m* can be defined as

$$\log M/M_r,$$

where M = molar concentration of a species, and M_r = reference molar

Table 14-9. Summary of application results.

Site	GHI
Low-level waste facility	2.7×10^{11}
High-level waste repository	4.4×10^{13}
Uranium ore deposit	3.0×10^{12}
Mercury ore deposit	1.3×10^{13}
Lead ore deposit	9.7×10^{12}

Source: From Smith *et al.* (1980).

concentration (i.e., 10^{-11}). Thus there is room for much variation in m factors, and qualitative terms such as $+m$ (great, small, etc.) are used when it is not possible to estimate a numerical value. Wachter and Kresan (1982) base much of their work on reports from the Oklo Natural Reactor (see Chapter 11) and uranium deposits of the Colorado Plateau as well as vein-type deposits. Some of the matrix information used by Wachter and Kresan (1982) are summarized in Table 14-10.

The $+m$ factors for total dissolved uranium given in Table 14-10 are suspect. First, a P_{CO_2} of 10^{-2} is unreasonably high for most waters. More important, if the mobility is so great (10^4–10^7) then one is hard pressed to explain the presence of U(VI)-bearing species, such as carnotite, tyuya-munite, uranophane, etc., in the natural environment. The author (Brookins, 1981e) has dated some U(VI) minerals by the U–Pb method and found them to give ages from 3 to 11 million years, and the degree of concordancy of the ^{238}U–^{206}Pb and ^{235}U–^{207}Pb suggests that these are formational ages largely unaffected by surface conditions. At pH = 7, the dominant form of U(VI) in solution is $UO_2(CO_3)_2 \cdot 2H_2O^{2-}$, not UO_2^{2+}. The very high P_{CO_2} values (Wachter and Kresan, 1982, p. 19) of 1 atm. are not reasonable for repository conditions, hence the high ($+$) m factors are also not relevant here. Positive $+m$ values for uranyl fluorides and sulfates, at low pHs, are to be expected, however. Immobility is reflected in the $-m$ values for all forms of carbon, pyrite, Fe^{2+}, etc. (Wachter and Kresan, 1982; pp. 18–19). This is to be expected because of the extremely high probability of reduction of U(VI) to U(IV) and, similarly, sorption of U(IV) by carbon, zeolites, clay minerals, pyroxene, biotite, and various oxides is also to be expected (Wachter and Kresan, 1982, p. 19).

Of extreme importance from the work of Wachter and Kresan (1982) is their conclusion that "geochemical parameters (especially oxidation/reduction potential) are apparently more potent mobilizers than geological hydrological conditions in many, if not most, geologic environments for most radioactive waste elements." They conclude (p. iv) that depth of a repository is not necessarily the factor be followed for radwaste disposal, as shallow to moderate depth in shale may be as geochemically containing as a much greater depth in other rock types.

A method to describe the *hazard from high-level radioactive waste* has been described by Gera (1975). In his approach, the toxicity of long-lived heavy radionuclides is described in LAI terms, that is, *l*imits of *a*nnual *i*ntake either by ingestion (LAI_{ing}) or inhalation (LAI_{inh}). Gera (1975) prefers the use of LAI values to other ways of reporting radiotoxicity of waste nuclides because the intake of the limit of any one radionuclide is equivalent to a 50-year dose commitment. In his approach, the LAI values per cubic centimeter of waste are tabulated for LAI_{ing} and LAI_{inh} for LWR and LMFBR heavy radionuclides from HLW. The LAI values have been calculated (Gera, 1975) based on known radiotoxicity of other isotopes of each element. Further, the assumption is made that the waste can be directly ingested or

Table 14–10. m-factors that affect uranium mobility.

Conditions	m	Reference
Given reducing conditions (Eh below 0.273 v at 25°C; uranium is in the +4 oxidation state. Concentration of the uranous ion is generally below 0.01 ppb except at pH values below 3 and above 7. Langmuir (1978) concludes that the extreme insolubility of uraninite and coffinite at normal groundwater pH values (4–8) makes uranium practically immobile in low-Eh environments.		Langmuir (1978)
at pH = 7, $[UO_2]$ = ~0.01 ppb or $10^{-11}M$ $[UO_2]$ at 25°C	0	Langmuir (1978)
at pH = 4, $[U]$ = ~0.0001 ppb or $10^{13}M$ $[UO_2]$	-2	Langmuir (1978)
at pH = 2, $[U]$ = ~0.01 ppb or $10^{-11}M$ $[UO_2]$	0	Langmuir (1978)
100-fold increase in uraninite solubility at pH = 6 and PO_4 = 0.1 ppm ($10^{-6}M$) and Eh exceeding +0.05 v.	+2	Langmuir (1978)
1000-fold increase in uraninite solubility with a rise of P_{CO_2} from atmospheric ($10^{-3.5}$ atm) to a groundwater value of 10^{-2} atm at Eh = −0.05 v, $[UO_2]$ = $10^{-7}M$.	+3 to +4	Langmuir (1978)
Uranous fluoride complexes are stable up to pH = 4 @ 25°C, at 2 ppm F and pH = 2. $[U]$ = 1 ppm or $10^{-5}M$ $[UO_2]$.	+6	Langmuir (1978)
Solubility of *uraninite* increases about 10-fold as temperature increases from 25°C to 100°C.	+1	Langmuir (1978)
Well-crystallized uraninite is much *less* soluble than amorphous UO_2 (pitchblende).	-2 to -6	Langmuir (1978)
Solubility of uraninite increases with temperature, with maximum solubility at 260°C in pure water.		Langmuir (1978)
at 25°C, $[UO_2]$ = ~0.01 ppb	0	
at 100°C, $[UO_2]$ = ~0.1 ppb	+1	
at 260°C, $[UO_2]$ = ~75 ppm	+5	

Table 14–10. (*continued*)

Conditions	m	Reference
Background values for uranium in streams is around 0.1 ppb. In uraniferous areas, the uranium concentration in waters can range from 1 to 400 ppb. (*m* factor is expressed relative to the concentration of uraninite in pure water at 25°C ($10^{-11}M$) because that represents "immobile" conditions.) Uranium is oxidized above 0.273 v.		Langmuir (1978)
[Carnotite] = $10^{-7}M$ at pH = 7 and $P_{CO_2} = 10^{-2}$ atm at 25°C w/0.1 ppm vanadium (10^{-6} vanadate) Solubility of carnotite increases above pH = 8 because of formation of carbonate complexes.	+4	Langmuir (1978)
[Autunite] = $10^{-4}M$ at pH = 7 and $P_{CO_2} = 10^{-2}$ atm $\Sigma PO_4 = 10^{-6}M$ (0.1 ppm); [Ca] = 80 ppm or $10^{-2.7}M$ and [K] = 39 ppm or $10^{-3}M$. terrain was more likely to have a [PO_4] of 0.1 ppm or more as compared to sedimentary rocks.	+7	Langmuir (1978)
[Uranyl hydroxide] in slightly alkaline and slightly acidic solutions is about $10^{-5}M$.	+6	Krauskopf (1967)
[Uranyl carbonate] complexes at pH = 7 with 0.01M [carbonate] is $10^{-4}M$ at 25°C or 10 times more solubility than in a carbonate free solution.	+7	Krauskopf (1967)
Uranyl carbonate complexes: [UO_2CO_2] = $10^{-3.5}M$ at pH = 6 and CO_2 = 1 atm	+7	Rich *et al.* (1977)
At CO_2 pressure of 10^{-2} atm and 25°C the uranyl carbonate complexes are the major species in pure water down to a pH = 5.		Langmuir (1978)
Uranyl sulfate complexes stable up to pH = 4 possibly significant up to pH = 7.	$+m$ weak	Langmuir (1978)
Carbon, carbonaceous matter and/or methane	$-m$ very strong	Langmuir (1978)

Sulfide (pyrite, H_2S)	$-m$	Langmuir (1978)
Fe^{+2} (biotite, augite)	strong	Langmuir (1978)
	$-m$	
	moderate to strong	
Presence of the above reducing agents will generally decrease the solubility to below the detection limit.		
Reduction of U due to metamorphic reactions involving the oxidation of ferrous ion and resultant reduction of uranyl ion.	$-m$	McMillan (1978)
Reduction of uranium during clay-mineral formation	$-m$	Brookins (1980a)
Sorption of 1 ppm uranium at pH = 5 to 8	0	Langmuir (1978)
at pH = 9 without CO_2	0	Langmuir (1978)
Presence of sulfate, halogens, or carbonate inhibits sorption.	$+m$	Langmuir (1978)
Humic acid concentrates uranium 10,000:1 at pH = ~5.	$-m$	Langmuir (1978)
	strong	Dyck (1978)
Sorption agents		
carbon	$-m$	Brookins (1980)
	strong	
zeolites	$-m$	Szalay (1967)
	strong	
clay	$-m$	Shiao et al. (1979)
	moderate	Beall et al. (1979)
augite and biotite	$-m$	Haire and Beall (1979)
	moderate	
oxides	$-m$	Allard et al. (1980)
	weak	

Source: From Wachter and Kresan (1982).

inhaled, but, as Gera (1975) points out, this assumption is ". . . far-fetched." Yet use of LAI values is, for the heavy radionuclides, a good indicator of relative toxicity of these isotopes. It must be emphasized, however, that no consideration is given to containment of wastes in a canister–sleeve–backfill–rock repository nor to the difficulty of getting the radionuclides into the food chain, hence these LAI values, too, are severely limited in value.

Most of the radionuclides listed by Gera (1975) are bone seekers, and thus the LAI totals are more or less legitimate if it is assumed that bone is a critical organ. The tables show some interesting data. At about 10^5 years of decay, the rate of slowing down the hazard potential is very slow: there is only a 10- to 20-fold drop in total LAI_{ing} between 10^5 and 5×10^6 years. After about 10^5 years, ^{226}Ra is the critical nuclide with Np isotopes next in importance. For inhalation, however, ^{239}Pu is the critical nuclide at 10^5 years, but within a few thousand years it is replaced in importance by Np series isotopes with ^{229}Th critical. After 10^6 years, Np series isotopes dominate the LAI_{inh} data.

Very long-term hazard of the heavy radionuclides is difficult to address, i.e., can containment be assured for times in excess of 5 million years, or, alternatively, is containment failture acceptable after, say, 100,000 years (Gera, 1975)?

To discuss the toxicity of the heavy radionuclides in HLW in a convenient perspective, it is reasonable to compare the HLW, after some set decay time, to naturally occurring radioactive ore. Uranium ore commonly contains from less than 0.1% U_3O_8 to 0.3% U_3O_8, although some ores do contain much higher amounts. For 1% hypothetical ore, Gera (1975) has calculated that the potential hazard is 1.2 LAI_{ing} and 23 LAI_{inh} (density for ore is 3 g/cm^3), and for thorium ore, the potential hazard is 0.15 LAI_{ing} and 9.3 LAI_{inh}. Yet for 1 cm^3 of pitchblende, the potential hazard is 220 LAI_{ing} and 4200 LAI_{inh}, which is comparable to both the LWR and LMFBR waste allowed to decay from under 10^3 years to a few thousand years for ingestion, although for inhalation, the waste is more hazardous than pitchblende after even a million years. Gera (1975) properly concludes that the prime consideration for hazard is the inhalation term in the long-term situation. Yet for buried waste, heavy radionuclides will enter the food chain by a complex pattern of dissolution, transport, fixation, and only the LAI_{ing} data are applicable as, despite its critical importance for fallout, etc., all radionuclides will be in solution (including colloids, etc.) during transport.

Gera (1975) concludes that, in the case where inhalation is negligible (i.e., buried waste), the HLW with time is comparable to naturally occurring uranium, some thorium, and some mixed uranium–thorium ores. In this view, it is interesting to note that in the NURE (National Uranium Resource Evaluation) carried out by the USERDA and USDOE from about 1975–1981, surface uranium enrichment over uranium ore bodies buried even at shallow depths was often negligible. In fact, sampling over known bodies in the Grants Mineral Belt, New Mexico, showed no U surface enrichment.

This means that insignificant (if any) amounts of uranium from the ore dissolved and were transported to the surface, and it must be remembered that the ore is in the rocks without benefit of container, backfill, or even, in some cases, geologic criteria generally believed favorable. Thus it can be assumed that in the case of encapsulated buried HLW, the likelihood of any heavy radionuclide reaching the surface in quantities that could pose a health hazard is probably negligible. This is especially true, since in the HLW Np-series, Pu-series species are much more important, relative to U, than in naturally occurring uranium ore, and only U shows moderate mobility in most naturally occurring water-bearing rocks (see discussion in Chapter 11). The Eh–pH diagrams of Np, Pu, Am (and probably Cm) argue for much less mobility under burial conditions than for U; hence if no U seepage from U sources is noted, then no seepage of Np, Pu, Am, and Cm would be expected either. In the case of Ra, U ore at or near the surface can often be identified by Ra haloes, although the Ra has not reached the surface. Radium is only moderately soluble in groundwater systems, and is commonly removed by sulfate mineral precipitation. Were even slight amounts of Ra to reach the surface over a U-ore body, then this would be readily detectable by radioactivity measurement, but, at least in the cases the author is familiar with, Ra accumulation over U ore is rare.

It is of further interest to compare radioactive wastes to naturally occurring ores. To do this, it is convenient to introduce the relative toxicity index (RTI), which is defined as (Tonnessen and Cohen, 1977):

$$RTI = HM_{waste}/HM_{comparative\ base},\qquad (14\text{-}9)$$

where, for practical purposes, the hazard measure (HM) of Tonnessen and Cohen (1977) may be replaced by the hazard index (HI) (see Smith et al., 1980) defined as:

$$HI = \sum_{i=1}^{n} \frac{Q_i/V}{RCG_i} \qquad (14\text{-}10)$$

where

Q_i = dissolved quantity of each radionuclide,

RCG_i = maximum permissible concentration of element in question,

V = volume.

This equation is useful for it allows a direct comparison of different radioactive wastes with common ore bodies of known percent concentrations of the key element. In Fig 14-1 the RTI values as a function of time for SURF, HLW, and fission products, all from light-water reactors (LWR), as well as values for ores of Hg, Se, Pb, Cd, Cr, Ag, As, and Ba are plotted. What is strikingly apparent is that after about 300 years the HLW and fission products show RTI values less than all the ores shown, and the SURF

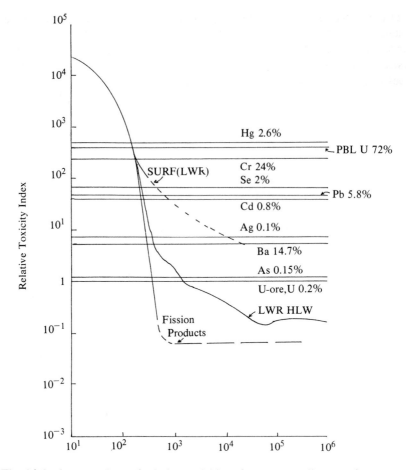

Fig. 14-1. A comparison of relative toxicities of some naturally occurring ores and nuclear wastes. Modified from Tonnessen and Cohen (1977).

is below most of the ores. Further, the SURF under chemically reducing conditions is immobile, and thus less likely to move into the ecosystem than ores of Ag, As, and Ba.

The interpretation of this comparison of various radioactive wastes with equal volumes of different ores is obvious. After a containment time of only a few hundred years the wastes are, in effect, much less hazardous than the naturally occurring ores. The reader is referred to Tonnessen and Cohen (1977), for further details on this problem.

Concluding Statements

There is no simple conclusion for this chapter. In terms of total hazard, many earth materials formed by natural processes are potentially much more hazardous than radwaste. Isolation of the radwaste by the total barrier system will result in essentially no measurable releases to the biosphere, hence any figure for total hazard or geotoxicity for the isolated radwaste is, basically, unrealistic, except to compare gross-hazard parameters for many types of naturally occurring and man-made concentrations of materials.

CHAPTER 15
Conclusions

Throughout this book most of the emphasis has been on rocks, and how radioactive elements, their daughter elements, and analog elements behave in rocks. There are abundant examples given in the preceding chapters that demonstrate and document the simple fact that many rocks will retain and retard the constituent elements of buried radwaste. The Oklo natural reactor of Gabon, and contact zone studies of the Eldora–Bryan stock and Alamosa River Stock, both in Colorado, all show the extreme immobility of radwaste–analog elements, even under elevated temperatures. Further, routine radiometric age determinations indicate that rocks also retain many elements of interest under essentially closed-system conditions for up to billions of years in some cases.

Repository sites under consideration in general appear to be well-suited for radwaste disposal, both from a generic sense and from a detailed examination of specific sites. The multibarrier system planned for buried radwaste is supported by numerous examples of geologic materials that show how well this system will work. The very long times required to leach radionuclides from the waste form, canister and sleeve materials, and engineered backfill, as well as the repository rocks proper, lead to the conclusion that the isotopic and chemical integrity of the buried radwaste will not be perturbed until after such time as there is no longer any threat to public health by release of radionuclides.

Subseabed disposal of radwaste, while not treated in depth in this book, also appears to warrant further investigation. In addition, uranium milling techniques and the understanding of mill tailings have improved greatly over the last few years so that few adverse environmental effects will result from these activities. The handling of low-level radioactive wastes, while cumbersome, is nevertheless straightforward, and, in general, favorable sites for shallow land burial are readily available.

The close working relationships of various Federal and State and other agencies or equivalent institutions has resulted in significant progress in addressing the managerial aspects of radwaste disposal, and this progress is supported by many geological and geochemical studies, especially in furthering our greater understanding of how radionuclides behave in the crust of the earth.

Rocks can and do retain the elements of concern of radwaste. The repository approach for radwaste isolation is sound. Future work, including both basic research and engineering development, will attest to the applicability of geomedia for successful radwaste isolation from the biosphere.

References

Aaron, W.S., C. Quinby, and Kobisk. E. H., 1979. Development of cermets for high-level radioactive waste fixation. In: T.D. Chikalla and J.E. Mendel, Eds., Ceramics in Nuclear Waste Management, CONF-790420. NTIS, Springfield, Virginia.

Abrego, L., and H.J. Rack 1981. The Slow Strain Rate Behavior of TiCode-12 (ASTM grade 12) in Aqueous Chloride Solutions. SAND80 = 1738C., paper presented at National Association of Corrosion Engineers, CORROSION/81 meeting, Toronto, Ont., Canada. April.

Adams, P.B. 1979. Relative leach behavior of waste glasses and naturally occurring glasses. In: T.D. Chikalla and J.E. Mendel, Eds., Ceramics in Nuclear Waste Management, CONF-790420. NTIS. Springfield, Virginia. p. 233–237.

Akatsu, E., R. Ono, K. Tsukuechi, and H. Uchiyama. 1965. Radiochemical study of adsorption behavior of inorganic ions on zirconium phosphate, silica gel and charcoal: Jour. Nucl. Sci. Tech., v. 2, p. 141–146.

Allard, B. 1979. Sorption of radionuclides on rock. In: American–Swedish Minisymposium on Geochemistry and Chemistry in Relation to Underground Disposal of Radioactive Waste. Berkeley, Calif., Oct. 11–13, 1979.

Allard, B., and G.W. Beall, 1978. Sorption of Americium on Geologic Media: DOE RHO-BWI-C-35.

Allard, B., H. Kipatsi, and J. Rydberg. 1977. Sorption of Long-Lived Radionuclides in Clay and Rock. Part 1: Determination of Distribution Coefficients. KBS Report 19:01-1977-10-10 (Swedish).

Allard, B., B. Torstenfelt, K. Anderson, and J. Rydberg. 1980. Possible retention of iodine in the ground. In: C.J.M. Northrup, Ed., The Scientific Basis for Nuclear Waste Management. v. II. Plenum Press, New York, p. 673–680.

Allard, B., B. Torstenfelt, and K. Anderson. 1981. Sorption studies of $H^{14}CO_3^-$ on some geologic media and concrete. In: J.G. Moore, Ed., The Scientific Basis for Nuclear Waste Management. v. III. Plenum Press, New York, p. 465–472.

Allard, B., G.W. Beall, and T. Krajewski. 1978. The sorption of actinides in igneous rocks: Nucl. Tech.

Allen, E.J. 1978. Criticality Analysis of Terminal Underground Nuclear Waste Storage. Union Carbide Corp. Report ORNL/TM-7405.

Ames, L.L. 1978. Characterization of rock samples. Controlled sample program publication No. 1. Battelle Pacific Northwest Laboratory Report PNL-2797/ UC-70.

Ames, L.L., and D. Rai. 1978. Radionuclide interactions with soil and rock media. Vol. 1: Processes Influencing Radionuclide Mobility and Retention; Element Chemistry and Geochemistry; Conclusions and Evaluation: E.P.A. Report 520/ 6-78-007A.

Anderson, R.Y. 1981. Deep-Seated Salt Dissolution in the Delaware Basin, Texas and New Mexico: New Mexico Bur. Mns. Min. Resources Spec. Paper v. 10, p. 133–146.

Apps, J.A. 1978. The Simulation of Reactions between Basalt and Groundwater: Progress Report for the Hanford Waste Isolation Project, Task II, High-Level Water Basalt Interactions. Lawrence Berkeley Laboratory Informal Report.

Apps, J.A., J. Lucas, A.K. Mathur, and L. Tsao, 1977. Theoretical and experimental evaluation of waste transport in selected rocks. In: 1977 Annual Report LBL Contract No. 45901AK, DOE Contract W-7405-ENG-48.

Baas Becking, L.G.M., I.R. Kaplan and D. Moore. 1960. Limits of the natural environment in terms of pH and oxidation-reduction potential: Jour. Geol., v. 68, p. 243–256.

Bailey, C.E., and I.W. Marine. 1980. Parametric Study of Geohydrologic Performance Characteristics for Geologic Waste Repositories. E. I. du Pont de Nemours & Co. Report DP-1555.

Barner, H.E., and R.V. Scheuerman. 1978. Handbook of Thermochemical Data for Compounds and Aqueous Species. Wiley–Interscience, New York.

Barney, G.S., and B.J. Wood. 1980. Identification of Key Radionuclides in a Nuclear Waste Repository in Basalt. RHO-BWI-ST-9, Rockwell Hanford Operations, Richland, Washington.

Barton, P.B., and B.J. Skinner. 1979. Sulfide mineral stabilities. In: H.L. Barnes, Ed., Geochemistry of Hydrothermal Ore Deposits. 2nd ed. Wiley–Interscience, New York, p. 278–403.

Beall, G.W. 1979. Sorption behavior of trivalent actinides and rare earths on clay minerals. In: S. Fried, Ed., Radioactive Waste in Geologic Storage, Am. Chem. Soc. Symp. Series 100, Sept. 1978, Miami Beach, Florida, p. 201–213.

Beall, G.W., G.D. O'Kelley and B. Allard. 1980. An autoradiographic study of actinide sorption on Climax Stock granite: Oak Ridge National Laboratory Report ORNL-5617, Oak Ridge, Tennessee.

Beasley, T.M., R. Carpenter, and C.D. Jennings. 1982. Plutonium, [241]Am and ^{137}Cs ratios, inventories, and vertical profiles in Washington and Oregon continental shelf sediments: Geochim. Cosmochim. Acta, v. 46, p. 1931–1946.

Behrenz, P., and K. Hannerz. 1978. Criticality in a Spent Fuel Repository in Wet Crystalline Rock. KBS-Teknisk Report 108 (Swedish).

Benson, L.V., and L.S. Teague. 1979. A Study of Rock–Water–Nuclear Waste Interactions in the Pasco Basin, Washington. DOE Report 2-7405-ENG-48.

Benson, L.V., C.L. Carnahan, J.A. Apps, C.A. Mouton, D.J. Corrigan, C.J. Frisch, and L.K. Shomura. 1980. Basalt Alteration and Basalt-Waste Interaction in the Pasco Basin of Washington State. DOE Report W8A-SBB-51760.

Bibler, N.E., and J.A. Kelley. 1978. Effect of Internal Alpha Radiation on Borosilicate Glass Containing Savannah River Plant Waste. E.I. du Pont de Nemours & Co. Report DP-1482, Aiken, South Carolina.

Bird, G.W., and V.J. Lopata. 1980. Solution interaction of nuclear waste anions with selected geologic materials. *In*: C.J.M. Northrup, Ed., The Scientific Bases for Nuclear Waste Management, v. II. Plenum Press, New York, p. 419–426.

Blanco, R.E. and A.L. Lotts. 1979a. High-Level Waste Program Progress Report for July 1, 1979 through September 30, 1979. Oak Ridge National Laboratory Report ORNL/TM-7118, Oak Ridge, Tennessee.

Blanco, R.E. and A.L. Lotts. 1979b. High-Level Waste Program Progress Report for October 1, 1978 through March 31, 1979. Oak Ridge National Laboratory Report ORNL/TM-6866, Oak Ridge, Tennessee.

Blanco, R.E., and A.L. Lotts. 1979c. High-Level Waste Program Progress Report for April 1, 1979 through June 30, 1979. Oak Ridge National Laboratory Report ORNL/TM-7013, Oak Ridge, Tennessee.

Bodine, M.W. 1978. Clay mineral assemblages from drill core of Ochoan evaporites, Eddy County: New Mexico Bur. Mins. Min. Resources Circ. v. 159, p. 21–32.

Boltwood, B.B. 1907. On the ultimate disintegration products of the radioactive elements: Amer. Jour. Sci., v. 23 (4), p. 77–88.

Bondietti, E.A., and C.W. Francis. 1979. Geologic migration potentials of technetium-99 and neptunium-237: Science, v. 203, p. 1337–1340.

Bonhomme, M., F. Weber, and R. Favre-Mercuret. 1965. Age par la methode rubidium strontium des sediments du bassin de Franceville, Republique Gabonaise: Bull. Serv. Carte Geol. Als. Lorr., v. 18, p. 243.

Bonhomme, M., J. Leclerc, and F. Weber. 1978. Etude radiochronologique complémentaire de la série du Francevillien et de son environment: IAEA Tech. Pub. v. 119, p. 19–24.

Boult, K.A., J.T. Dalton, A.R. Hall, A. Hough, and J.A.C. Marples 1978. The Leaching of Radioactive Waste Storage Glasses. AERE R-91880, Atomic Energy Research Establishment, Harwell, Oxfordshire, OX11 ORA England.

Bradley, D.J., C.O. Harvey, and R.P. Turcotte. 1979. Leaching of Actinides and Technetium from Simulated High-Level Waste Glass, Pacific Northwest Laboratory Report PNL-3152, Richland, Washington.

Bradshaw, R.L. and W.C. McClain. 1971. Results and Analysis, Project Salt Vault: A Demonstration of the Disposal of High-Activity Solidified Wastes in Underground Salt Mines. ORNL-4555, Oak Ridge National Laboratory, p. 161–256.

Braithwaite, J.W. 1979. Brine chemistry effects on the durability of a simulated nuclear waste glass. *In*: G.J. McCarthy, Ed., The Scientific Basis for Nuclear Waste Management, v. II. Plenum Press, New York, p. 99.

Bresetti, M., F. Girardi, S. Orlowski, and P. Venet. 1980. Radioactive waste disposal in geological formations: Research activities of the Commission of European Committees. *In*: C.J.M. Northrup, Ed.), The Scientific Basis for Nuclear Waste Management. v. II. Plenum Press, New York, p. 31–38.

Brookins, D.G. 1976. The Grants Mineral Belt, New Mexico: Comments on the Coffinite–Uraninite Relationship, Clay Mineral Reactions, and Pyrite Formation. New Mexico Bur. Mns. Min. Resources Special Paper No. 6, p. 158–166.

Brookins, D.G. 1977. Uranium deposits of the Grants Mineral Belt, geochemical constraints: Rocky Mountain Geol. Assoc. Guidebook, v. 9, p. 337–352.

Brookins, D.G. 1978a. Application of Eh–pH Diagrams to Problems of Retention and/or Migration of Fissiogenic Elements at Oklo. IAEA Tech. Pub., v. 119, p. 243–265.

Brookins, D.G. 1978b. Eh–pH diagrams for elements from Z=40 to Z=52: Application to the Oklo natural reactor: Chem. Geol., v. 23, p. 324–342.

Brookins, D.G. 1978c. Geochemical Constraints on Accumulation of Actinide Critical Masses from Stored Nuclear Waste in Natural Rock Repositories. DOE EY-76-C-06-1830; ONWI Report 17.

Brookins, D.G. 1978d. Oklo reactor reanalyzed: Geotimes, v. 23, p. 26–28.

Brookins, D.G. 1978e. Possible Accumulations of Critical Masses of U, Pu, Np, Am and Cm from Stored Fuel Rods in Natural Rock Repositories: Battelle Memorial Inst.–Office of Nucl. Waste Isolation, ONWI-17.

Brookins, D.G. 1978f. Retention of transuranic and actinide elements and bismuth at the Oklo natural reactor, Gabon: Application of Eh–pH diagrams: Chem. Geol., v. 23, p. 309–323.

Brookins, D.G. 1979a. Thermodynamic considerations underlying the migration of radionuclides in geomedia: Oklo and other examples. Basis for Nuclear Management. v. I. *In*: G.J. McCarthy, Ed., The Scientific Plenum Press, New York, p. 355–366.

Brookins, D.G. 1979b. Uranium Deposits of the Grants, New Mexico mineral belt (II). DOE Report BFEC-GJO-029E.

Brookins, D.G. 1980a. Clay minerals suitable for overpack in waste repositories: evidence from uranium deposits: *In*: C.J.M. Northrup, Ed., The Scientific Basis for Nuclear Waste Management v. II. Plenum Press, New York, p. 479–486.

Brookins, D.G. 1980b. Near-field reactions in evaporites considered for waste repositories: Proc. Jour. Ariz.–Nev. Acad. Sci., p. 43.

Brookins, D.G. 1980c. Syngenetic model for Some Early Proterozoic Uranium Deposits: evidence from Oklo: Int. Uranium Sym. on the Pine Creek Geosyncline: CSIRO Inst. Earth Resources and IAEA p. 709–719.

Brookins, D.G. 1980d. Use of evaporite materials for K–Ar and Rb–Sr geochronology: evidence from bedded Permian evaporites, southeastern New Mexico, USA: Die Naturwissenschaften, v. 67, p. 604.

Brookins, D.G. 1981a. Alkali and alkaline earth studies at Oklo: *In*: J.G. Moore,

Ed., The Scientific Basis for Nuclear Waste Management. v. III. Plenum Press, New York, p. 275–282.

Brookins, D.G. 1981b. Geochronologic studies near the WIPP site, southeastern New Mexico: summary and interpretation: New Mexico Bur. Mns. Min. Resources Special Papers v. 10, p. 147–152.

Brookins, D.G. 1981c. Geochemical study of a lamprophyre dike near the WIPP Site: In: J.G. Moore, Ed., The Scientific Basis for Welfare Waste Management. v. III, Plenum Press, New York, p. 307–314.

Brookins, D.G. 1981d. Periods of mineralization in Grants Mineral Belt, New Mexico: in Sym. Grants Uranium Region, New Mexico Bur. Mns. Min. Resources Mem., v. 38, p. 52–57.

Brookins, D.G. 1981e. U–Pb ages for U(VI) hydrosilicates, Grants, New Mexico: Isochron/West, v. 32, p. 25–27.

Brookins, D.G. 1982a. Geochemistry of clay minerals for uranium exploration in the Grants Mineral Belt, New Mexico: Mineralium Deposita, v. 17, p. 37–53.

Brookins, D.G. 1982b. Geochemistry of the Dakota Formation of northwestern New Mexico: relevance to radioactive waste studies: Nucl. Tech., v. 59, p. 420–428.

Brookins, D.G., Ed. 1983. The Scientific Basis for Nuclear Waste Management, Vol. VI, Elsevier Sci. Pub. Co., New York.

Brookins, D.G., and W.H. Dennen. 1964. Trace element variations across some igneous contacts: Trans. Kansas Acad. Sci., v. 67, p. 70–91.

Brookins, D.G., and A.W. Laughlin. 1983. Rb–Sr geochronological investigations of Precambrian samples from drill holes GT-1, GT-2, EE-1 and EE-2, Los Alamos dry hot rock program, Fenton Hill, NM: Jour. Volc. Geoth. Res, v. 14, p. 18–35.

Brookins, D.G., M.J. Lee, B. Mukhopadhyay, and S.L. Bolivar. 1976. Search for fission produced Rb, Sr, Cs and Ba at Oklo: IAEA Sym. 204, p. 401–414.

Brookins, D.G., J.K. Register, M.E. Register, and S.J. Lambert. 1980. Long-term stability of evaporite minerals: geochronological evidence: In: C.J.M. Northrup, Ed., The Scientific Basis for Nuclear Waste Management. v. II. Plenum Press, New York, p. 427–436.

Brookins, D.G., M.S. Abashian, L.H. Cohen, and H.A. Wollenberg. 1981a. Radwaste Storage in Crystalline Rocks: A Natural Analog. IAEA SM-257, Extended synopses, p. 77–79.

Brookins, D.G., L.H. Cohen, and H.A. Wollenberg. 1981b. Geochemistry of a contact metamorphosed zone: Implications for radwaste disposal in crystalline rocks: In Proc. Sym. on Uncertainties Associated with the Regulation of the Geologic Disposal of High Level Radioactive Waste: Oak Ridge National Laboratory–NRC, NUREG/CP-0022, CONF-810372, p. 349–354.

Brookins, D.G., with J.A. Wethington and J.F. Merklin. 1981b. Exploration, reserve estimation, mining, milling, conversion, and properties of uranium: Kansas State University Learning Module on the Nuclear Fuel Cycle.

Brookins, D.G., M.S. Abashian, L.H. Cohen, A.E. Williams, H.A. Wollenberg, and S. Flexser. 1982a. Suitability of crystalline rocks for radwaste storage: I.

Investigations of two plutons, Colorado: Geol. Soc. Amer. Prog. w. Abs., v. 14, p. 305.

Brookins, D.G., M.S. Abashian, L.H. Cohen, and H.A. Wollenberg. 1982b. A natural analogue for storage of radwaste in crystalline rocks: *In*: S.V. Topp, Ed., 4th Int. Sym. on the Sci. Basis for Nuc. Wste. Mngt., North-Holland Press, Amsterdam.

Brookins, D.G., M.S. Abashian, L.H. Cohen, A.E. Williams, H.A. Wollenberg, and S. Flexser. 1983, Natural analogues: Alamosa River monzonite intrusive into tuffaceous and andesitic rocks. *In*: D. Brookins, Ed. The Scientific Basis for Nuclear Waste Management, v. VI, Elsevier Sci. Publ. Co., New York, p. 299–306.

Brookins, D.G., H.A. Vogler, and J.J. Cohen. 1983. Use of poisoned land/inland seas for low level radioactive waste disposal: *In*: D.G. Brookins, Ed. The Scientific Basis for Nuclear Waste Management. v. VI. Elsevier Scientific Pub. Co., New York.

Burkholder, H.C., and G.E. Koester. 1975. Nuclear Waste Management and Transportation Quarterly Progress Report. BNWL-1913, Battelle Pacific Northwest Laboratories.

Calzia, J.P., and W.L. Hiss 1978. Igneous rocks in northern Delaware basin, New Mexico and Texas. *In*: Geology and Mineral Deposits of Ochoan Rocks in Delaware Basin and Adjacent Areas, compiled by G.S. Austin, New Mexico Bureau of Mines and Mineral Resources, Circular 159, p. 39–45.

Carnahan, C.L., L.V. Benson, and J.A. Apps. 1978. Preliminary simulations of mass transfer. *In*: Basalt Alteration and Basalt–Waste Interaction in the Pasco Basin of Washington State. Lawrence Berkeley Laboratory Report LBL-8532. 7–15.

Card, T.R., and G. Jansen. 1975. Solubility of Elements in U.S. Western Desert Groundwater and Comparison with Chemical and Radiological Concentration Limits for Drinking Water. U.S.E.R.D.A., BNWL-B-378.

Carr, J., C.R. Delannoy, W. Pardue, M. Pobereskin, and J. Waddell. 1979. The Disposal of Spent Nuclear Fuel. ONWI-59.

Cater, F.W. 1972. Salt anticlines within the Paradox Basin. *In*: Geologic Atlas of Rocky Mountains. Rocky Mountain Assoc. Geol., p. 137–167.

Cerny, P., 1979. Pollucite and its alteration in geologic occurrences and in deep-burial radioactive waste disposal. *In*: G.J. McCarthy, Ed., The Scientific Basis for Nuclear Waste Management. v. I. Plenum Press, New York, p. 231–236.

Chayes, F. 1972. Silica saturation in Cenozoic basalt: Philos. Trans. Roy. Soc. London-A. v. 271, p. 285–296.

Christensen, A.B., J.A. Del Debbio, J.D. Knecht, and J.E. Tanner. 1981. Loading and leachage of krypton immobilized in zeolites and glass. *In*: J.G. Moore, Ed., The Scientific Basis for Nuclear Waste Management v. III. Plenum Press, New York, p. 267–274.

Claiborne, H.C. 1976. Thermal Analysis of a Fuel Cladding Repository Pilot Plant in Salt. ORNL/TM-5221.

Cohen, B.L., and H.N. Jow. 1978. A generic hazard evaluation of low-level waste burial grounds: Nucl. Tech., v. 41, p. 381–388.

Cohen, J.J., and C.F. Smith. 1981. Validation of predictive models for geologic disposal of radioactive waste via natural analogs. *In*: D. Kocher, Ed., Proceedings of the Symposium on Uncertainties Associated with the Regulation of the Geologic Disposal of High-Level Radioactive Waste. NRC-ORNL Sym., NUREG/CP-0022, CONF-810372.

Cohen, J.J., C.F. Smith, and T.E. McKone. 1977. Determination of Performance Criteria for High-Level Solidified Nuclear Waste. Lawrence Livermore National Laboratory Report NUREG-0279.

Conservation Foundation. 1981. Toward a National Policy for Managing Low Level Radioactive Waste. Conservation Foundation, Washington, D.C..

Coons, W.E., E.L., Moore, M.J. Smith, and J.D. Kaser. 1980. The Functions of an Engineered Barrier System for a Nuclear Waste Repository in Basalt. DOE Report RHO-BWI-LD-23.

Cornman, W.R. comp. 1980. Composite Quarterly Technical Report on Long-Term High-Level Waste Technology, July–September 1979. E.I. du Pont de Nemours & Co., DP-27-157-3, Aiken, South Carolina.

Cornwall, H.R. 1972. Geology and mineral deposits of southern Ny County, Nevada: Nev. Bur. Mns. Geol. Bull. v. 77.

Dalrymple, G.B., and M.A. Lanphere. 1969. Potassium–Argon Dating. W.H. Freeman & Co., San Francisco.

De Donder, T., and P. Van Rysselberghe. 1936. Thermodynamic Theory of Affinity: A Book of Principles. Stanford Univ. Press, Stanford, Cal..

DeLaeter, J.R., J.K.R. Rosman, and C.L. Smith. 1980. The Oklo natural reactor: Cumulative fission yields and retentivity of the symmetric mass region fission products: Earth Plan. Sci. Lett. v. 50, p. 239–246.

Désiré, B., M. Hussenois, and R. Guillaument. 1969. Determination of the first hydrolysis constant of americium, berkelium, and californium: C.R. Acad. Sci. Ser. C., v. 269, p. 448–480.

DOE. 1980. Management of Commercially Generated Radioactive Waste: Final Environmental Impact Statement. DOE/EIS-0046F, 3 vols.

DOE. 1979. Technology for Commercial Radioactive Waste Management. DOE/ET-0028, UC-70, 5 vols.

DOE. 1980. In the Matter of Proposed Rulemaking on the Storage and Disposal of Nuclear Waste. DOE/NE-0007.

DOE. 1980. Low-Level Radioactive Waste Policy Act Report (Response to Public Law 96-573). DOE/NE-0015, Dist. Cat. UC-70.

DOE. 1982. National Plan for Siting High-Level Radioactive Waste Repositories and Environmental Assessment. DOE/NWTS-4.

Dosch, R.G. 1979a. Assessment of Potential Radionuclide Transport in Site-Specific Geologic Formations. Sandia National Laboratories Report SAND 79-2468.

Dosch, R.G. 1979b. Radionuclide migration studies associated with the WIPP site in S. New Mexico. *In*: G.J. McCarthy, Ed. The Scientific Basis for Nuclear Waste Management, v. I. Plenum Press, p. 395–398.

Dosch, R.G., and A.W. Lynch. 1980. Radionuclide transport in a dolomite aquifer. *In*: C.J.M. Northrup, Ed., The Scientific Basis for Nuclear Waste Management. v. II. Plenum Press, New York, p. 617–624.

Drever, J.I. 1982. The Geochemistry of Natural Waters. Prentice-Hall, Englewood Cliffs, New Jersey, 388 pp.

Drozd, R.J., C.M. Hohenberg, and C.J. Morgan. 1974. Heavy rare gases from Rabbit Lake (Canada) and the Oklo mine (Gabon): Natural spontaneous chain reactions in old uranium deposits: Earth Plan. Sci. Lett., v. 23, 28–37.

Dyck, W. 1978. The mobility and conclusions of U and its decay products in temperate surficial environments. In: M.M. Kimberley, Ed., Short Course in Uranium Deposits: Their Mineralogy and Origin, Mineralogical Association of Canada, Toronto, Ontario, Canada.

EPA. 1976. A study of waste generation, treatment and disposal in the metals mining industry. Midwest Res. Inst., EPA Report 68-01-2665, Oct. 1976.

ERDA. 1976. Alternatives for Managing Waste From Reactors and Post-Fission Operations on the LWR fuel cycle, v. 1–5. U.S. Energy Research and Development Administration Report 76–43.

Ewing, R.C. 1979. Natural glasses: Analogues for radioactive waste forms. In: G.J. McCarthy, Ed., The Scientific Basis for Nuclear Waste Management. v. I. Plenum Press, New York, p. 57–68.

Faure, G. 1977. Principles of Isotope Geology. Wiley–Interscience, New York.

Feates, F. and H.J. Richards. 1983. United Kingdom regulatory procedures for radioactive wastes: In: D. Brookins, Ed. The Scientific Basis for Nuclear Waste Management. v. VI. Elsevier Sci. Pub. Co., New York, in press.

Federal Register. 1981. Disposal of high-level radioactive wastes in geologic repositories: licensing procedures: Federal Register, v. 46, n. 37.

Fleischer, R.L., and O.G. Raabe. 1978. Effects of recoiling alpha-emitting nuclei: Mechanisms for uranium-series disequilibrium: EOS Trans. Amer. Geophys. Un., v. 59, p. 389.

Flynn, K.F., R.E. Barletta, L.J. Jardine, and M.J. Steindler 1979. Trans. Am. Nucl. Soc., v. 32, 394.

Folsom, T.R., and A.C. Vine. 1957. On the tagging of water masses for the study of physical processes in the oceans: Nat. Acad. Sci. Wash., Publ. No. 551, p. 121–132.

Ford, Bacon, and Davis. 1977. Determination of a Radioactive Waste Classification System, Lawrence Livermore National Laboratory Report PO 6320703.

Frejacques, C., C. Blain, C. Devilliers, R. Hagemann, and J.C. Ruffenbach. 1976. Conclusions tirées de l'étude de la migration des produits de fission. In: IAEA, The Oklo Phenomenon, p. 509–526.

Fuger, J., and F. Oetting. 1976. The Chemical Thermodynamics of Actinide Elements and Compounds. Part 2: The Actinide Aqueous Ions: IAEA STI/PUB/424/2.

Fyfe, W.S. 1983. Suitability of natural media for radwaste storage. In: D. Brookins, Ed., The Scientific Basis for Nuclear Waste Management. v. VI. Elsevier Sci. Pub. Co., New York, in press.

Gancarz, A. 1978. U–Pb age (2.05×10^9 years) of the Oklo uranium deposit. In IAEA. Natural Fission Reactors. Tech. Commun. 119, p. 513–520.

Gancarz, A.J., G.A. Cowan, D.B. Curtis, and W.J. Maeck. 1979. ^{99}Tc, Pb, and Ru Migration around the Oklo Natural Fission Reactor. Mat. Res. Soc. Sym. G., Scientific Basis for Nuclear Waste Management, Ext. Abs., p. 59–60.

Gancarz, A., G. Cowan, D. Curtis, and W. Maeck 1980. Tc-99, Pb and Ru migration around the Oklo natural fission reactors. In: C.J.M. Northrup, Ed., The Scientific Basis for Nuclear Waste Management, v. II. Plenum Press, New York, p. 601–608.

Garrels, R.M. 1959. Mineral Equilibria, Harper & Brothers, New York.

Garrels, R.M. and C.G. Christ. 1965. Solutions, Minerals and Equilibria, Harper & Row, New York.

Gauthier-Lafaye, F. 1977. Oklo et les gisements d'uranium du Francevillien: aspects tectonique et metallogenique. Ph.D. Thesis, University of Strasbourg.

GE. 1977. Chart of the Nuclides, 12th Ed., General Electric Co.

Geffroy, J. 1975. Étude microscopiques des minérais uraniferais d'Oklo. In: IAEA, The Oklo Phenomenon, p. 133–152.

Gera, F. 1975. Geochemical Behavior of Long-Lived Radioactive Wastes. Oak Ridge National Laboratory Report ORNL-TM-4481. Oak Ridge, Tennessee.

Germanov, A.I., S.G. Batulin, G.A. Volkov, A.K. Lisitsin, and V.S. Serebrennikov. 1958. Some regularities of uranium distribution in underground waters. In: Proc. U.N. Intern. Conf. Peaceful Uses At. Energy, 2nd Conf., Geneva, v. 2, p. 161–177.

Glass, R.S. 1981. Effects of Radiation on the Chemical Environment Surrounding Waste Canisters in Proposed Repository Sites and Possible Effects on the Corrosion Process SAND81-1677, Sandia National Laboratories, Albuquerque, New Mexico.

Godbee, H.W., and D.S. Joy. 1974. Assessment of the loss of radioactive isotopes from radioactive waste solids to the environment. Part I: Background and theory: Oak Ridge National Laboratory Report ORNL-TM-4333, Oak Ridge, Tennessee.

Goldberg, E.D. 1965. Minor elements in sea water: In: J.P. Riley and G. Skirrow, Eds., Chemical Oceanography, Academic Press, New York.

Guber, W., M. Hussain, L. Kahl, G. Ondracek, and J. Saidl. 1979. Preparation and characterization of an improved high level radioactive waste (HAW) borosilicate glass. In: G.J. McCarthy, Ed., The Scientific Basis of Nuclear Waste Management, v. 1. Plenum Press, New York, p. 37.

Haaker, R.F., and R.C. Ewing. 1980. Uranium and thorium minerals: Natural analogues for radioactive waste forms. In: C.J.M. Northrup, Ed., The Scientific Basis for Nuclear Waste Management. v. II. Plenum Press, New York, p. 281–288.

Haaker, R.F. and R.C. Ewing 1981. Natural analogues for crystaline radioactive waste forms, Part II: Non-Actinide phases. In: J.G. Moore, Ed., The Scientific Basis for Nuclear Waste Management, v. III. Plenum Press, New York, p. 299–306.

Haire, R.G., and G.W. Beall 1979. Consequences of radiation from sorbed transplutonium elements on clays selected for waste isolation. In: S. Fried, Ed.,

Radioactive Waste in Geologic Storage, Am. Chem. Soc. Symp. Series 100, Sept. 1978, Miami Beach, Florida, p. 291–295.

Hale, W.E., L.S. Hughes, and E.R. Cox. 1954. Possible Improvement of Quality of Water of the Pecos River by Diversion of Brine of Malaga Bend, Eddy County, New Mexico. Report of Pecos River Comm., New Mexico and Texas, with U.S. Geol. Surv. Water Resources Div.

Hanson, G.N., and P.W. Gast. 1967. Kinetic studies in contact metamorphic zones: Geochim. Cosmochim. Acta, v. 31, p. 1119–1154.

Harshman, E.N. 1972. Geology and Uranium Deposits, Shirley Basin Area, Wyoming. U.S. Geol. Surv. Prof. Paper 745, 82 p.

Hart, S.R., G.L. Davis, R.H. Steiger, and G.R. Tilton. 1968. A comparison of the isotopic mineral age variations and petrologic changes induced by contact metamorphism. In: E.I. Hamilton and R.D. Farquhar, Eds., Radiometric Dating for Geologists, Wiley–Interscience, New York, p. 73–110.

Hench, L.L. 1977. Leaching of glass. In: Ceramic and Glass Radioactive Waste Forms, CONF-770102. NTIS, Springfield, Virginia.

Hinga, K.R. 1982. Ocean research conducted for the subseabed disposal program: EOS Trans.–Amer. Geophys. Un., v. 63, p. 802.

Hiss, W.L. 1975. Stratigraphy and ground water hydrology of the Capitan Aquifer, southeastern New Mexico and western Texas: Unpubl. Ph.D. Thesis, University of Colorado.

Hite, R.J. 1960. Stratigraphy of the Saline Facies of the Paradox Member of the Hermosa Formation of Southeastern Utah and Southwestern Colorado. Four Corners Geologic Society Third Field Conf., p. 86–89.

Hite, R.J. and S.W. Lohman, 1973. Geologic appraisal of Paradox Basin salt deposits for waste emplacement: U.S. Geol. Surv. Open File Report.

Hite, R.J., O.B. Rauo, J.A. Peterson, and C.M. Molenaar. 1972. Carbonate and evaporite facies of the Paradox Basin: The Mountain Geol., v. 9, p. 301–310.

Holliger, P., C. Devillers, G. Retali. 1978. Evaluation des temperatures neutroniques dans les zones de reaction d'Oklo par l'etude des rapports isotopiques. In: IAEA. National Fission Reactors, p. 553–568.

Hostetler, P.B. and R.M. Garrels. 1962. Transportation and precipitation of uranium and vanadium at low temperatures, with special reference to sandstone type uranium deposits: Econ. Geol., v. 57, p. 137.

Hurley, P.M., J.M. Hunt, W.H. Pinson, and H.W. Fairbairn. 1963. K–Ar age values on the clay fractions in dated shales: Geochim. Cosmochim. Acta, v. 27, p. 279–284.

IAEA. 1962. Basic Safety Standards for Radiological Protection. IAEA Safety Series No. 9.

IAEA. 1975. The Oklo Phenomenon. IAEA Symp. v. 204.

IAEA. 1978. Natural fission reactors. IAEA Tech. Commun. 119.

Jardine, L.J., and M.J. Steindler. 1978. A Review of Metal-Matrix Encapsulation of Solidified Radioactive High-Level Waste. Argonne National Laboratory, ANL-78-19, Argonne, Illinois.

Jenks, G.H. 1975. Gamma-Irradiation Effects in Geologic Formations of Interest in Waste Disposal: A Review and Analysis of Available Information and Suggestions for Additional Experimentsation. Oak Ridge National Laboratory Report ORNL-TM-4827, Oak Ridge, Tennessee.

Jockwer, N. 1981. Laboratory investigations of water content within rock salt and its behavior in a temperature field of disposed high-level waste. *In*: J.G. Moore, Ed., The Scientific Basis for Nuclear Waste Management. v. III. Plenum Press, New York, p. 35–42.

Johnson, K.D.B., and J.A.C. Marples. 1979. Glass and Ceramics for Immobilization of Radioactive Wastes for Disposal. Atomic Energy Research Establishment, AERE-R 9417, Harwell, England.

Johnstone, J.K. 1980. Radiation effects on tuff. *In*: J.K. Johnstone and K. Wolfsberg, Eds. Evaluation of Tuff as a Medium for Nuclear Waste Repository: Interum Status Report on the Properties of Tuff. Sandia National Laboratory Report SAND-1464, p. 75–76.

Johnstone, J.K., and K. Wolfsberg, Eds. 1980, Evaluation of Tuff as a Medium for Nuclear Waste Repository: Interim Status Report on the Properties of Tuff. Sandia National Laboratory Report SAND80-1464.

Jones, C.L. 1973. Salt Deposits of the Los Medanos Area, Eddy and Lea Counties, New Mexico. U.S. Geol. Surv. Open File Report 4339-7, p. 29–30.

Katayama, Y.B., D.J. Bradley, and C.O. Harvey. 1980. Status Report on LWR Spent Fuel IAEA Leach Tests. Pacific Northwest Laboratory Report PNL-3173 Richland, Washington.

Keller, C. 1971. The chemistry of the transuranium elements: Kernchemie in Einzeldarstellungen: Verlag Chemie Gmbh., v. 3.

Kharaka, Y.K., and F.A.F. Berry. 1973. Simultaneous flow of water and solutes through geologic membranes. I. Experimental investigations: Geochim. Cosmochim. Acta, v. 37, p. 2577–2604.

Kibbey, A.H., and H.W. Godbee. 1979. Physicochemical characterization of solidification agents used and products formed with radioactive wastes at LWR nuclear power plants. *In*: G.J. McCarthy, Ed., The Scientific Basis of Nuclear Waste Management. v. I. Plenum Press, New York, p. 495–498.

Klement, A.W., C.R. Miller, R.P. Minx, and B. Shleien. 1972. Estimates of Ionizing Radiation Doses in the United States 1600–2000. EPA Report ORP-OSD 72-1.

Klingsberg, C., and J. Duguid. 1980. Status of Technology for Isolating High-Level Radioactive Wastes in Geologic Repositories. EPA Report TIC11207.

Kocher, D. 1981. Proceedings of the Symposium on Uncertainties Associated with the Regulation of the Geologic Disposal of High-Level Radioactive Waste. NRC–ORNL Sym. NUREG/CP-0022 CONF-810372.

Komarneni, S., and R. Roy. 1980. Superoverpack: Tailor-made mixtures of zeolites and clays. *In*: C.J.M. Northrup, Ed., The Scientific Basis for Nuclear Waste Management. v. II. Plenum Press, New York, p. 411–418.

Komarneni, S., B.E. Scheetz, G.J. McCarthy, and W.E. Coons. 1980. Hydrothermal Interactions of Cesium and Strontium Phases from Spent Unprocessed

Fuel with Basalt Phases and Basalt. Rockwell International Report RHO-BNI-C-70.

Krauskopf, K.B. 1979. Introduction to Geochemistry. 2nd edn., McGraw–Hill, New York.

Krauskopf, K. B. 1967. Introduction to Geochemistry. McGraw-Hill, New York, 721 pp.

Krestov, G.A. 1972. Thermochemistry of compounds of rare earths and actinide elements: AEC transl. 7505.

Kupfer, M.J. 1979. Vitrification of Hanford Radioactive Defense Wastes. Rockwell Hanford Operations Report RHO-SA-89, Richland, Washington.

Kupfer, M.J., D.M. Strachan, and W.W. Schulz, 1979. Technical challenges in the immobilization of Hanford defense waste. *In*: G.J. McCarthy, Ed., The Scientific Basis for Nuclear Waste Management. v. I. Plenum Press, New York.

Lambert, S. J. 1983. Evaporite dissolution relevant to the WIPP site, northern Delaware basin, southeastern New Mexico. *In*: D. G. Brookins, Ed., The Scientific Basis for Nuclear Waste Management, v. VI. Elsevier, New York, p. 291–298.

Langmuir, D. 1978. Uranium solution-mineral equilibria at low temperatures with applications to sedimentary ore deposits: Geochim. Cosmochim. Acta, v. 42, p. 547–568.

Lanza, F., N. Jacquet-Francillon, and J.A.C. Marples. 1980. Methodology of Leach Testing of Borosilicate Glasses in Water. Presented at First European Community Conference on Radioactive Waste Management and Disposal, Luxembourg.

Lappin, A.R. 1980. Thermal conductivity. *In*: J.K. Johnstone and K. Wolfsberg, Eds., Evaluation of Tuff as a Medium for Nuclear Waste Repository: Interum Status Report on the Properties of Tuff. Sandia National Laboratory Report SAND-1464, p. 51–63.

Lappin, A.R., R.G. Van Buskirk, D.O. Enniss, S.W. Butters, F.M. Prater, C.B. Muller, and J.L. Bergosh. 1981. Thermal Conductivity, Bulk Properties, and Thermal Stratigraphy of Silicic Tuffs from the Upper Portion of Hole USW-G1, Yucca Mountain, Nye County, Nevada. Sandia National Laboratory Report SAND81-1873.

Latimer, W.L. 1961. Oxidation Potentials, 2nd edn., Prentice–Hall, Princeton, N.J.

Laul, J.C., and J.J. Papike. 1982. Chemical migration by contact metamorphism between granite and silt carbonate systems: IAEA Symp. 257, p. 603–614.

Leach, D.L., K.P. Puchlik, and R.K. Glanzman. 1980. Geochemical exploration for uranium in plavas: Jour. Geochem. Explor., v. 13, p. 251–284.

Ledbetter, J.O., W.R. Kaiser, and E.A. Ripperger. 1975. Radioactive Waste Management by Burial in Salt Domes. Univ. Texas, Austin, Eng. Mech. Res. Lab. Report AEC Contract AT 40-1-4039.

Lee, M.J. 1976. Geochemistry of the Sedimentary Uranium Deposits of the Grants

Mineral Belt, Southern San Juan Basin, New Mexico. Ph.D. Thesis, University of New Mexico.

Levin, G.B. 1981. Low level radioactive waste management in the U.S.: A proving ground: Nucl. News, v. 24, p. 72–76.

Lipman, P. 1975. Evolution of the Platoro Caldera Complex, and Related Volcanic Rocks, Southeastern San Juan Mountains, Colorado. U.S. Geol. Surv. Prof. Paper 852.

Lippolt, H.J., and I. Raczek. 1979a. Cretaceous Rb–Sr total rock ages of Permian salt rocks: Die Naturwissenschaften, v. 66, p. 422–423.

Lippolt, H.J., and I. Raczek. 1979b. Rinneite-dating of episodic events potash salt deposits: Jour. Geophys., v. 46, p. 225–228.

Lisitsin, A.K. 1969. Conditions of molybdenum and selenium deposition in exogenous epigeonetic uranium deposits: Lithol. Min. Res. no. 5, p. 541–548.

Loehr, C.A. 1979. Mineralogical and geochemical effects of basaltic dike intrusion into evaporite sequences near Carlsbad, New Mexico: unpub. M.S. Thesis, New Mexico Institute of Mining Technology.

Lokken, R.O. 1979. A Review of Radioactive Waste Immobilization in Concrete. Pacific Northwest Laboratory Report PNL-2654, Richland, Washington.

Lomenick, T.F., and R.L. Bradshaw. 1969. Deformation of rock salt in openings mined for disposal of radioactive wastes: Rock Mech., v. 1, p. 5–30.

Long, P.E., and N.J. Davidson. 1981. Lithology of the Grande Ronde Basalt with emphasis on the Umtanum and McCoy Canyon flows. *In* C.W. Myers and S.M. Price, Eds., Subsurface Geology of the Cold Creek Syncline. Rockwell International Report RHO-BWL-ST-14, p. 5-1–5-55.

Long, P. E. and R. D. Landon. 1981. Stratigraphy of the Grande Ronde Basalt. *In:* M. W. Myers, and S. M. Price, Eds., Subsurface Geology of the Cold Creek Syncline. Rockwell International Report RHO-BW1-ST-14, p. 4-1–4-43.

Lutze, W., Ed. 1982. The Scientific Basis for Nuclear Waste Management. v. V. Elsevier Sci. Pub. Co., New York.

Mackenzie, F.T., and R.M. Garrels. 1971. Evolution of Sedimentary Rocks. W.W. Norton & Co., New York.

Maeck, W.J., K.E. Apt, and G.A. Cowan. 1978a. The measurement of ruthenium in uranium ores and ^{238}U spontaneous fission yield. *In:* Natural Fission Reactors. IAEA Tech. Comm. 119, p. 521–540.

Maeck, W.J., K.E. Apt, and G.A. Cowan. 1978b. A possible uranium-ruthenium method for the measurement of ore age. *In:* Natural Fission Reactors. IAEA Tech. Comm. 119, p. 535–542.

Maini, Y.N.T. 1972. Theoretical and Field Consideration on the Determination of in situ Hydraulic Parameters in Fractured Rock. Proc. Sym. Percolation through Fissured Rock, Stuttgart, p. TI-El-TI-E8.

Marin, B., and T. Kikindai. 1969. Etude comparée de l'hydrolyse de l'europium et de l'americium en milieus chlorue par électrophorése sur papier: C.R. Acad. Sci., Ser. C., v. 268, p. 1–4.

Markos, G. 1979. Geochemical Mobility and Transfer of Contaminants in Uranium Mill Tailings. Symosium on Uranium Mill Tailings Management, Fort Collins, CO, p. 55–69.

Martin, J.B., M.J. Bell, and M.R. Knapp 1981. Uncertainties in the regulation and licensing of a high-level waste repository. *In*: Proceedings of the Symposium on Uncertainties Associated with the Regulation of the Geologic Disposal of High-Level Radioactive Waste, Gatlinburg, Tennessee, March 9–13, 1981, NUREG/CP-0022, CONF-810372, p. 3–16.

McCarthy, G.J. 1977. High-level ceramics: materials considerations, process simulation, and product characterization: Nucl. Tech., v. 32, p. 92–105.

McCarthy, G.J. 1979a. Ceramics from defense high-level wastes: Nucl. Tech., Letter to the Editor, v. 10.

McCarthy, G.J., Ed. 1979b. The Scientific Basis for Nuclear Waste Management v. I. Plenum Press, New York.

McCarthy, G.J., and M.T. Davidson. 1975. Ceramic nuclear forms. I. Crystal chemistry and phase formation: Ceramic Bull., v. 54 (g), p. 782–786.

McCarthy, G.J., S. Komarneni, B.E. Scheetz, and W.B. White. 1978a. Hydrothermal reactivity of simulated nuclear waste forms and water-catalzyed waste–rock interactions. *In*: G.J. McCarthy, Ed., The Scientific Basis for Nuclear Waste Management, v. I. Plenum Press, New York.

McCarthy, G.J., B.E. Scheetz, S. Komarneni, and D.K. Smith. 1978b. Reaction of Water with a Simulated High-Level Nuclear Waste Glass at 300°C, 300 bars. DOE Report RHO-BWI-C-35.

McLennan, S.M., and S.R. Taylor. 1979. Rare earth elements mobility associated with uranium mineralization: Nature, v. 282, p. 247–249.

McMillan, R.H. 1978. Origin of stratiform uranium Deposits. *In*: M.M. Kimberley, Ed., Short Course in Uranium Deposits: Their Mineralogy and Origin, Mineralogical Association of Canada, Toronto, Ontario, Canada, p. 199.

McVay, G.L. 1979. Gamma-Irradiation Effects on Leaching Characteristics of Waste Containment Materials: Pacific Northwest Laboratory Report PNL-SA-7992, Richland, Washington, pp. 83–91.

McVay, G.L., D.J. Bradley, and J.F. Kircher. 1981. Elemental Releases from Glass and Spent Fuel: Pacific Northwest Lab., Battelle Memorial Inst. Report ONWI-275.

Mendel, J.E. 1978. The Storage and Disposal of Radioactive Waste as Glass in Canisters. Pacific Northwest Laboratory Report PNL-2764, Richland, Washington.

Mendel, J.E., R.D. Nelson, R.P. Turcotte, W.J. Gray, W.D. Merz, F.P. Roberts, W.J. Wever, J.H. Westsik, D.E. Clark. 1981a. A State-of-the-Art Review of Materials Properties of Nuclear Waste Forms. Battelle Pacific Northwest Laboratories Report PNL-3802-UC-70.

Mendel, J.E., J.H. Westsik, and D.E. Clark. 1981b. Chemical durability. *In*: A State of the Art Review of Materials Properties of Nuclear Waste Forms. Battelle Pacific Northwest Laboratories Report PNL-3802-UC-70, p. 4.1–4.69.

Mercer, J.W., and B.R. Orr. 1977. Review and Analysis of Hydrogeologic

Conditions near the Site of a Potential Nuclear Waste Laboratory, Eddy and Lea Counties, New Mexico. U.S. Geol. Surv. Open File Report p. 77–123.

Merritt, W.F. 1977. High-level waste glass: field leach test: Nucl. Tech. v. 32, p. 88–89.

Merz, M.D. 1981. Physical properties. *In*: A State of the Art Review of Materials Properties of Nuclear Waste Forms. Battelle Pacific Northwest Laboratory Report PNL-3802-UC-70, p. 2.1–2.5.

Moak, D.P. 1981. Waste Package Materials Screening and Selection. Battelle Mem. Inst. Report ONWI-312, Columbus, Ohio.

Moody, J.B. 1982 Radionuclide Migration/Retardation: Research and Development Technology Status Report. Battelle Mem. Inst. Report ONWI-321, Columbus, Ohio.

Moore, J.G., Ed. 1981. The Scientific Basis for Nuclear Waste Management. v. III. Plenum Press, New York.

Moore, J.G., et al. 1975. Development of Cementitious Grouts for the Incorporation of Radioactive Wastes. Part 1. Leach Studies. Oak Ridge National Laboratory Report ORNL-4962, Oak Ridge Tennessee.

Moore, J.G., H.W. Godbee, A.H. Kibbey, and D.S. Joy 1975. Development of Cementitious Grouts for the Incorporation of Radioactive Wastes. U.S.D.O.E. Rep. ORNL-4962, Oak Ridge National Laboratory, Oak Ridge, Tennessee, p. 116 (available from NTIS).

Morgan, M.T., J.G. Moore, H.E. Devaney, G.C. Rogers, C. Williams, and E. Newman 1979. The disposal of Iodine-129. *In*: G.J. McCarthy, Ed., The Scientific Basis for Nuclear Waste Management, v. I. Plenum Press, New York, p. 453–459.

Murphy, E.S., and G.M. Holter. 1980. Technology, safety and costs of decommisioning a reference low level waste burial ground: Battelle Pacific Northwest Laboratory Report NUREG/CR-0570, v. I.

Naudet, R. 1974. Summary report on the Oklo Phenomenon: Fr. Cent. Energie Atom., Rept, Bull, Inf. Sci. Tech., v. 193., p. 7–85.

Naudet, R. 1978. Les réacteurs d'Oklo: Cinq ans d'exploration du site. *In*: Natural Fission Reactors IAEA Tech. Pub. 119, p. 3–18.

Noble, D.C. 1972. Some observations on the Cenozoic volcano-tectonic evolution of the Great Basin, western United States: Earth Plan. Sci. Lett., v. 17, p. 142–150.

Neretnieks, I., 1977. Retardation of Escaping Nuclides from a Final Depository. KBS Report No. 130, Stockholm, Sweden.

Northrup, C.J.M., Ed. 1980. The Scientific Basis for Nuclear Waste Management, v. II: Plenum Press, New York.

Nowak, E.J., 1979. The Backfill as an Engineered Barrier for Nuclear Waste Management. SAND79-0990C, Sandia National Laboratories, Albuquerque, New Mexico.

Nowak, E.J. 1980a. Radionuclide Sorption and Migration Studies of Getters for Backfill Barriers: Sandia National Laboratories Report SAND-79-1110.

Nowak, E.J. 1980b. The Backfill Barrier as a Component in a Multiple Barrier Nuclear Waste Isolation System. Sandia National Laboratories Report SAND79-1305.

Nowak, E.J. 1980c. The backfill as an engineered barrier for nuclear waste management. In: C.J.M. Northrup, Ed., The Scientific Basis for Nuclear Waste Management. v. II. Plenum Press, New York.

Nowak, E.J. 1982. The diffusion of Cs(I) and Sr(II) in liquid-saturated beds of backfill materials. Sandia National Laboratories Report SAND82-0750.

NUREG. 1979. Determination of Performance Criteria for High-Level Solidified Nuclear Waste. Lawrence Livermore National Laboratory Report NUREG-0279.

NUREG. 1980. Final Generic Environmental Impact Statement on Uranium Milling. 3 Vols. U.S. Nuclear Regulatory Commission Project M-25, NUREG-0706.

Obert, L., and W.I. Duvall. 1966. Rock Mechanics and the Design of Structures in Rocks. John H. Clay & Sons, Pubs., Inc. NY.

O'Hare, P.A.G., and H.R. Hoekstra, 1975. Thermodynamics of formation of cesium and rubidium uranates at elevated temperatures: Jour. Chem. Thermo., v. 7, p. 931–838.

Olsson, W.A., and L.W. Teufel. 1980. Mechanical properties of tuff. In: J.K. Johnstone and J. Wolfsberg, Eds., Evaluation of Tuff as a Medium for Nuclear Waste Repository: Interum Status Report on the Properties of Tuff. Sandia National Laboratories Report SAND80-1464, p. 64–75.

ONWI. 1980. Regional Environmental Characterization Report for the Paradox Bedded Salt Region and Surrounding Territory. Battelle Mem. Inst. Report ONWI-68.

ONWI. 1981. Geologic Evaluation of Gulf Coast Salt Domes: Overall Assessment of the Gulf Interior Region. ONWI-106.

ONWI. 1982. Environmental Characterization Report for the Paradox Basin Study Region Utah Study Areas. Battelle Mem. Inst. Report ONWI-144.

ORNL. 1980. Radiological Surveys at inactive uranium-mill sites: Oak Ridge National Laboratories Reports ORNL/TM-5251 and ORNL-5447–ORNL-5465.

People's Republic of China. 1980. Health survey in high background radiation areas in China: Science, v. 209, p. 877–880.

Pettijohn, F.I. 1975. Sedimentary Rocks, 3rd. edn., Harper & Row, New York.

Pickett, G.R. 1968. Properties of the Rocky Mountain Arsenal Disposal Reservoir and their relation to Derby Earthquakes: Colo. School Mns. Quart., v. 63, p. 73–100.

Pinder, G.F., and W.G. Gray. 1977. Finite Element Simulation in Surface and Subsurface Hydrology. Academic Press, New York.

Pitman, S. G., 1980. Investigation of Susceptibility of titanium-grade 2 and titanium-grade 12 to Environmental Cracking in a Simulated Basalt Repository Environment. PNL-3915, Battelle Pacific Northwest Laboratory, Richland, Washington, August.

Plodinec, M.J., and J.R. Wiley. 1979. Evaluation of Glass as a Matrix for Solidifying Savannah River Plant Waste: Properties of Glasses Containing Li_2O. Savannah River Laboratory, DP-1498, Aiken, South Carolina.

Polzer, W.L. 1971. Solubility of Pu in Soil/Water Environments: A theoretical Study. In: Proc. Rocky Flats Sym. Safety in Pu Handling. ORNL, AEC Report Conf. 710401, p. 412–429.

Pusch, R. 1977. Required Physical and Mechanical Properties of Buffer Masses. KBS Teknisk Rapport 33, Stockholm (Swedish).

Pusch, R., and A. Bergstrom 1981. Highly compacted bentonite—a self-healing substance for nuclear waste isolation. In: J.G. Moore, Ed., The Scientific Basis for Nuclear Waste Management, v. III. Plenum Press, New York, p. 553–560.

Quinby, T.C. 1978. Method of Producing Homogeneous Mixed Metal Oxide and Metal–Metal Oxide Mixtures. United States Patent 4,072,501, U.S. Patent Office, Washington, D.C.

Rafalsky, R.P. 1958. The experimental investigation of the conditions of uranium transport and deposition by hydrothermal solutions. In: Proc. Sec. U.N. Int. Conf. on Peaceful Uses of Atomic Energy. v. 2. p. 432–444.

Rai, D., and R.J. Serne, 1978. Solid Phases and Solution Species of Different Elements in Geologic Environments. DOE PNL-2651.

Rai, D., R.J. Serne, and D.A. Moore. 1980. Solubility of plutonium compounds and their behavior in soils: Soil Sci. Soc. Am. J., v. 44, p. 490–495.

Ramspott, L.D., and J.A. Howard. 1975. Principal rock types and test areas at Nevada Test Site (map). In: Borg et al., Eds., Information Pertinent to the Migration of Radionuclides in Groundwater at the Nevada Test Site. I. Review and Analysis of Existing Information Lawrence Livermore Laboratory Report UCRL52078.

Rankin, W.N., J.A. Kelley. 1978. Microstructures and leachability of vitrified radioactive wastes: Nucl. Tech., v. 41.

Register, J.K., and D.G. Brookins. 1980. Rb-Sr isochron age of evaporite minerals from the Salado formation (Late Permian), southeastern New Mexico: Isochron/West, n. 29, p. 39–41.

Register, J.K., D.G. Brookins, M.E. Register, and S.J. Lambert. 1980. Clay mineral–brine interactions during evaporite formation: Lanthanide distribution in WIPP samples. In: C.J.M. Northrup, Ed., The Scientific Basis for Nuclear Waste Management. v. II. Plenum Press, New York, p. 445–452.

Rich, R.A., H.D. Holland, and U. Petersen 1977. Hydrothermal Uranium Deposits. Elsevier Scientific Pub. Co., New York.

Ringwood, A.E., S.E. Kesson, N.G. Ware, W. Hibbesson, and A. Major 1979. Immobilization of high level nuclear wastes in SYNROC: Nature, 278, 219.

Ringwood, A.E., S.E. Kesson, and N.G. Ware. 1980a. Immobilization of U.S. defense nuclear wastes using SYNROC processes. In: C.J.M. Northrup, Ed., The Scientific Basis for Nuclear Waste Management v. II. Plenum Press, New York, p. 265–272.

Ringwood, A.E., V.M. Oversby, and W. Sinclair. 1980b. The effects of radiation damage on SYNROC. *In*: C.J.M. Northrup, Ed., The Scientific Basis for Nuclear Waste Management v. II, Plenum Press, New York, p. 273–280.

Ritzma, H.R., and H.H. Doelling. 1969. Mineral resources, San Juan County, Utah and adjacent areas: Utah Geol. Min. Surv. Special Studies 24.

Robertson, J.B. 1980. Shallow land burial of low level radioactive wastes in the USA: geohydrologic and nuclide migration studies. *In*: Underground Disposal of Radioactive Wastes, v. II, IAEC Symp. p. 253–275.

Robie, R.A., B.S. Hemingway, and J.R. Fisher. 1978. Thermodynamic properties of minerals and related substances at $290.15°K$ and one bar (10^5 pascals) pressure and at higher temperatures: U.S. Geol. Surv. Bull., v. 1452, 456 p.

Robinson, T.W., and G.W. Lang. 1938. Geology and Ground Water Conditions of the Pecos River Valley in the Vicinity of Laguna Grande de la Sal, New Mexico, with Special Reference to the Salt Content of the River Water. N.M. State Engineer 12th and 13th Bienn. Reports, p. 77–100.

Robinson, B.P. 1962. Ion-Exchange Minerals and Disposal of Radioactive Wastes— A Survey of Literature. U.S. Geol. Surv. Water Supply Paper 1616.

Roeder, E. and H.E. Belkin. 1979. Application of studies of fluid inclusions in Permian Salado salt, New Mexico, to problems of siting the waste isolation pilot plant. *In*: G.J. McCarthy, Ed., The Scientific Basis for Nuclear Waste Management. v. I. Plenum Press, New York, p. 313–322.

Roeder, E., and H.E. Belkin. 1980. Thermal gradient migration of fluid inclusions in single crystals of salt from the waste isolation pilot plant site (WIPP). *In*: C.J.M. Northrup, Ed., The Scientific Basis for Nuclear Waste Management v. II. Plenum Press, New York, p. 453–462.

Ross, W.A. 1975. *In*: J.L. McElroy, Ed., Research and Development Activities Waste Fixation Program, April through June 1975. Pacific Northwest Laboratory Report BNWL-1932, Richland, Washington.

Ross, W.A., C.O. Harvey, R.O. Lokken, R.P. May, F.P. Roberts, C.L. Timmerman, R.L Treat, and J.H. Westsik, Jr. 1982. A comparative assessment of TRU waste forms and immobilization processes. *In*: D.G. Brookins, Ed., The Scientific Basis for Nuclear Waste Management, v. VI. Elsevier, New York, p. 497–504.

Rudolph, G., P. Vejmelka, and R. Koster. 1981. Leach and corrosion tests under normal and accident conditions on cement products from simulated intermediate level evaporator concentrates. *In*: J.G. Moore, Ed., The Scientific Basis for Nuclear Waste Management. v. III, Plenum Press, New York, p. 339–346.

Rusin, J.M., W.J. Gray, and J.W. Wald. 1979. Multibarrier Waste Forms. Part II. Characterization and Evaluations. Pacific Northwest Laboratory Report PNL-2668-2, Richland, Washington.

Salander, C., and P. Zuhlke. 1979. The concept for the treatment and disposal of radioactive wastes in the German Ertsorgungszeutrum. CONF-790420, p. 13–16.

SAND 1978. Geological Characterization Report, Waste Isolation Pilot Plant

(WIPP) Site, Southeastern New Mexico. *In*: D.W. Powers, S.J. Lambert, S.E. Shaffer, L.R. Hill, W.D. Weart, Eds., DOE Report AT(29-1)-789; 2 vols.

Sauer, H.I. and F.R. Brand. 1971. Geographic patterns in the risk of dying. *In*: H.L. Cannon and H.C. Hopps, Eds., Environmental Geochemistry in Health and Disease Geol. Soc. Amer. Mem. 123, p. 131–150.

Scheetz, B.E., W.P. Freeborn, J. Pepin, and W.B. White 1982. The system $SrMoO_4$-$BaMoO_4$-$CaMoO_4$: Compatibility relations, the implications for supercalcine ceramics. *In*: D.G. Brookins, Ed., The Scientific Basis for Nuclear Waste Management, v. VI, Elsevier, New York, p. 155–162.

Schilling, J.A. 1973. K-Ar dates on Permian potash minerals from southeastern New Mexico: Isochron/West, n. 6, p. 37–38.

Schulz, W.W. 1976. The Chemistry of Americium. ORNL, ERDA Report TID 26971, p. 47–121.

Schumm, R.H., D.D. Wagman, S. Bailey, W.H. Evans, and V.B. Parker. 1973. Selected Values of Chemical Thermodynamic Properties: Tables for the Lanthanide (Rare Earth) Elements. Nat. Bur. Standards Tech. Note 270-7.

Sergeyeva, E.I., A.A. Nikitin, I.L. Khodakovskiy, and G.B. Naumov. 1972. Experimental investigations of equilibria in the system UO_3-CO_2-H_2O in the $25°$-$200°C$ temperature range: Geochem. Internat., v. 9, p. 900–910.

Shalinets, A.B., and A.V. Stepanov. 1972. Investigation of complex formation of the trivalent actinide and lanthanide elements by the method of electromigration. XVII. hydrolysis: Soviet Radiochem., v. 14, p. 290–293.

Shea, M.E. 1982. Uranium migration at Marysvale, Utah: a natural analog for radioactive waste isolation: Unpub. M.S. Thesis, University of California, Riverside.

Shiao, S.Y. 1979. Ion-exchange equilibrium between montmorillonite and solutions of moderate-to-high ionic strength. *In*: S. Fried, Ed., Radioactive Waste in Geologic Storage, Am. Chem. Soc. Symp. Series 100, Sept. 1978, Miami Beach, Florida, p. 297–324.

Simmons, E.C., and C.E. Hedge. 1978. Minor element and Sr isotopic geochemistry of Tertiary stocks, Colorado Mineral Belt: Contr. Min. Petrol., v. 67, p. 379–396.

Simmons, J.H., P.B. Macedo, A. Barkatt, and T.A. Ltovitz 1979. Fixation of radioactive waste in high silica glasses: Nature, v. 278, p. 729–731.

Simpson, D.R. 1980. Desiccant Materials Screening for Backfill in a Salt Repository. ONWI-214, Office of Nuclear Waste Isolation, Battelle Memorial Institute, Columbus, Ohio.

Smith, C.F., and J.J. Cohen. 1981. Models and criteria for waste repository performance: *In*: Kocher (Ed.), p. 209–214.

Smith, C.F., J.J. Cohen, and T.E. McKone, 1980. A hazard index for underground toxic material: Lawrence Livermore National Laboratory Report UCRL-52889.

Smith, J.M., T.W. Fowler, and A.S. Goldin. 1981. Mathematical models for estimating population health effects from the disposal of high-level radioactive waste: *In*: D. Kocher, Ed., Proceedings of the Symposium on Uncertainties Associated with the Regulation of the Geologic Disposal of the High-Level Radioactive Waste. NRC-ORNL Symposium, NUREG/CP-0022 CONF 810372.

Smyth, J.R. and M.L. Sykes. 1980. Zeolite stability studies: *In*: J.K. Johnstone and K. Wolfsberg, Eds., Evaluation of Tuff as a Medium for Nuclear Waste Repository: Interum Status Report on the Properties of Tuff. Sandia National Laboratory Report SAND80-1464, p. 37–38.

Spitsyn, V.I., and V.D. Balukova 1973. Issledovanija vzaimodejstive othodov, s material ami plastov- khranilish i razrabotka metodov podgotovki othodov d zahoroneniu. Trudy koferentsii SEV in Kolobzege, tom II, Poland, Oct. 1972, izd. Warsaw.

Spitsyn, V.I., A. A. Minaev, and V. D. Balukova, 1973. Physical and Chemical Factors Governing the Disposal of High-Activity Liquid Wastes into Deep Geologic Formations: Symposium on Management of Radioactive Waste from Fuel Reprocessing, Paris, France, p. 987–1001.

Spitsyn, V.I., and V.D. Balukova. 1979. The scientific basis for, and experience with, underground storage of liquid radioactive wastes in the USSR: *In*: G.H. McCarthy, Ed., The Scientific Basis for Nuclear Waste Management. v. 1. Plenum Press, New York, p. 237–248.

Spitsyn, V.I., A.A. Minaev, L.I. Barsova, P.Ya. Glazunov, and V.N. Vetchkanov. 1982. The study of the interaction of the components of imitator glasses of glassified radioactive wastes with various surrounding rocks. *In*: S.V. Topp, Ed., The Scientific Basis for Nuclear Waste Management, v. IV, Elsevier, New York, p. 9–14.

Steward, D.C., and W.C. Bentley. 1954. Analysis of uranium in sea water: Science, v. 120, p. 50–53.

Stommel, H. 1957. The abyssal circulation of the oceans: Nature, v. 180, p. 4587.

Stone, J.A. 1977. Evaluation of Concrete as a Matrix for Solidification of Savannah River Plant Waste. DP-1448, Savannah River Laboratory, Aiken, South Carolina.

Storch, S.N., and B.E. Prince. 1979. Assumptions and Ground Rules used in Nuclear Waste Projections and Source Term Data. ONWI-24.

Sykes, M.L., and J.R. Smyth. 1980. Location-specific studies, Yucca Mountain. *In*: J.K. Johnstone and K. Wolfsberg, Eds., Evaluation of Tuff as a Medium for Nuclear Waste Repository: Interum Status Report on the Properties of Tuff. Scandia National Laboratory Report SAND80-1464, p. 39–48.

Szalay, A. 1967. The role of humic acids in the geochemistry of uranium and their possible role in the geochemistry of other cations," *In*: Vinogradov, Ed., Chemistry of the Earth's Crust, Israel Prog. of Sci. Transl., Jerusalem, p. 456–471.

Tamura, T. 1972. Sorption phenomena significant in radioactive waste disposal. *In*: Underground Waste Management and Environmental Implications. Amer. Assoc. Petrol. Geol. Special Publication.

10CFR61 1981. Title 10, Code of Federal Regulations, Chapter 61. Licensing Requirements for Land Disposal Facilities for Radioactive Wastes.

Theis, C.V., and A.N. Sayre. 1942. Geology and Groundwater in (U.S.) National Resources Planning Board, 1942. Pecos River Joint Investigation–Reports of the participating agencies: U.S. Gov't. Printing Office, Washington, D.C. p. 27–38.

Tokarev, A.N. and A.V. Scherbakov. 1960. Radiohydrology. Moscow, 1956. AEC transl. 4100, 1960.

Tonneson, K.A., and J.J. Cohen. 1977. Survey of Naturally Occurring Hazardous Materials in Deep Geologic Formations: A Perspective on the Relative Hazard of Deep Burial of Nuclear Wastes. Lawrence Livermore National Laboratory Report UCRL-52199.

Topp, S.V. Ed. 1982. The Scientific Basis for Nuclear Waste Management, v. IV, Elsevier, New York.

Tremba, E.L. 1969. Isotope geochemistry of strontium in carbonate and evaporite rocks of marine origin: Unpub. Ph.D. thesis, Ohio State University.

U.S. Environmental Protection Agency 1975. Interim Status Report: Development of Radiological Monitoring Programs for Waste Burial Sites. Contract AT (49-1)-3759.

Usiglio, J. 1849. Analyse de l'eau de la Mediterranée sur les cotes de France: Annalen der Chemie, v. 27, pp. 92–107, 172–191.

Vogler, H.A., and D.G. Brookins. 1982. Playas and Naturally Poisoned Areas of Nevada and Western Utah. Lawrence Livermore. National Laboratory Report, in press.

Wachter, B.G., and P.L. Kresan. 1982. Availability of Geotoxic Material: Univ. Arizona Nuc. Fuel Cycle Res. Program; subcon. Lawrence Livermore National Laboratory Report 1879901.

Wald, J.W., and J.H. Westsik. 1979. Devitrification and leaching effects in HLW glass-comparison of simulated and fully radioactive waste glass. In: T.D. Chikalla and J.E. Mendel, Eds. Ceramics in Nuclear Waste Management, CONF-790420, NTIS, Springfield, Virginia.

Wang, R., and Y.B. Katayama. 1981. Probable leaching mechanisms for UO_2 and spent fuel. In: G.H. McCarthy, Ed., The Scientific Basis for Nuclear Waste Management. v. 3. Plenum Press, New York, p. 379–386.

Warner, D.L., and D.H. Orcutt. 1973. Industrial wastewater injection wells in United States—Status of use and regulations. In: Second International Symposium on Underground Waste Management and Artificial Recharge Amer. Assoc. Petrol. Geol., New Orleans p. 687–697.

Weber, F., and M. Bonhomme. 1975. Données radiochronologiques nouvelles sur le Francevillien et son environment: IAEA Symposium 204, p. 17–36.

Weber, R. 1969. Une série précambrienne du Gabon: Le Francevillien, sédimentologie, géochimie, relations avec les gites minéraux. Thesis, University of Strasbourg, Strasbourg.

Wedepohl, K.H. 1967. Geochemistry. Holt, Rinehart & Winston, Inc., New York.

Wengerd, S.A., and M.L. Matheney. 1958. Pennsylvanian system of the Four Corners Region. Amer. Assoc. Petrol. Bull., v. 42, p. 2048–2106.

Westerman, R.E. 1980. Investigation of Metallic Ceramics, and Polymeric Materials for Engineered Barrier Applications in Nuclear Waste Packages. PNL-3484, Pacific Northwest Laboratory, Richland, Washington.

Westermark, T. 1980. Persistent Genotoxic Wastes—An Attempt at a Risk Assessment. Talk at the conference on Chemical Pathways in the Environment, Paris, Sept. 22–26, 1980, sponsored by the Societe de Chimie Physique.

Westsik, J.H., Jr., and C.O. Harvey. 1981. High-Temperature Leaching of a Simulated High-Level Waste Glass. PNL-3172, Pacific Northwest Laboratory, Richland, Washington.

Westsik, J.H., Jr., and R.P. Turcotte. 1978. Hydrothermal Reactions of Nuclear Waste Solids—A Preliminary Study PNL-2752, Pacific Northwest Laboratory, Richland, Washington.

White, A.F., and H.C. Claassen. 1979. Dissolution kinetics of silicate rocks, application to solute modeling. *In*: E.A. Jenne, Ed., Chemical Modeling in Aqueous Systems. Chapter 22, Amer. Chem. Soc. Sym. Series 93, Amer. Chem. Soc., Columbus, Ohio.

Wilkening, M.H., and D.E. Watkins. 1976. Exchange and ^{222}Rn concentrations in the Carlsbad Caverns: Health Physics, v. 31, p. 139–145.

Williams, A.E. 1980. Investigation of oxygen-18 depletion of igneous rocks and ancient meteoric–hydrothermal circulation in the Alamosa River Stock region, Colorado. Unpub. Ph.D. thesis, Brown University.

Winslow, C.D. 1981. The Sorption of Cesium and Strontium from Concentrated Brines by Backfill Barrier Materials. Sandia National Laboratory Report SAND80-2046.

Wolfsberg, K. 1980. Sorptive properties of tuff and nuclide transport and retardation. *In*: J.K. Johnstone and K. Wolfsberg, Eds., Evaluation of Tuff as a Medium for Nuclear Waste Repository: Interum Status Report on the Properties of Tuff. Sandia National Laboratory Report SAND80-1464, p. 39–48.

Index